T0257965

Handbook of Atmospheric Aerosols

Handbook of Atmospheric Aerosols

Edited by **Luke McCoy**

New York

Published by Callisto Reference,
106 Park Avenue, Suite 200,
New York, NY 10016, USA
www.callistoreference.com

Handbook of Atmospheric Aerosols
Edited by Luke McCoy

© 2015 Callisto Reference

International Standard Book Number: 978-1-63239-369-2 (Hardback)

This book contains information obtained from authentic and highly regarded sources. Copyright for all individual chapters remain with the respective authors as indicated. A wide variety of references are listed. Permission and sources are indicated; for detailed attributions, please refer to the permissions page. Reasonable efforts have been made to publish reliable data and information, but the authors, editors and publisher cannot assume any responsibility for the validity of all materials or the consequences of their use.

The publisher's policy is to use permanent paper from mills that operate a sustainable forestry policy. Furthermore, the publisher ensures that the text paper and cover boards used have met acceptable environmental accreditation standards.

Trademark Notice: Registered trademark of products or corporate names are used only for explanation and identification without intent to infringe.

Printed in the United States of America.

Contents

Preface

This book discusses the characteristics of atmospheric aerosols and their effects on climatic conditions. Ground-based, air-borne and satellite data have been collected and analyzed. It also includes extensive information on the organic and inorganic elements of atmospheric aerosol. The author explicates how aerosol particles are chemically and physically processed, temporally and spatially distributed, emitted, formed and transported. This book caters to the need of researchers to resolve several issues related to the complex interaction between atmospheric aerosols and climatology.

This book is a result of research of several months to collate the most relevant data in the field.

When I was approached with the idea of this book and the proposal to edit it, I was overwhelmed. It gave me an opportunity to reach out to all those who share a common interest with me in this field. I had 3 main parameters for editing this text:

1. Accuracy – The data and information provided in this book should be up-to-date and valuable to the readers.

2. Structure – The data must be presented in a structured format for easy understanding and better grasping of the readers.

3. Universal Approach – This book not only targets students but also experts and innovators in the field, thus my aim was to present topics which are of use to all.

Thus, it took me a couple of months to finish the editing of this book.

I would like to make a special mention of my publisher who considered me worthy of this opportunity and also supported me throughout the editing process. I would also like to thank the editing team at the back-end who extended their help whenever required.

Editor

Aerosols Regional Characteristics

Variations in the Aerosol Optical Depth Above the Russia from the Data Obtained at the Russian Actinometric Network in 1976–2010 Years

Inna Plakhina

Additional information is available at the end of the chapter

1. Introduction

Investigation results of the atmospheric aerosol over the Russia territory are of great interest for the ecology and climate developments. The regularities of spatial and temporal variations in the Aerosol Optical Depth (AOD) and Air Turbidity Factor (T) can be received by the Russian actinometric network data (RussianHydrometeorologicalResearchCenter). Our analysis will be based on the "Atmosphere Transparency" special-purpose database created at the Voeikov Main Geophysical Observatory (MGO) on the basis of observational actinometric data. Author has many years cooperation with MGO in the region of the processing and analysis of these observation data. The relationship between the increases in the global surface air temperature and in the atmospheric content of greenhouse gases has been proven. The warming over the past 50 years has mainly been related to human activities (IPCC, Climate Change 2001, 2007). Along with the anthropogenic factor, climate is affected by such natural factors as variations in the solar constant, cyclic interactions between the atmosphere and the ocean, and atmospheric aerosol; these factors are pronounced within time intervals of several years to several decades. The sign of aerosol forcing may be different: the stratospheric aerosol layer causes the reflection of solar radiation incident upon the atmospheric upper boundary and, thus, decreases the warming of the underlying air layers. For example, the sulfate aerosol which formed in the stratosphere after the Pinatubo eruption (June 1991) caused "short" (in 1993) global cooling. Tropospheric aerosol can increase or decrease the surface air temperature, and its influence on the ecological state of the air is well understood (Isaev, 2001). Therefore, monitoring the atmospheric aerosol component is important and necessary now from the standpoints of its climatic forcing and ecology. The study of current spatiotemporal variations in the atmospheric aerosol component is of scientific interest and presents a problem. Current ground-based networks of monitoring (in particular, AERONET) are the results of such interest (Holben et al., 1998). There are eight AERONET stations in Russia; seven of them are located in Siberia [8]. The maps, which show a global distribution of the sources of different anthropogenic, natural, organic, mineral, marine, and volcanic) aerosols arriving in the atmosphere, the total aerosol optical depth in the atmospheric thickness according to model data (IPCC, Climate Change 2001) and the aerosol optical depth according to satellite (MODIS) monitoring (IPCC, Climate Change 2007), show Russia as a territory of decreasing aerosol optical depth (AOD) going from south to north. At the same time, Russia occupies the entire northeastern part of Eurasia (30°E –180°E; 50°N – 80°N) and includes different climatic zones which differ in water content, air temperature, cloudiness, solar radiation flux incident upon the land surface, underlying surface, and air-mass circulation. In addition, the density of population and the degree of industrialization of different Russian regions are very inhomogeneous in space. In the studies (Plakhina et al., 2007, 2009) we have shown that an analysis of the AOD of a vertical atmospheric column can be made on the basis of observational data obtained at the Russian actinometric network, in particular, on the basis of data on the integral atmosphere transparency (P), because P variations are, to a great extent, determined by the aerosol component of the attenuation of direct solar radiation; other components of the attenuation (water vapor and other gases) have little effect on its time variations. Thus, on the basis of data on the homogeneous (calibrated against a single standard and obtained with a unified method) observational series of direct solar-radiation fluxes at the land surface and estimates of the integral (total and aerosol) transparency, it is possible to analyze variations in the AOD of a vertical atmosphere. Now we continue this analysis on the basis of an extended database (the number of stations -- 53, and the period of observations – 1976 -2010 years. Now we present the character of multiyear seasonal variations in AOD, the simplest statistical parameters (means, extrema, and variation coefficients) of spatial variations in AOD annual means, the "purification" of the atmosphere from aerosol over the past 15 years (1995-2010 y.y.). Also we compare the effects of the two natural factors (the global factor—the powerful volcanic eruptions in the latter half of the 20th century which resulted in the formation of a stratospheric aerosol layer— and the regional tropospheric factor—for example, the arrival of aerosol in the atmosphere due to tundra and forest fires) on AOD.

2. Russian actinometric network data

Fig. 1 gives a map showing the location of 53 actinometric stations of the Russian network (Makhotkina et al., 2005, 2007; Luts'ko et al., 2001)for which the AODs of vertical atmospheric columns were estimated for a wavelength of 0.55 µ from the measured fluxes of direct solar radiation at land surface. These stations cover a large part of Russia and are located outside the zones of direct local anthropogenic sources of industrial and municipal aerosol emissions (suburbs, rural areas, uplands, etc.). In other words, the considered spatiotemporal variations in AOD are formed under the influence of natural factors: the advection of air masses from the regions with an increased or decreased aerosol load, volcanic eruptions, and forest and tundra fires. In analyzing the 1976–2010 observational data, our goal was to obtain an averaged pattern of the spatial distribution of atmospheric aerosol over Russia and to compare this pattern with that of the global aerosol distribution which is presented in the IPCC third (modeling) and fourth (satellite data, MODIS) reports (IPCC, Climate Change 2001, 2007). In this case, the estimates obtained with our method supplement the international data on the model approximations and satellite monitoring of AOD. The advantages of our estimates are the great length of the series of actinometric observations under consideration (35 years), the universal methods of measurements and data treatment for all the stations, and the vast coverage area of Russia's large territory.

Figure 1. Layout of 53 actinometric stations whose data will be analyzed in the chapter. It is possible that the list of the observation stations will be increased up to 80 for the special estimations.

3. Empirical data and analysis procedure

The special-purpose Atmosphere Transparency database formed at the Main Geophysical Observatory makes it possible to analyze both the integral and aerosol transparencies of the atmosphere. The stations given in Fig. 1 were selected with consideration for the quality and completeness of the instrumental series. The integral air transparency :

$$P = (S/S_0)^{1/2} \tag{1}$$

Where S is the direct solar radiation to the normal-to-flux surface, reduced to the average distance between the Earth and the Sun and a solar altitude of 30°; S_0 is the solar constant equal to 1.367 kW/m². The Linke turbidity factor is unambiguously correlated with P:

$$T = \lg P / \lg P_i = (\lg S_0 - \lg S) / (\lg S_0 - \lg S_i) = -\lg P / 0.0433 \tag{2}$$

The AOD of the vertical atmosphere was calculated with a method specially developed and used at the MoscowStateUniversity meteorological observatory (Abakumova et al., 2006) with consideration for its limitations and errors:

$$AOD = \{lnS - [0.1886w^{(-0.1830)} + (0.8799w^{(-0.0094)} - 1)/\sinh]\}/\{0.8129w^{(-0.0021)} - 1 + (0.4347w^{(-0.0321)} - 1)/\sinh\} \quad (3)$$

An index of the Angström spectral attenuation which depends on the size distribution of particles and the coefficient of particle reflection—is assumed to be equal to 1; Sis the direct solar radiation reduced to the average distance between the Earth and the Sun, W/m^2; and w is the water content of the atmosphere, g/cm^2. The conditions of observations at the stations, as a rule, correspond to the weather of an anticyclonic type (clear or slightly cloudy) when the Sun is not blocked by clouds.

4. Spatial variations in aerosol optical depth AOD

Table 1 gives the multiyear means and extrema of the annual values of AOD and the standard deviations from these means, which are averaged over the all 53 stations under consideration (pointed in Fig.1) for the two periods. It is seen that the AOD mean over all the stations and the entire observation period is equal to 0.14 and varies from 0.29 to 0.07, which is in good agreement with the spatial range of the AOD variations obtained from the satellite and model data (for the Russian region) that are given in the IPCC third and fourth reports (0.30–0.05).

Period	AOD		σ	Trend of AOD variations inover 10 years
	Mean	0.14	0.04	-0.02
1976 – 2010	Maximum	0.29		+0.02
	Minimum	0.07		-0.05
	Mean	0.12	0.04	-0.01
1995 – 2010	Maximum	0.22		+0.05
	Minimum	0.05		-0,06

Table 1. Multiyear means, maxima, minima, and standard deviations of the annual means of AOD over all stations in absolute units.

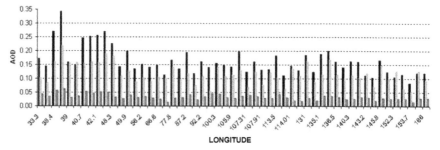

LONGITUDE

Figure 2. Statistics of the annual means of AOD for each of the stations: the ratio of the AOD means (black) over the period 1976–1994 and the AOD means (grey) over the period 1995–2010 and the standard deviations (red) in the series of the annual values of AOD for each of the stations.

Variations in the Aerosol Optical Depth Above the Russia from the Data Obtained at the Russian
Actinometric Network in 1976–2010 Years

7

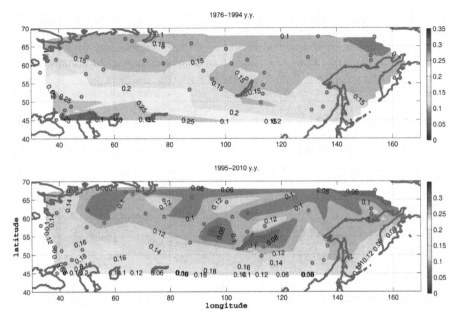

Figure 3. Spatial distributions of the multiyear means of AOD over the observation periods 1976–1994
(upper part) and 1995–2010 (lower part).

The annual values of AOD for each of the stations multiyear means over 1976–1994 years,
over 1995–2010 years, their standard deviations are given in Fig.2. Each column of the
diagram corresponds to the longitude positionof the station in accordance with Table 1. The
means of the AOD characteristics are corresponds to the multiyear (1976–1994 years) annual
means of AOD (black), grey corresponds to the multiyear (1995–2010 years) annual means,
red corresponds to standard deviations of the annual values of AOD from its mean for each
station. A spatial distribution of AOD is shown in more detail in the maps (Fig. 3) drawn by
interpolating the data obtained at 53 stations to Russia's territory. For this interpolation, the
technologies of the MATLAB 7.5.0. program package were used: there are options to create a
uniform grid for the entire region, onto which the given functions $Z= F(x,y)$were projected,
where x and $y$are the latitude and longitude, respectively, for each of 53 observation points,
and Z is the AOD mean. In addition, a bilinear interpolation of data was performed. Under
bi-cubic and bi-square interpolations, the results, in principle, do not differ from those given
in Fig. 3. The spatial distribution of the AOD means over the 35 - year period is in a good
agreement with the results of modeling a spatial atmospheric-aerosol distribution, which are
given in the IPCC third report (IPCC, Climate Change 2007). The model described in this
report takes into account aerosols of different origins anthropogenic and natural sulfates,
organic particles, soot, mineral aerosol of natural origin, and marine saline particles) which
have certain specific properties of distribution over the globe, and it yields a decrease in
AOD over Eurasia from the southern to the northern latitudes in the presence of areas with
increased atmospheric turbidity over southern Europe, the Middle East, southeastern Asia,

Ukraine, and Kazakhstan. Fig. 3 shows that the AOD over Russia decreases from the southwest to the northeast. The increased values of aerosol haziness in the southeast and southwest are most likely caused by an advective arrival of air masses from the regions with high aerosol content in the atmosphere: from Ukraine and Kazakhstan in the southwest and from southeastern Asia and China in the southeast. Fig. 3 (upper part) shows the localizations of regional tropospheric aerosol sources (western and eastern Siberia and Primorskii Krai). In the last 15 years (Fig. 3, lower part), in the absence of powerful volcanic eruptions and under conditions the atmosphere being purified of the stratospheric aerosol layer, the sources of aerosol arriving in the troposphere have become more pronounced. In addition, in the last decade, the AOD has noticeably increased for a few stations in the Far East, which is probably due to increased volcanic activity on Kamchatka .

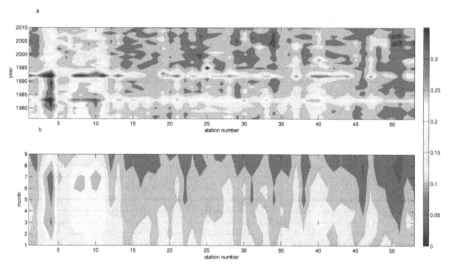

Figure 4. Spatiotemporal variations in AOD: (a) multiyear variations in the annual values of AOD for all 53 stations under consideration and (b) mean seasonal variations in AOD for all 53 stations under consideration.

The spatiotemporal inhomogeneities of the AOD annual values clearly reflect their causes (Fig. 4a): the peaks of the volcanic eruptions (El Chichon, 1982, and Pinatubo, 1991) and the tundra fires of the last decade in eastern Siberia, the frequency and intensity of which have increased due to climate changes. Fig. 4b shows variations in the mean annual cycle of AOD. The features of the AOD mean annual cycle for each concrete station are formed under the influence of seasonal variations in the character of air-mass transport to a given point from regions with different aerosol contents (synoptic processes) and seasonal variations in air temperature, humidity, and in the state of the underlying surface, in combination with an industrial load of some regions. The AOD maxima are, as a rule, observed in April and July–August, but the summer maximum is more pronounced at stations (N° 4, 8, 9, 10, and 11) located in the south of European Russia. First of all, this is related to the fact that, in

Variations in the Aerosol Optical Depth Above the Russia from the Data Obtained at the Russian
Actinometric Network in 1976–2010 Years

9

summer, tropical air masses dominate here which are characterized by high contents of moisture and aerosol. The spring maximum is caused by snow cover melting and the replacement of the dominating arctic air masses by temperate or tropical air masses.

5. Time variations in aerosol optical depth AOD

Fig. 5a gives some examples of time variations in the annual means of AOD for stations with negative and positive trends. In Fig. 7b, the examples of the time trends of the AOD annual values are supplemented by the corresponding variations in the flux of direct solar radiation (for the Sun's height h= 30°), which reach 100 W/m2 over the course of 35 years (3 W/m2 per year); estimates were obtained for two stations with the maximum and minimum means of AOD. Thus, the influence of a decreased aerosol load on the flux of direct solar radiation incident upon the land surface under clear skies is empirically estimated. For total radiation, this influence is less pronounced. And our estimate of the rate of a decrease in direct solar radiation does not contradict the satellite data (IPCC, Climate Change 2007) on the rate of a decrease in the flux of the total reflected (upward) solar radiation (–0.18 ± 0.11) W/m2 per year (the ISCCP project) and (–0.13 ± 0.08) W/m2 per year (the ERBS project)) over the course of 1984–1999 and the assumption made in (IPCC, Climate Change 2007), that this is caused by a global decrease in stratospheric aerosol (the so-called phenomenon of "aerosol dimming").

At most observation sites, the atmosphere was purified of aerosol within the period under consideration. On the whole, for Russia, the trend of AOD variations is negative (Fig. 6); the absolute value of the trend (over 10 years) varies from (–0.05) to (+ 0.02) and increases generally from the south-west to the north-east of Russia. The mean of the relative trend accounts for (–14%) over 10 years, its maximum is 21% over 10 years, and its minimum is (–35%) over 10 years at a determination coefficient of no more than 0. 5. (See also Table 1). It is evident that, in this case, a decrease in the AOD mean must be observed during the last 15 years of the whole region. The largest negative trends are observed at the Solyanka station (in the south of the Krasnoyarsk Krai), in Chita (Transbaikalia), Khabarovsk (Primorskii Krai), and in the south of European Russia. The combination of the two factors—global purification of the atmosphere from transformed volcanic aerosol and decreased anthropogenic forcing—forms the negative trends in these regions. Positive trends are observed in Arkhangelsk and the Far East (Kamchatka and Okhotsk), and almost zero trends are observed in western (station nos. 18, 19, and 20) Siberia. The positive (Arkhangelsk) and decreased negative (the indicated Siberian stations) trends may be caused by increased industrial emissions in these regions, an increase in the number and intensity of fires, and comparatively low-power volcanic eruptions (for example, in Kamchatka). The estimates of the AOD trends and integral transparency obtained by other authors (for example, Ohmura, 2006) were compared with our estimates earlier in (Plakhina et al., 2007). This comparison shows an agreement with the results presented in this paper.

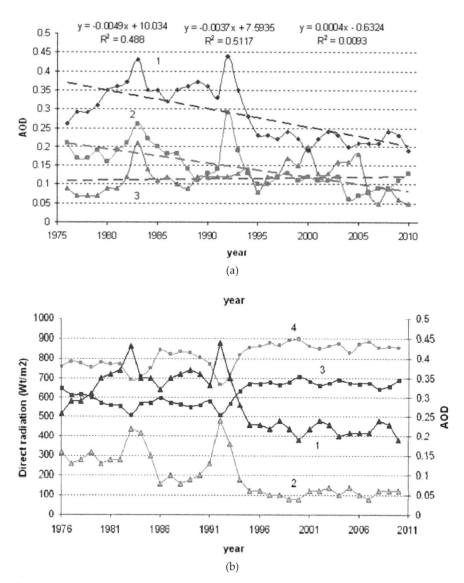

(a)

(b)

Figure 5. Time variations in the annual values of AOD and in the flux of direct solar radiation for the Sun's height 30°: (a) multiyear variations in the annual values of AOD for three stations (Krasnodar (1), Chita (2), and Okhotsk (3)) and (b) multiyear variations in the annual values of AOD and in the annual mean of direct solar radiation flux at the Sun's height 30° for the two stations with the maximum and minimum means of AOD. For both graphs, the period under analysis is 1976– 2010. Krasnodar(1 corresponds to AOD and 3corresponds to direct radiation), Solyanka (2 corresponds to AOD and 4 corresponds to direct radiation)

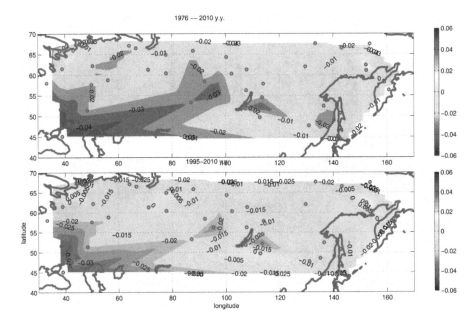

Figure 6. Spatial distributions of the multiyear variability of AOD: trends of the time variations over the period 1976-2010 years (in absolute values over 10 years) and trends of the time variations over the period 1995-2010 years (in absolute values over 10 years)

6. Effects of the volcano eruptions

6.1. Influence of the volcano eruptionson AOD

Fig. 7 gives a "long" (45 years) series of annual means of AOD for the Ust' Vym station (62.2°N, 50.4°E), which demonstrates a characteristic multiyear trend of variations in the annual values of AOD and its response to stratospheric disturbances. The four powerful volcanic episodes— Agung (8°S, 116°E, 1963), Fuego (14°N, 91°W, 1974), El Chichon (17°N, 93°W, 1982), and Pinatubo (15°N, 120°W, 1991)—are clearly pronounced and quantitatively estimated. In particular, the maximum effect observed a year after the eruptions is 100% (in deviations from the multiyear norm); throughout the year, its attenuation occurs with the dissipation and transformation of the stratospheric aerosol layer. A decrease in the AOD values for 1995–2006 is also clearly manifested. Such a character of multiyear variations in the annual values of AOD is characteristic of most stations and is, to a great extent, determined by the four powerful volcanic eruptions in the latter half of the 20th century, because seasonal and local disturbances caused by the effects of tropospheric aerosol, when annually averaged, become leveled and have almost no influence on the distribution of the multiyear values of AOD.

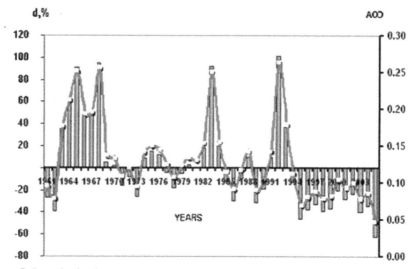

Figure 7. Example of multiyear variations in the annual means of AOD (red) and their deviations from the averaged (blue),

$$d = 100\% * (AOD_i - AOD_m) / AOD_m.$$

6.2. Influence of the volcano eruptions on the turbidity factor (T)

From the data of 80 observation stations over the Russia the special analysis of the turbidity factor (T) have been fulfilled: time variations during 1976-2010 y.y. and during 1994-2010 y.y. have been estimated. For the 9 regions over all Russia territory long-term trends for the characteristics of the integral atmospheric transparency have described. For all regions during 1976-2010 y.y. negative T and AODvariations tendencies exist; during 1994-2010 y.y. negative T2 иAOT variations tendencies remain at the same level as during 1994-2009 y.y. practically for all Russia regions. So, for the most part of Russia territory the conditions of the relatively high atmospheric transparency (in 1994-2010 y.y. – 17 years) remain as well as the atmospheric transparency increase within this 17 years time interval remain. Comparatively stable, longterm and intensive variations (increase) take place in post-volcanic periods: 1) for El Chichon eruption (1982 year, April) – from the last 1982 year to October of 1983 year; 2) for Pinatubo (1991 year, June) – from the September of 1991 year to July of 1993 year. Anomalies of the mean month values of the T2 during these "post volcanic" period after the eruptions of El Chichon and Pinatubo are presented in Table 2.

Estimations of the volcano contribution into the multiyear mean values (for the months and year) of the factor turbidity and aerosol optical depth during 1976 – 2005 years period and durings the so called "stable" 1976-2005 years period (without 1982, 1983, 1991,1992, 1993 years) are pointed in Table 3. It is obvious that effects, connected with eruptions lead to increase of the multiyear mean values equal 3% (from 1% - to 7%) for Tand equal 7% (from 2% - to 12%) for AOD.

Region	$\Delta T_2\%=100*(T_i-T_m)/T_m$		$\Delta T_2=(T_i-T_m)/\sigma$	
	El Chichon	Pinatubo	El Chichon	Pinatubo
North of EPR	20	32	1.9	3.0
Central part of EPR	18	30	1.8	3.1
Southof EPR	14	20	1.8	2.6
Ural	26	23	2.6	2.3
West Siberia	19	35	1.7	3.1
North–east of APR	19	35	1.7	3.2
Central part of APR	19	27	2.0	2.8
Southof EPR	22	38	1.8	3.0
Far East of the Russia	15	36	1.5	3.4

Table 2. Anomalies of the mean month values of the T during post volcanic period after the eruptions of El Chichon and Pinatubo.

$$T_{(1976\text{-}2005)} / T_{stab\ (1976\text{-}2005)}$$

Month	1	2	3	4	5	6	7	8	9	10	11	12	Year
North of EPR		1,03	1,03	1,04	1,03	1,03	1,02	1,02	1,03	1,02	1,04		1,03
Central part ofEPR	1,04	1,05	1,04	1,04	1,03	1,02	1,02	1,02	1,01	1,02	1,02	1,05	1,03
Southof EPR	1,04	1,04	1,03	1,03	1,02	1,02	1,01	1,01	1,01	1,01	1,03	1,04	1,02
Ural	1,05	1,06	1,06	1,04	1,03	1,03	1,02	1,02	1,03	1,02	1,05	1,05	1,04
West Siberia	1,03	1,05	1,04	1,03	1,04	1,03	1,02	1,03	1,04	1,03	1,02	1,02	1,03
North–east ofAPR	1,07	1,07	1,04	1,03	1,03	1,03	1,03	1,02	1,02	1,03	1,05	1,07	1,04
Central part ofAPR	1,04	1,02	1,02	1,02	1,02	0,98	0,98	1,02	1,01	1,00	1,06	1,04	1,02
Southof EPR	1,06	1,04	1,04	1,04	1,03	1,02	1,02	1,02	1,02	1,02	1,03	1,03	1,03
Far East of the Russia	1,06	1,05	1,04	1,03	1,02	1,02	1,01	1,01	1,01	1,02	1,03	1,05	1,03

$$AOD_{(1976\text{-}2005)} / AOD_{stab.(1976\text{-}2005)}$$

Month	1	2	3	4	5	6	7	8	9	10	11	12	Year
North of EPR		1,11	1,10	1,10	1,08	1,07	1,07	1,06	1,06	1,06	1,14		1,07
Central part of EPR	1,06	1,07	1,08	1,09	1,10	1,08	1,08	1,07	1,06	1,02	1,09	1,10	1,07
Southof EPR	1,10	1,09	1,06	1,07	1,05	1,04	1,03	1,02	1,03	1,04	1,10	1,14	1,06
Ural	1,08	1,12	1,14	1,11	1,09	1,08	1,07	1,08	1,14	1,08	1,07	1,08	1,06
West Siberia	1,05	1,10	1,11	1,09	1,11	1,10	1,08	1,06	1,08	1,13	1,07	1,08	1,08
North–east of APR	1,03	1,13	1,11	1,08	1,09	1,08	1,08	1,07	1,09	1,11	1,13	1,02	1,09
Central part of APR	1,08	1,07	1,08	1,09	1,10	1,08	1,08	1,07	1,06	1,05	1,13	1,11	1,06
Southof EPR	1,09	1,08	1,09	1,10	1,08	1,05	1,05	1,05	1,05	1,04	1,06	1,06	1,07
Far East of the Russia	1,12	1,11	1,09	1,06	1,06	1,07	1,06	1,03	1,04	1,04	1,08	1,12	1,07

Table 3. Estimation of the volcano contribution into the multiyearmean values of the factor turbidity and aerosol optical depth during 1976 - 2005 years period; EPR -- the European Part of Russia; APR -- theAsian Part of Russia.

The examples of the long-term time variations for Tand AOD in the differentRussia regions:North,Central part and South of theEuropean Part of the Russia andRussianFar East are presented in Fig. 8.

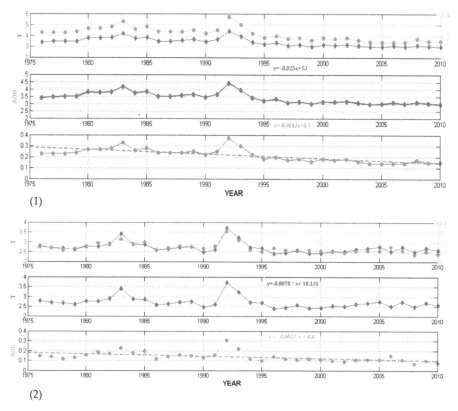

Figure 8. Examples of the long-term time variations for T(blue)and AOD (green) in the different Russia regions: South of theEuropean Part of the Russia (1)and North part (2).

7. Fires above the European Part of Russia (EPR) under conditions of abnormal summer of 2010

The spatial variations in the air turbidity factor according to ground-based measurement data from 18 solar radiometry stations within the territory (40°–70° N, 30°–60° E) in summer 2010. We have shown earlier (Makhotkina et al., 2005; Plakhina et al., 2007, 2009, 2010) that the spatial distribution of the aerosol optical depth (AOD) over the territory of Russia averaged over more than 30 years corresponds to the model of global atmospheric aerosol distribution over Eurasia and the satellite AOD monitoringresults, presented in the 3rd and 4th IPCC reports; it shows a decrease in the aerosol turbidity from southwest to northeast.

The events of summer 2010 (abnormal heat and forest and peat fires) evidently changed both the average values of air turbidity and the character of its spatial variations. Therefore, our estimates are of interest in the analysis (All Russia Meeting, 2010) of the situation on the European Part of Russia (EPR) in summer 2010. Fig. 9 presents the coordinates of solar radiometry stations on the EPR (Luts'ko et al., 2001); data from it were used in this work. The long-term annual average (over a "post-volcanic" period of 1994–2009 years) values T_{post} for summer months and the corresponding monthly values T_{2010} for 2010 are given in Table 4, along with the monthly average maxima of T and the relative difference (%) $D = (T_{2010} - T_{post})/T_{post}$. As it is seen, the average July and August T in 2010 and in the "postvolcanic" period differ by –6% and +4%, respectively (the differences D vary from – 28% to +11% of the average value for a certain station in June and from – 22% to +25% in July). The value of $D = (T_{2010} - T_{post})/T_{post}$ is 14% in August (for the region) and varies from – 11% to +48% for certain stations.

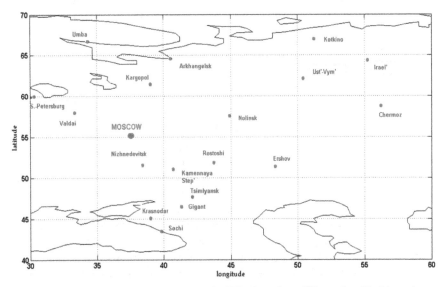

Figure 9. Layout of 18 actinometric stations on the EPR whose data will be analyzed in this section.

Spatial variations in $T$are shown in Fig. 10. To interpolate the data of the stations to the whole region under study , we also used features of the MATLAB package, i.e., the option for creating a homogeneous grid for the EPR region under study, the option of bilinear (horizontal and vertical) interpolation of data from 18 stations to the territory (40°–70° N, 30°–60° E), and the projection of the function $T= F(\phi, \lambda)$ (where $\phi$and $\lambda$are the longitude and latitude, respectively, for each of the observational points) to the grid. The spatial distribution of the mean T_{post}(for June, July, and August) for the "postvolcanic" period corresponds to the results obtained earlier (Plakhina et al., 2009) for the long-term annual average AOD. In this period, T_{post}quasi -monotonically decreased from southwest to

Period	Month	Mean (maximum)	D	Standard deviation in the series of monthly average values for different stations
1994 – 2009 years	June	3.0 (3.9)		13%
	July	3.2 (4.2)		13%
	August	3.2 (4.3)		14%
2010 year	June (165)	2.95 (4)	-6%	18%
	July (250)	3.42 (4.1)	+4%	19%
	August (125)	3.73 (5.3)	14%	21%

Table 4. Long-term monthly average values of the turbidity factor T (1994-2009 years) and the corresponding values for summer 2010 along with the regional maximum values of mean *T*. The number of daily average values of *T* used in the averaging is mentioned in the parenthesis in the second column.

northeast; the regions of localization of regional tropospheric aerosol sources are invisible (except for Archangelsk). The June–July average values of *T* at the Archangelsk station have been increased during the "postvolcanic" period: a local (and/or regional) atmospheric aerosol source is traceable; it can be both frequent natural forest fires and anthropogenic industrial factors in this Russian region. The pattern differed significantly before 2010. In June, the spatial variations in T were close to distributions of Tpost with a certain northward shift of the regions of maximum transparency (T= 2 – 2.5) with a decrease in means for June (Table 2) throughout the region in comparison with the "postvolcanic" period. In July, the monotonicity in a decrease in the turbidity was obviously disturbed in the northeast direction. A south-to-north "tongue" of increased values of the turbidity factor is observed (T= 3.5 – 4.0). Finally, in August, an epicenter (closed region) of anomalous air turbidity (T= 4.5 – 5.5) was formed within the region 48°– 55° N and 37°– 42° E, which is located to the south of Moscow and covering the Moscow region by its periphery (T= 4.0 – 4.5). This pattern resulted from the action of the blocking anticyclone, which prevented air mass ingress from the west, provided for closed air circulation in the EPR, and a favored temperature rise over the EPR and a rapid increase in the forest fire area. Fire aerosols accumulated in the atmosphere through this period. This process was the most pronounced in the 1st decade of August. Our pattern of spatial distribution of *T* in August 2010, obtained from ground based measurements of the direct solar radiation flux, is in a good agreement with the map of AOD distribution in the EPR (within the region 50°– 65° N, 30°– 55° E) in the 1st decade of August presented in (Sitnov, 2010).

Thus, we have ascertained the peculiarities of spatial variations in the air turbidity factor in summer 2010 in comparison with the long-term average spatial variations, which have been manifested in both distribution character and the value of the anomalies of the turbidity factor.

8. Changes in integral and aerosol atmospheric turbidity in Trans-Baikal and Central Siberia

The turbidity factor T and atmospheric aerosol optical thickness AOD for the wavelength 0.55 μ is used in this section as atmospheric transparency characteristics. The series T and AOT were analyzed for the 1976–2010 years for the stations presented in Fig.11.

From Fig. 12, 13 and Table 5 it is evident that AOD and T have the apparent seasonal dependence: maximal values of the aerosol turbidity are observed in spring (mean excess above the year ones is 25%), maximal values of the integral turbidity are observed in summer (mean excess above the year ones is 10%). At the same time the structure of the spatial distribution is similar in April and in July as for AOD so for T. The sources, formed AOD and T spatial distribution: prevailing air circulations, bearing of the aerosol and water vapor rich air masses (and vice versa), "constant" local aerosol sources, antropogenic or natural (for example, the forest or peat fires).

From Fig. 14, 15 it is evident that in July it is exist the constant district in Trans-Baikal region with high AOD (may be the fires); in April a structure of the AOD field is formed by the air masses arriving from the south directions. But in April a structure of the T is not so apparent. In Fig.16 the variation of the month averaged values of FQ - fires quantity (aircraft data) and factor of integral turbidity T are showed. It is observed that FQ have seasonal variations with the maximum value in May, at the same time maximum for T is observed in July. We can also see the growth of T, connected with the El Chichon eruption in 1982 - 1983 years. In Fig.17 the wavelet spectra of the fire quantity (FQ) and factor of integral turbidity (T) series demonstrate the oscillations by 12 moths period (in both series) up to 1982-83 years (1982 y. - volcano eruption). Then the oscillations by 6 - 3 month periods are exist in the FQ series, which are connected with the variations of the FQ - fire quantity. Colorbar represents normalized variances.

In Fig.18 the variation of the month averaged values of fires quantity FQ (aircraft data) and factor of integral turbidity T are showed during 1992 -2009 years. It is also observed that the FQ have seasonal variations with the maximum value in May, at the same time maximum for T is observed in July. We can also see the decrease of T, connected with the Pinatubo eruption during 1992-1993 years. In Fig.19 the wavelet spectra of the fire quantity (FQ) and factor of integral turbidity (T) series demonstrate the oscillations by 12 moths period (in both series) from 1997 year up to 2005 year (in FQ series) and up to 2006 year (in T series) . Then the oscillations by 7- 3 monhs are exist in the FQ series, which may be also connected with the variations of the fire quantity. Colorbar represents normalized variances.

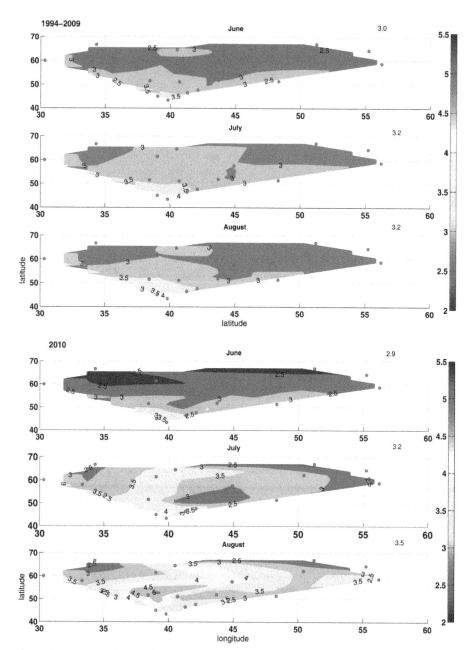

Figure 10. Spatial distribution of mean values of the turbidity factor T for June, July, and August in 1994 – 2009 years (top part) and in summer 2010 year (lower part).

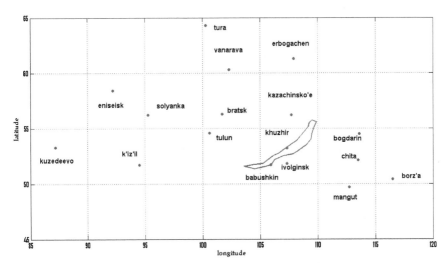

Figure 11. Layout of 17 actinometric stations whose data were analyzed in this section to investigate integral and aerosol atmospheric turbidity variability in Trans-Baikal and Central Siberia.

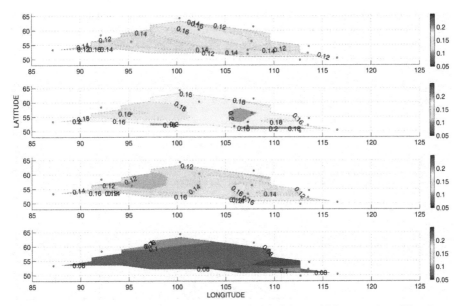

Figure 12. Spatialdistribution of AOD in Trans-Baikal and Central Siberia region of Russia for the year and season AOD values:year,April,July and October values(from the top to the bottom), averaging period 1976-2010 years (14 stations).

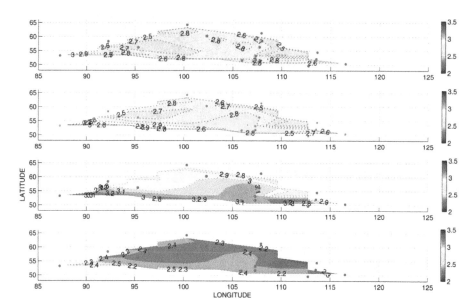

Figure 13. Spatial distribution of T in Trans-Baikal and Central Siberia region of Russia for the year and season AOD values: year, April, July and October values (from the top to the bottom), averaging period 1976-2010 years (14 stations).

	YEAR	APRIL	JULY	OCTOBER
AOD	0.14	0.19	0.14	0.09
variation coefficient %	13	16	15	16
turbidity factor T	2.72	2.76	3.02	2.39
variation coefficient %	36	43	44	66

Table 5. Multiyear mean values of the T turbidity factors of the atmospheric aerosol optical thickness (AOD) and their variation coefficients (the period of averaging is 1976–2010)

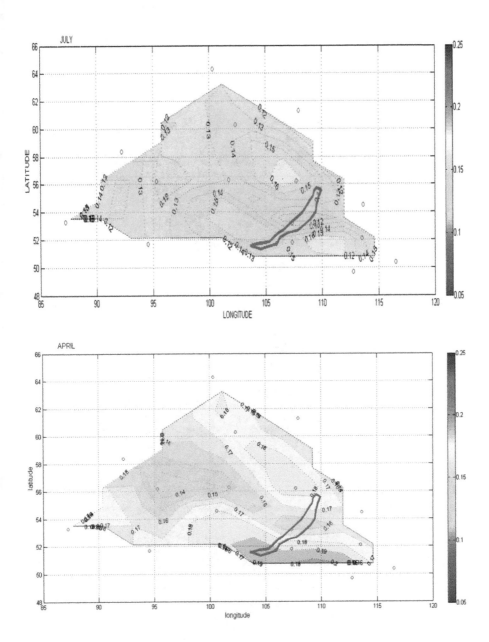

Figure 14. Spatial distribution of AOD in Trans-Baikal and Central Siberia region of RF for the July
(top) and April (bottom) values, averaging period 1976-2010 years, 17 stations.

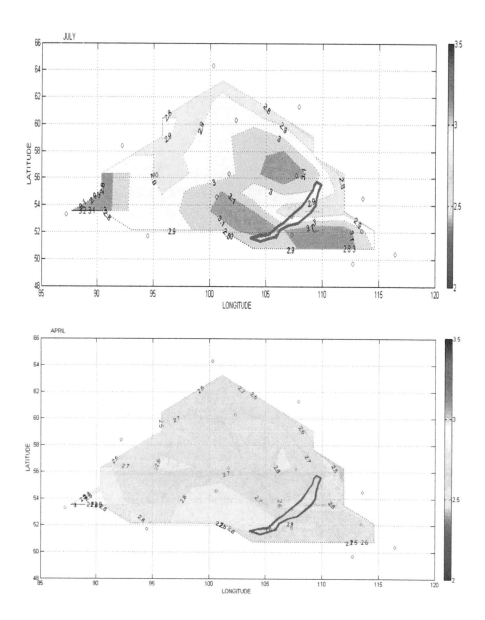

Figure 15. Spatial distribution of T in Trans-Baikal and Central Siberia region of RF for the July (top) and April (bottom) values, averaging period 1976-2010 years, 17 stations.

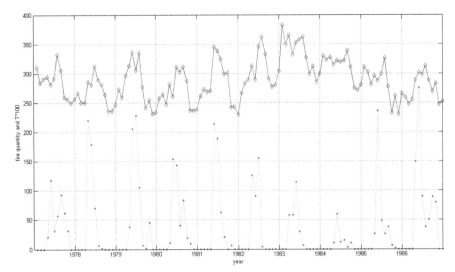

Figure 16. Fire quantity (FQ) (dotted line) and factor of integral turbidity T (firm line) during 1977 - 1986 years in Trans-Baikal region.

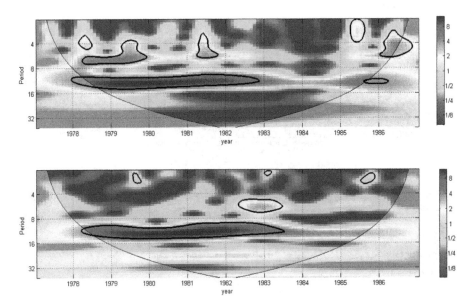

Figure 17. A result of the one-dimensional wavelet transformation of the two signals: fire quantity FQ (top) and factor of integral turbidity T (bottom) during 1977 -1986 years in Trans-Baikal region.

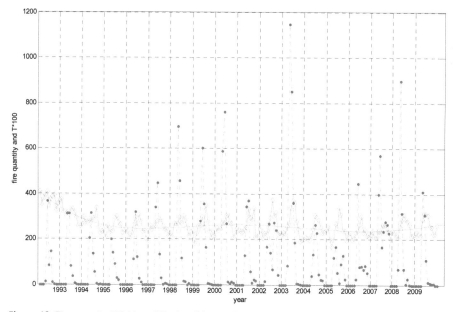

Figure 18. Fire quantity FQ (dotted line) and factor of integral turbidity T (firm line) during 1992 - 2009 years in Trans-Baikal region.

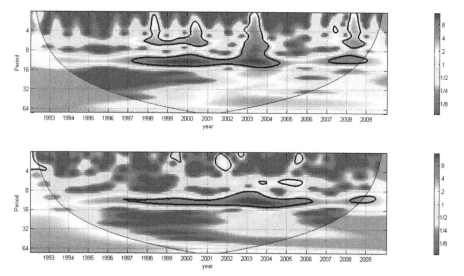

Figure 19. A result of the one-dimensional wavelet transformation of the two signals: fire quantity FQ (top) and factor of integral turbidity T (bottom) during 1992 - 2009 years in Trans-Baikal region.

Variations in the Aerosol Optical Depth Above the Russia from the Data Obtained at the Russian
Actinometric Network in 1976–2010 Years

25

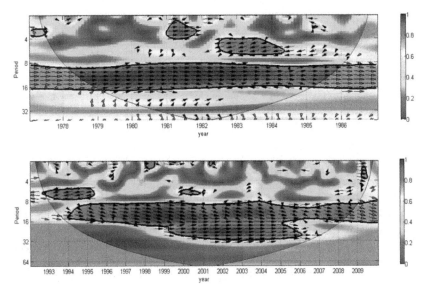

Figure 20. A result of the WTC transformation of the two signals: fire quantity FQ and factor of integral turbidity T during 1987- 1986 years (top) and 1992 – 2009 years (bottom) in Trans-Baikal region. Units of the colorbar are wavelet squared coherencies. An arrow, pointing from left to right signifies in-phase, and an arrow, pointing upward means that T series lags FQ series by 90° (phase angle is 270°).

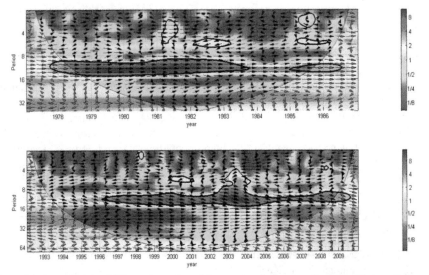

Figure 21. A result of the XWT transformation of the two signals: fire quantity FQ and factor of integral turbidity T during 1987- 1986 years (top) and during 1992 – 2009 years (bottom) in Trans-Baikal region. Units of the colorbar are wavelet squared coherencies. An arrow, pointing from left to right signifies in-phase, and an arrow, pointing upward means that T series lags FQ series by 90° (phase angle is 270°).

Notice:

The continuous WT (wavelet transformation) expands the time series into time frequency space. Cross wavelet transform (XWT) finds regions in time frequency space where the time series show high common power. Wavelet coherence (WTC) finds regions in time frequency space where the two time series co-vary (but does not necessarily have high power); See

(Grinsted et al., 2004; Jevrejeva, 2003)

9. Conclusions

The application of the AOD and T estimationtechnique for processing the results of observations at the Russia actinometric network stations allows obtaining qualitatively new and detailed information on the level of aerosol content of the atmosphere in separate regions and in Russia as a whole. Our analysis has made it possible to formulate the following conclusions about the spatiotemporal distribution of AOD over Russia. The spatial distribution of the AOD values averaged over the 35-year period underconsideration generally corresponds to the model of global aerosol distribution over Eurasia, which is represented in the IPCC third and fourth reports. This is manifested in the AOD decrease from the southwest to the northeast in the presence of regions with continuous increasing aerosol turbidity in southwestern and southeastern Russia. Against this background, regions with increased troposphere aerosol loads are pronounced that have been more noticeable under the global purification of the atmosphere from the stratospheric aerosol layer that started in 1995. These troposphere sources are related either to an anthropogenic load (cities in southern Russia, western Siberia, and Primorskii Krai) or to forest and tundra fires in Siberia, in particular, at the Tura station in the Evenki Area. One more cause of the decreased transparency in the atmosphere over eastern Russia, which is manifested in the annual means of AOD, is the volcanoes of Kamchatka. On the whole, for Russia, the trends of multiyear variations have been negative in the last decades. However, there are stations at which the AOD trends are positive; this is particularly true of the stations of Kamchatka and the Far East.We have ascertained the peculiarities of spatial variations in the air turbidity factor in summer 2010 year in comparison with the long-term average spatial variations, which have been manifested in both distribution character and the value of the anomalies of the turbidity factor at ETR. Also we have ascertained the peculiarities of spatial and seasonal variations in the air turbidity factor T and aerosol optical depth AOD in Trans-Baikal and Central Siberia region of RF for the year and different season

The analysis of AOD variationsduring the last 35 years shows the following concrete estimations:

1. Total averaged AOD over all stations and the whole period under study (0.14) is very close to the annual mean global AOD value (0.14) calculated from the ECHAM-HAM model and to the estimates obtained from satellite data (0.16);
2. Maximum annual mean AOD (0.29) is reached in Krasnodar (№ 4) and the minimum one (0.07) is observed at the "aerosol pure" station of Srednekolymsk (№ 51). The

averaged special-time changes (standard deviations from the all year-averaged values of AOD for all stations) are 0.04 and are equal to mean space changes (standard deviations of mean AOD over all stations) which are found to be 0.03;

3. Minimum annual AOD values for European and Asian parts of Russia are, respectively: 0.03 and 0.02;

4. "Purification" of the atmosphere from aerosol is caused by the absence of large volcanic eruptions and by industrial "calm" conditions during the last decades. The mean AOD for the last 15 years $\{[AOD_{(1976-1994)} - AOD_{(1995-2010)}] / AOD_{(1976-2010)}\}100\% = 28\%$ is lower than in the preceding 19 years. Negative tendencies are almost similar for remote and urban (as well as for rural) stations; they are less pronounced in the fall than in spring and summer;

5. Local effect of the AOD increase due to volcanic eruptions can reach 100%, while the average effects, within our consideration period, are of some percents.

Author details

Inna Plakhina
Oboukhov Institute of Atmospheric Physics, RussianAcademy of Science, Russia

Acknowledgement

The study was supported by the Russian Foundation for Basic Research, project № 10-05-01086. I thank Academician G. S. Golitsyn for helpful remarks; I am helpful to my close colleague Makhotkina E.L. for herhelp in preparing and analysis of the experimental data.

10. References

Abakumova, G.; Gorbarenko, E.; Chubarova, N. (2006). Estimation of Determination Accuracy of Atmosphere Aerosol Optical Thickness and Moisture Content by Data of Standart Observations on the Base of Comparison with Measurements by Solar Photometer SIMEL, Proc. of the Intern. Symp. of SNG Countries on Atmospheric Radiation MSAR-2006, 27–30 June 2006 (St.-Petersb. Gos. Univ., St.-Petersburg, 2006), pp. 43–44

All Russia Meeting on the Status of the Air Basin in Moscow and European Russia under Extreme Weather Conditions in Summer 2010. (2010) http://www.ifaran.ru/messaging/forum/

Grinsted, A.; Moore, J.; Jenrejeva, S. (2004). Application of the cross wavelet transform and wavelet coherence to geophysical time series, *Nonlinear Processes Geophys.*, 11, 561-566 (or http:www.pol.ac.uk/research/ waveletcoherence/)

Holben, B.; Eck, T.; Slutsker, I. et al. (1998). AERONET: a Federated Instrument Network and Data Archive for Aerosol Characterization, *Rem. Sens. Envir.*66, 1–16

IPCC, Climate Change (2001). Working Group I, *Contribution to the Intergovernmental Panel on Climate Change. 3rd Assessment Report Climate Change 2001: the PhysicalSciense Basis* (Cambridge Univ., UK, New York); http://www.grida.no/climate/ipcc_tar/wg1/166.html

IPCC, Climate Change (2007). Working Group I, *Contribution to the Intergovernmental Panel on ClimateChange. 4th Assessment Report of Climate Change: ThePhysical Science Basis* (Cambridge Univ., UK, New York), Ch. 2, pp. 130–234

Isaev, A.A. (2001). *Ecological Climatology* (Naucka. Mir, Moscow, 2001)

Jevrejeva, S., Moore, J., Grinsted, A. (2003). Influence of the Arctic Oscillation and El Nino-Southern Oscillation (ENSO) on ice conditions in the Baltic Sea: The wavelet approach, J. Geophys. Res., 108(D21), 4677, doi:10.1029/2003JD003417, 2003

Luts'ko, L.; Makhotkina, E.; Klevantsova, V. (2001). The Development of Actinometric Observations, *Current Studies of the Main Geophysical Observatory: A Jubilee Collection* , Gidrometeoizdat, St. Petersburg, pp. 184–202 [in Russian]

Makhotkina, E.; Plakhina, I.; Lukin, A. (2005). Some Feature of Atmospheric Turbidity Change over the Russian Territory in the Last Quarter of the 20th Century . *Russian Meteorology and Hydrology*, No. 1, pp. 28 – 36, ISSN 0130 – 2906

Makhotkina, E.; Lukin, A. Plakhina, I. (2007). Monitoring of Integral Atmosphere Transparency. *Proc. of the All-Russ. Conf. on Development of Monitoring System of Atmosphere Structure (RSMSA)* (Max Press, Moscow, 2007), p. 104

Makhotkina, E.; Plakhina, I.; Lukin, A. (2010). Changes in Integral and Aerosol Atmospheric Turbidity in Trans-Baikal and Central Siberia. *Russian Meteorology and Hydrology, Vol.35*, No. 1, pp. 34 – 46, ISSN 1068 – 3739

Ohmura, A. (2006). Observed Long-term Variations of Solar Irradiance at the Earth Surface. *Space Sceince Reviews*, Vol.125, No 1- 4, pp.111-128

Plakhina, I.; Makhotkina, E.; Pankratova, N. (2007). Variations of Aerosol Optical Thickness of the Atmosphere in Russia in 1976-2003. *Russian Meteorology and Hydrology*, Vol. 32, No. 2, pp. 85 – 92, ISSN 1068 – 3739

Plakhina, I.; Pankratova, N., Makhotkina, E. (2009). Variations in the Aerosol Optical Depth from the Data Obtained at the Russian Actinometric Network in 1976-2006. *Izvestiya, Atmospheric and Oceanic Physics*, Vol. 45, No. 4, pp. 456 – 466, ISSN 0001 – 4338

Plakhina, I.; Pankratova, N., Makhotkina, E. (2011). Spatial Variations in the Air Turbidity Factor above the European part of Russia under Conditions of Abnormal Summer of 2010. *Izvestiya, Atmospheric and Oceanic Physics*, Vol. 47, No. 6, pp. 708 – 713, ISSN 0001 – 4338

Sitnov, S. (2010). The Results of Satellite Monitoring of the Content of Trace Gases in the Atmospheric and Optical Characteristics of Aerosol over the European Territory of Russia in April–September 2010, in AllRussia Meeting on the Problem of the State of the Air Basin in Moscow and European Part of Russia under the Extreme Weather Conditions in Summer 2010 pp. 26–27, http://www.ifaran.ru/messaging/forum/

Aerosol Characteristics over the Indo-Gangetic Basin: Implications to Regional Climate

A.K. Srivastava, Sagnik Dey and S.N. Tripathi

Additional information is available at the end of the chapter

1. Introduction

The climatic and environmental effects of atmospheric aerosols are the critical issues in global science community because aerosols, derived from variety of natural and man-made (or anthropogenic) emission sources, are well known to affect the air quality, human health and radiation budget [1]. While comparing the third and fourth assessment report of the Intergovernmental Panel on Climate Change (IPCC) as shown in Figure 1, the level of scientific understanding for the role of green house gases (GHGs) in projected temperature changes is higher relative to that of aerosols [2,3]. This is because of inadequate measurements of aerosols, their microphysical and optical properties and poor understanding of their role in the Earth's radiation budget.

Aerosols influence the Earth's climate directly by scattering and absorbing the solar and terrestrial radiations and indirectly by modifying the cloud macro- and micro-physical properties [4]. The direct and indirect effects of atmospheric aerosols are shown in the schematics in Figure 2a and 2b, respectively. Varity of aerosols present in the atmosphere from natural and anthropogenic emission sources can influence our Earth's atmosphere directly by absorbing/scattering the incoming solar radiations (Figure 2a). It can also absorb and re-radiated the outgoing radiations emitted from the Earth. On the other hand, aerosols indirectly affect the climate system by acting as cloud condensation nuclei (CCN) and ice nuclei (IN) and thereby modify the cloud properties and their impacts depending upon the environment like polluted or un-polluted regions (Figure 2b). In a recent study, reported in [5], they have investigated the indirect aerosol effect during the successive contrasting monsoon seasons over Indian subcontinent. However, a different study reported in [6] was carried out to investigate the intensity and the spatial extent of the indirect effect over the Indo-Gangetic Basin (IGB) region.

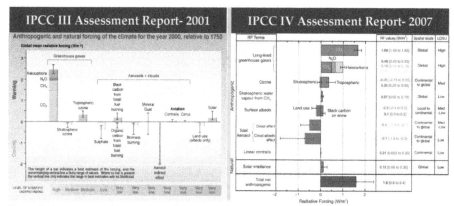

Figure 1. A comparison between III and IV assessment reports of IPCC for 2001 and 2007 (*Adopted from [2,3]*).

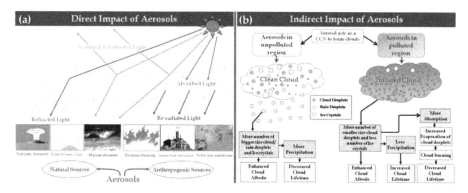

Figure 2. Schematics showing (a) Direct and (b) Indirect impacts of atmospheric aerosols.

The uncertainty in quantifying these impacts have no doubt improved over the years due to assimilation of observations (especially after global observations of aerosols by EOS-Terra started in 1999), but not up to the desired level, particularly at regional scale [3]. The Indian subcontinent is one such region, where heterogeneity in aerosol optical and microphysical properties over a wide range of spatial and temporal scales continues to hinder in improving the estimates of aerosol-induced climate forcing. Thus, it is important to improve aerosol characterization with high spatio-temporal resolutions; particularly over the IGB region, which supports nearly 70% of the country's population and is one of the highly polluted regions in the world. The problem is also critical due to lack of adequate long-term measurements of aerosol properties and large uncertainty in emission factors, leading to poor representation of aerosol distribution by General Circular Model (GCM) [7].

Although, aerosol properties have been measured at many sites in India in continuous and campaign modes (the *in-situ* observations have summarized in [8]) in the last two decades, only few of them have fairly long-term data of aerosol microphysical properties [9-11]. Satellites are proved to be a good tool to understand the broad spatio-temporal characteristics of aerosols and associated effects from global to local scales [12-15]; but they are unable to provide an in-depth view of aerosol properties on a local scale and pose higher uncertainties as compared to the ground-based instruments [16]. In the IGB, aerosols of natural and anthropogenic origins mix with each other during dust loading season [17,18]. As a result, aerosol properties change, leading to even larger uncertainty in satellite retrievals [14,19] as this is not considered in the aerosol retrieval algorithm [20]. National Aeronautics and Space Agency (NASA) has setup ground-based aerosol monitoring network under the Aerosol Robotic Network (AERONET) program [21], in which automatic sun/sky radiometers are deployed at various places around the world. As per India, particularly in the northern part, is concerned, the routine measurements of aerosols under this network were started initially by the deployment of the sun/sky radiometer at Kanpur over the IGB region in year 2001 [22]. At a later stage, it was deployed at other places in the IGB, considering the region as crucial for aerosol measurements where significant aerosol loads of pollution mostly from the combustion of biomass, bio-fuel/fossil fuel emissions and the transported mineral dust have led to one of the largest regional TOA energy losses worldwide (For Example see [11,23-25]). The measurements by individual instruments can particularly be useful for the quantification of the regional impact of aerosols on the radiative energy balance. Provided a parallel individuation of aerosol types is possible, these local studies can also help to reduce the uncertainties on the effect of individual aerosol species, which is necessary because the direct radiative forcing by individual aerosol species is less certain than the total direct radiative forcing by all aerosols [3]. The complex mixture of aerosols over the IGB has been evaluated in the literature over the last decade starting with Indian Ocean Experiment (INDOEX) [26], and the research has been continued with the AERONET data [27 and references therein].

Understanding and quantifying the aerosol effects are important in the IGB region due to several pathways have been hypothesized to explain the possible impacts of aerosols on the regional hydrological cycle. The region is of great research interest due to its unique topography surrounded by the Himalayas to the north, moderate hills to the south, Thar Desert and Arabian Sea in the west, and Bay of Bengal to the east. The IGB region is dominated by the urban/industrial aerosols [28-30], which demonstrate significant seasonal variability based on the complex mixture of these aerosols with the naturally produced aerosols, particularly during the pre-monsoon and monsoon seasons. In addition to the urban-industrial pollution, desert dust is one of the other major natural sources of the aerosols over the Ganga basin [24,27,31-36], transported frequently from the neighboring desert regions, mostly during pre-monsoon periods. Dust storms are often experienced in northern and northwestern part of India, including over different parts of the IGB region during the pre-monsoon season, when dust aerosols are transported by southwesterly summer winds from the western Thar Desert [37,38]. High dust loading over the IGB region during the pre-monsoon period has been established by remote sensing data [39,40]. These dust storms apparently deposited silty materials in the downwind directions, as observed on the quartzite ridges in the Delhi area [41]. The wind also carries heavy metals to the IGB during the summer season [42] along with the dusts, causing severe air pollution and degradation in the visibility. On contrary, the spatial distribution of aerosols (in terms of AOD) during the winter season also revealed high aerosol loading over the IGB and its outflow to the northern Bay of Bengal due to high anthropogenic emission sources, which was observed by satellite [8,12,13,15] and ground-based measurements [22,43-46].

2. Factors affecting aerosol characteristics over IGB

Complex nature of aerosols over the IGB is mainly because the region is very diverse in topography, population distribution, meteorology and emission sources. Figure 3 shows unique topography of the IGB region, surrounded by the variety of aerosol emission sources and thus making it hotspot for aerosol research.

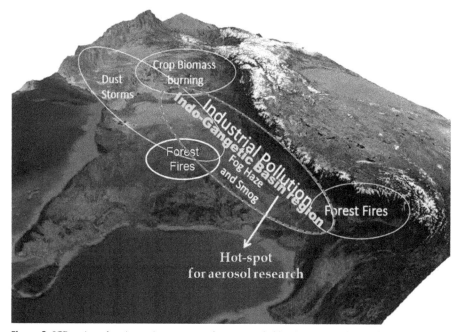

Figure 3. IGB region, showing unique topography surrounded by the variety of aerosol emission sources (Adopted after modification from personal presentation of William K. M. Lau on Aerosol, Monsoon Rainfall Variability and Climate Change).

2.1. Topography

The IGB region, world's most populated river basin having more than 700 million populations, stretches from Pakistan in the west to Bangladesh in the east, encompassing most of the northern part of India. The region is bounded by the Himalayas to the north, and by Vindhyan and Satpura range of mountains in the south. Due to its unique topography, this region can be summarized as a type of region, where, both anthropogenic and natural, aerosols show distinct seasonal characteristics and mixing [13,17,22,25,28,30,47]. General seasonal abundance shows that the winter months are dominated by the fine-mode aerosols, produced by various anthropogenic sources from the IGB region, and pre-monsoon or summer months are dominated by the coarse-mode mineral dust, primarily from the Thar Desert region in the western Rajasthan and its frequent transportation over the IGB region. Further details regarding geography, climate, regional sources and emissions of these aerosols over the IGB as well as over the other Indian region, however, can be found in [15]. This region also provides favorable climate for the agricultural activities due to its fertile soils and abundant water supply from the southwest monsoon and the rivers originating from the Himalayan glaciers such as the Ganges. Consequently, the cultivable land forms a major fraction of the total geographical area in the IGP region (~76%) as compared to the rest of India (~50%) (http://dacnet.nic.in/).

2.2. Synoptic conditions and aerosol characteristics over IGB

Synoptic meteorology (e.g. wind pattern, air temperature and specific humidity) over the IGB region along with its surroundings is shown in Figure 4 for (a) winter and (b) summer seasons for the period of 2007-2008. The European Centre for Medium-Range Weather Forecasts (ECMWF) reanalysis monthly data of weather parameters such as wind, air temperature and specific humidity at 850 hPa pressure level were used to study the synoptic meteorological conditions over the region. In both the figures, winds are shown with arrows pointing towards the wind direction, where length of arrows defines the magnitude of wind speed (in ms^{-1}), line contour represents air temperature (in ^{o}C) and shaded color contour represents specific humidity (in kg kg^{-1}), showing in dark blue color for low and red color for high magnitude of specific humidity. Results reveal that the IGB region during the winter period is relatively drier than during the summer. The persistence of low temperature and the westerlies (with low intensity) can be seen over the region during the winter whereas during the summer, relatively high temperature with intense southwesterly winds was observed to dominate. These winds are found to pass through arid regions of the western India (particularly from the Thar Desert) and bring dry air masses over the region [24,27,48].

The general aerosol characteristics over the entire IGB region are shown in Figure 5 as mean AOD values at 550 nm for (a) winter and (b) summer seasons for the period of 2007-2008 in color codes, obtained from the Moderate Resolution Imaging Spectroradiometer (MODIS). Large spatial heterogeneity in AOD can readily be noticed over the IGB region during both winter and summer periods, which has also been confirmed through various ground-based measurements, discussed in the later section of this chapter. Relatively large magnitude of AOD was observed throughout the IGB region during the summer, which is mainly due to frequent occurrence of dust storms over the Thar Desert region that caused large amount of dust particles to be transported over the station (showing the highest AOD). However, large

AOD during the winter is confined mostly over the eastern part of IGB. During the winter, the IGB region is often enveloped by thick fog and haze [49]. The prevailing winds over the region are westerly to northwesterly with relatively low wind speeds (<5 ms-1) as compared to the summer (as can be seen in Figure 4a) and the eastern parts of the IGB are impacted by a localized area of strong subsidence in winter [8,12,13,50]. These conditions tend to trap the pollution at low altitudes and responsible for the higher AOD along the eastern part of IGB. Results obtained, although, include the impacts of aerosol emissions from various natural and anthropogenic sources and the prevailing meteorology over the region, it also encourages to further investigate the plausible causes and impacts over the radiation budget as well as on weather and climate.

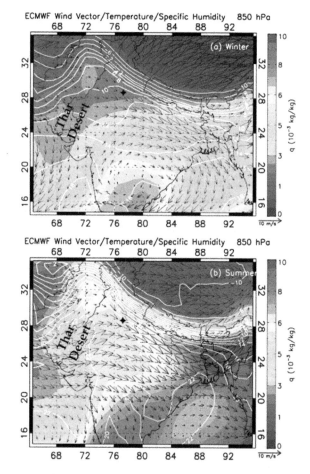

Figure 4. Synoptic meteorological conditions over the entire IGB region derived from ECMWF at 850 hPa pressure level during (a) winter and (b) summer periods (*Adopted from [46]*).

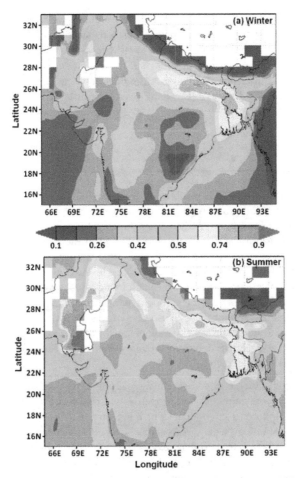

Figure 5. General aerosol characteristics in terms of AOD (550 nm) over the entire IGB region derived from MODIS during (a) winter and (b) summer periods (*Adopted from [46]*).

2.3. Emission sources

The IGB region, apart from being a major source region for aerosols, is bordered by densely populated and industrialized areas on the west and eastern sides from where different aerosol species such as mineral dust, soot, nitrate, sulfate particles and organics are produced and transported to this region and thus making it an aerosol hotspot, as can also be seen in Figure 3. The region itself has both, rural and urban population and various kinds of emission sources such as natural and industrial. In rural areas, bio-fuels burning such as wood, dung cake and crop waste, predominantly contribute to the major aerosols loading [51]. However, in urban areas, aerosol emissions from fossil fuels burning such as coal, petrol and diesel oil dominate [52,53]. Large fluxes of absorbing aerosol emissions (black carbon and inorganic oxidized matter, which is mostly fly ash from coal-based power plants and particles from open burning of crop waste/forest-fires) were reported over the IGB [51]. Apart from the dust emissions from the Thar Desert, predominantly during the pre-monsoon months, the influence of emissions from the forest-fires and open burning of crop waste from the central India were also found over IGB during these months as biomass aerosol contribution [15,27].

3. Measurements

3.1. Ground-based

Aerosol measurements in the Indian sub-continent started as early as the 1960s, when [54] studied Angstrom turbidity from solar radiance measurements. Later, a multi-wavelength radiometer was developed by the Indian Space Research Organization (ISRO) to monitor spectral AOD at Trivandrum in the year 1985 [55] and in the same year, aerosol vertical distribution measurement by ground-based lidar was initiated at Pune [56]. Further, NASA has setup ground-based aerosol monitoring network in India under the Aerosol Robotic Network (AERONET) program [21], in which automatic sun/sky radiometers are deployed at various places, particularly in the northern part of India, including the Himalayan foothills. The routine measurements of aerosols under this network were started initially by the deployment of the automatic sun/sky radiometer at Kanpur over the IGB region in year 2001 [22]. At a later stage, it was deployed at other places in the IGB, considering the region as crucial for aerosol measurements [36].

Using ground-based radiometric measurements, [22] have reported for the first time the seasonal characteristics of aerosol optical properties and the spectral behavior of AODs over Kanpur, an urban-industrial city, situated in the central part of the IGB. They showed pronounced seasonal influence of various aerosol properties, with maximum dust loading during the pre-monsoon season. The increase of pollution has a direct impact on climatic conditions, especially the increase of haze, fog, and cloudy conditions, which decrease the visibility especially during the winter season. On the other hand, *in-situ* aerosol measurements by Central Pollution Control Board (CPCB) have also showed very high annual average concentrations (>210 $\mu g/m^3$, in the critical range compared to the air quality standard in India) of particulate matter of diameter less than 10 μm (PM_{10}) in the atmosphere of the major cities of the Ganga basin like Delhi, Kanpur and Kolkata (http://www.cpcb.nic.in). These high PM_{10} concentrations provide an opportunity for SO_4 formation on the particulate surface, leading to very high concentration of sulfate aerosols in the atmosphere, which is the case observed over the IGB and reported in [29]. Several studies indicate strong seasonal variability in aerosol loading and changes in aerosol properties over the IGB [14,22,25,31,33-36,48,57,58]. In the recent studies, [23] and [24] have demonstrated the distribution of aerosols and associated optical and radiative properties in the IGB region and its further expansion to the foothills of Himalayas during the pre-monsoon period. The pre-monsoon period is of particular interest because this is the key period when locally generated and regionally transported aerosol loading peaks over the IGB region and spread up to the foothills of Himalayas [23,35,59-62], which has been linked to influence the monsoon circulation in India [63,64]. Significant gradient in the magnitude of most of the aerosol characteristics was observed over the IGB, which may be due to the gradual changes in weather parameters and/or emission sources. Such gradient is, ultimately, found to impact the Earth-atmosphere system by negative radiative forcing, thus causing cooling, at the surface, and positive aerosol forcing, thus causing heating in the atmosphere for the study period. Such gradient in heating rate raises several climatic issues, and is needed to be answered on the basis of longer period investigations at several stations to improve the scientific understanding of the regional climate in inter-annual as well as intra-seasonal scale.

The first simultaneous measurements of chemical composition (carbonaceous and inorganic species) and optical properties (absorption coefficient and mass absorption efficiency) of ambient aerosols ($PM_{2.5}$ and PM_{10}) have been recently reported in [58] at an urban site (Kanpur) in the IGB region. The study provides important information on the temporal variability in the abundance of organic matter and mineral dust over the IGB region, which has large implications to the large temporal variability in the atmospheric radiative forcing due to these aerosols. Based on the measured aerosol chemical composition, other studies have been carried out to understand the characteristics of anthropogenic aerosols and their quantification to the total radiative forcing over the IGB region, which are limited only at Kanpur [11] and Delhi [25]. Figure 6 shows seasonal variability of optical properties of composite aerosols estimated over Delhi (a typical urban station at the western part of the IGB near to the Thar Desert) during the winter, summer and post-monsoon seasons; however, the same for anthropogenic aerosols are shown in Figure 7. The anthropogenic components were found to be contributing ~72% to the composite aerosol optical depth ($AOD_{0.5}$ ~0.84) at Delhi. The contribution was found to be more during the winter (~84%) and post-monsoon (~78%) periods and less during the summer (~58%). On the other hand, mean SSA for composite aerosols was found to be ~0.70 (ranging from 0.63 to 0.79). However, SSA for anthropogenic aerosols was found to be slightly less (by ~1%) than that for composite aerosols, and the difference may be due to the mixing of natural dusts with anthropogenic aerosols in the region (for composite aerosols).

The resultant atmospheric forcing due to composite and anthropogenic aerosols at Delhi is shown in Figure 8. The anthropogenic contributions to the composite aerosol were found to be ~93%, 54%, and 88%, respectively during the winter, summer, and post-monsoon seasons (with a mean contribution of ~75%). However, the anthropogenic fraction of ~73% is responsible for the composite aerosol atmospheric heating rate (2.42 ± 0.72 Kday^{-1}) at Delhi. On the other hand at Kanpur, another typical urban station at central part of the IGB region, the heating rate due to anthropogenic aerosols was reported to be ~65% to the heating due to composite aerosols [11]. Relatively higher heating rate at Delhi may be caused by the large contribution of transported mineral dust aerosols due to the proximity of the station to the Thar Desert region as compare to Kanpur and their probable mixing with the other absorbing aerosol species like black carbon (BC) [27].

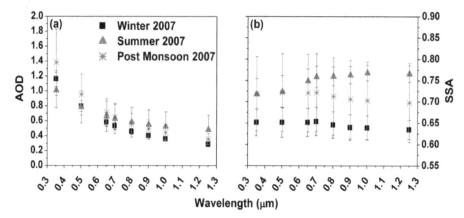

Figure 6. Seasonal mean spectral variation of (a) AOD and (b) SSA for composite aerosols over Delhi (*Adopted from [25]*).

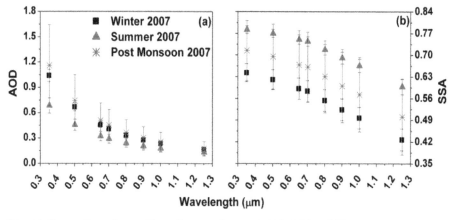

Figure 7. Same as figure 6, except for anthropogenic aerosols (*Adopted from [25]*).

Large atmospheric heating rate of the order of more than 2 Kday^{-1} is quite significant. Moreover, the large surface cooling due to negative forcing at the surface and strong heating due to positive forcing in the atmosphere, particularly for the anthropogenic aerosols, can strongly affect the atmospheric dynamics over the region. The warmer atmosphere close to the surface (due to high atmospheric absorption) and the colder surface during winter and post-monsoon periods over Delhi would create low-level inversions and strengthen the boundary layer stability [11], which restrict the mixing and dispersion of aerosols into the atmosphere. On the other hand, during summer, the observed large heating in the atmosphere, which is probably due to the mixing of anthropogenic aerosols with abundance of natural dusts, may supply excess energy to be trapped in the atmosphere during dry season and can have significant impact on regional climate and monsoon circulation systems [65,66]. Anomalous atmospheric heating due to absorbing aerosols (dust and BC) over the northern part of India during pre-monsoon season has been reported in [63]. A comparative study of aerosol direct radiative forcing was made with the available estimates from the literatures at various regions and given in Table 1. Various regions, characterized by different kinds of aerosol sources and prevailing meteorological conditions, are associated with different values of aerosol forcing and have provided an understanding of the aerosol radiative effect on regional scales, which are significantly different from the global mean radiative effect (indicating slightly cooling of the atmosphere).

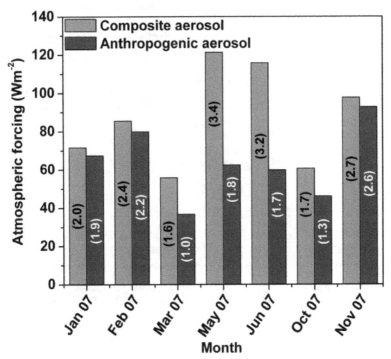

Figure 8. Monthly variation of atmospheric forcing for composite and anthropogenic aerosols over Delhi. The corresponding heating rate values for respective aerosols are given in the parenthesis (*Adopted from [25]*).

| Location | Type of Location | Period | Aerosol DRF (Wm⁻²) at | | References |
			Surface	Atmosphere	
New Delhi	Urban	Annual	-67	71	[57]
Kanpur	Urban	Annual	-32	28	[11]
Ahmedabad	Urban	Annual	-49	44	[131]
New Delhi	Urban	Jan-Nov 2007	-79	87	[25]

Location	Type of Location	Period	Aerosol DRF (Wm^{-2}) at		References
			Surface	Atmosphere	
New Delhi	Urban	Mar-Jun 2006	-77	80	[48]
Kanpur	Urban	Apr-Jun 2009	-26.1 to -29.2	19.5 to 16.1	[24]
Gandhi College	Rural	Apr-Jun 2009	-29.7 to -31.9	20.9 to 16.6	[24]
New Delhi	Urban	Winter (Dec 2004)	-66	67	[132]
Hissar	Urban	Winter (Dec 2004)	-10 (before fog) -20 to -25 (during fog)	15 (before fog) 25 to 40 (during fog)	[128]
Pune	Urban	Nov-Apr 2001 and 2002	-33	33	[133]
Hyderabad	Urban	Jan-May 2003	-33	42	[134]
Bangalore	Urban	Oct-Dec 2001	-23	28	[135]
Central India (multiple stations)	Urban and Rural	Winter (Feb 2004)	-15 to -40	16 to 29	[72]
Nainital	Rural (high-altitude)	Winter (Dec 2004)	-4.2	0.7	[77]
Nainital	Rural (high-altitude)	July 2006-May 2007	-14	14	[62]
Kathmandu	Urban (high-altitude)	Winter 2003	-25	25	[136]
Chennai	Urban	Feb-Mar	-19	13	[137]
Arabian Sea	Polluted Marine	Mar-Apr	-27	15	[70]
Bay of Bengal	Polluted Marine	Mar	-27	23	[138]
Indian Ocean	Polluted Urban	Feb-Mar	-29	19	[139]
Global mean	Natural and Anthropogenic			-0.5±0.4	[3]

Table 1. Aerosol direct radiative forcing (DRF) at different locations in India.

Due to high level of anthropogenic emissions, aerosol distribution in terms of type and loading undergo strong variability associated with the episodic yet strong influence of dust transport and biomass burning during the pre-monsoon period [23]. Dust was found to be one of the major components of aerosol composition (apart from other species) over the region [32], which significantly affects the region during pre-monsoon period due to enhanced surface convection activities [24,25,31,33-36,48,67], and thus essential to quantify its contribution over the region. Aerosol composition was measured with the chemical analysis method over IGB during different periods of time as reported in [44,52,53]. Retrieval of columnar black carbon and organic carbon has been carried out over IGB using the AERONET data [60,68]. Moreover, [7] have integrated AERONET and Cloud-Aerosol

Lidar with Orthogonal Polarization (CALIOP) data into atmospheric GCM to infer aerosol types at two AERONET sites in the IGB. However, in a recent study, [27] have discriminated the major aerosol types over the IGB region during pre-monsoon period using multi-year AERONET measured aerosol products associated with the size of aerosols (mainly fine mode fraction, FMF) and radiation absorptivity (mainly single scattering albedo, SSA). Figure 9a shows density plot of SSA versus FMF at Kanpur (KNP, a typical urban AERONET site over the central IGB region) and Gandhi College (GC, a typical rural AERONET site over the central IGB region) for different aerosol types. High dust enriched aerosols (i.e. polluted dust, PD) were found to contribute more over the central IGB station at Kanpur (~62%) as compared to the eastern IGB station at Gandhi College (~31%) whereas vice-versa was observed for polluted continental (PC) aerosols, which contain high anthropogenic and less dust aerosols. Contributions of carbonaceous particles having high absorbing (mostly black carbon, MBC) and low absorbing (mostly organic carbon, MOC) aerosols were found to be 11% and 10%, respectively at Gandhi College, which was ~46% and 62% higher than the observed contributions at Kanpur; however, very less contribution of non-absorbing (NA) aerosols was observed only at Gandhi College (2%). The mean SSA and FMF based on cluster analysis of daily-averaged data at Kanpur and Gandhi College, associated with the different aerosol categories is also shown in Figure 9b. The horizontal and vertical lines indicate the standard deviations of SSA and FMF from their respective means, indicating the variability of these parameters for different aerosol types. Although similar magnitude of SSA was observed for PD, PC and MBC type aerosols, they are further distinguished based on FMF thresholds following Lee et al. (2010), i.e. FMF<0.4 indicates dominantly coarse-mode and hence is assigned to PD aerosols, FMF>0.6 indicates dominantly fine-mode and hence is assigned to MBC aerosols, and PC aerosols are considered for 0.4≤FMF≤0.6. MOC and NA type aerosols have similar FMF, but higher scattering relative to the other aerosol types.

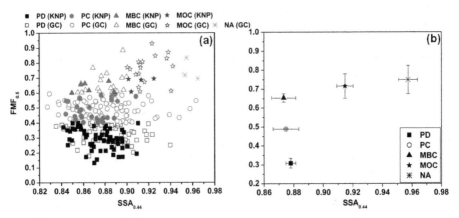

Figure 9. (a) Density plot and corresponding (b) cluster plot of AERONET-derived SSA vs. FMF for two stations over the IGB region (showing different aerosol types) during pre-monsoon period (*Adopted from [27]*).

Spectral information of SSA for each aerosol type was also shown and discussed in [27], which clearly discriminates the dominance of natural dust (SSA increases with increasing wavelength) with anthropogenic aerosols (SSA decreases with increasing wavelength) at Kanpur and Gandhi College over the IGB. As expected, SSA for PD and PC aerosols was found to have spectrally increased, suggesting relative importance of dust. PD has higher spectral trend relative to PC due to larger fraction of dust, which was found to be dominated at Kanpur as compared to Gandhi College. On the contrary to PD and PC type aerosols at Gandhi College, SSA for NA aerosols was found to have spectrally decreased with relatively larger magnitude at all the wavelengths. However, relatively less spectral dependence in SSA was seen for MBC and MOC aerosols, which shows slight decrease in spectral SSA at Gandhi College and opposite at Kanpur.

Further, the absorption aerosol optical depth (AOD$_{abs}$) at different wavelengths (λ) can be obtained, suggested in [69] as

$$\text{AOD}_{abs}\left(\lambda\right) = \left[1- \text{SSA}\left(\lambda\right)\right] \times \text{AOD}\left(\lambda\right) \tag{1}$$

The absorption Ångström exponent (AAE) has been computed as negative of the slope of fitted line of the natural logarithm of AODabs vs. natural logarithm of the respective wavelengths and used to substantiate the inferred aerosol types over IGB, as shown in Figure 10. The magnitude of AAE near to 1.0 (marked by dotted line in Figure 10) represents a theoretical AAE value for black carbon as reported in [69]. AAE values for PD and PC aerosol types were found to be 1.70 and 1.43, respectively at Kanpur and 1.30 and 1.18, respectively at Gandhi College. However, for MBC and MOC type aerosols, AAE values were relatively higher at Kanpur (~20%) than at Gandhi College, where values were found to be closer to the theoretical AAE value for black carbon (i.e. AAE≈1.0), thus indicating the presence of fresh BC at Gandhi College, which can be expected from the potential source of combustion of fossil fuel and biomass burning used for domestic purposes. On the other hand, aged BC or mixed BC can be expected at Kanpur (mostly from biomass burning and urban/industrial sources), which is favorable scenario during the summer periods [17-36]. The estimated AAE values over the IGB thus suggest relative dominance of absorbing type aerosols over the central part of IGB (due to dominant dust mixed with other absorbing aerosols) as compared to the eastern part during pre-monsoon period.

Apart from these continuous measurements, various field campaigns have also been conducted regionally to study and improve the aerosol remote sensing measurements as well as provide data for atmospheric prediction over the past decade. Campaigns conducted in or near India, which used space-based, airborne, and surface-based instrumentations to observe high aerosol loading over the Indian subcontinent and the surrounding Oceanic regions, included the Indian Ocean Experiment (INDOEX) reported in [26], Arabian Sea Monsoon Experiment (ARMEX) reported in [70], Indian Space Research Organization Geosphere Biosphere Programme (ISRO-GBP) Land Campaign reported in [43-45]. The first phase of Land Campaign (LC-I) was conducted during February to March 2004, to understand the spatial distribution of aerosols and trace gases over central/peninsular India. The details of these campaigns and the major findings have been reported in literatures in [71-74]. As a continuation of this experiment, second phase of land campaign (LC-II) was conducted during December 2004, to characterize the regional aerosol properties and trace gases across the entire IGB region including the Himalayan foothills. This phase of the campaign provided a comprehensive database on the optical, microphysical and chemical properties of aerosols over the IGB and the foothills of Himalayas and reported in [44,45,75-80]. All these studies showed the persistence of high aerosol loading (in terms of high AOD) and black carbon mass concentrations over the region.

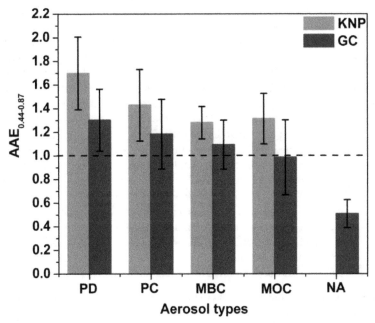

Figure 10. Mean AAE values for each aerosol type at Kanpur and Gandhi College over the IGB (*Adopted from [27]*).

Further, an International TIGERZ experiment was conducted by the NASA AERONET group within the IGB region around the industrial city of Kanpur during the pre-monsoon period [36]. The major objectives of TIGERZ include the spatial and temporal characterization of columnar aerosol optical, microphysical and absorption properties; the identification of aerosol particle types/mixtures; and the validation of remotely sensed aerosol properties from satellites. In a recent past, Ramanathan group from the Scripps Institution of Oceanography, University of California, USA conducted a field measurement (from November 2009–September 2010), called *Project Surya*, in a rural area over the IGB region. Studies were focused on to establish the role of both solid biomass based cooking in traditional stoves and diesel vehicles in contributing to high BC and organic carbon (OC), and solar absorption [81,82]. In continuation to this, [83] have studied the link between local scale aerosol properties and column averaged regional aerosol optical properties and atmospheric radiative forcing.

Apart from ground-based aerosol measurements, vertical distribution of aerosols were carried out for the first time over Kanpur in the IGB region during the winter and summer of the year 2005, reported in [84-85]. Vertical measurements of aerosols up to 1.5 km provided useful information during the winter because aerosols were mostly confined to the boundary layer; however, during summer, aerosols get convected and reached up to the higher altitudes. The Integrated Campaign for Aerosols, gases, and Radiation Budget (ICARB) was initiated to address these issues with multi-institutional, multi-instrumental, multi-platform field campaign, where integrated observations and measurements of aerosols with special emphasis on black carbon, radiation and trace gases along with other complementary measurements on boundary layers and meteorological parameters were made simultaneously [86]. The ICARB was conducted during February-May period of 2006 as an integrated campaign, comprising three segments namely the land, ocean and air, to assess the regional radiative impact of aerosols and trace gases, and to quantify the effect of the long-range transport of aerosols and trace gases over the Indian mainland, Arabian Sea, Bay of Bengal and the tropical Indian Ocean. The details of this campaign and the major findings have been reported in different literatures in [86-90]. ICARB was covered only the eastern part of the IGB (Bhubaneswar) [91] while focusing mostly on the peninsular India and surrounding oceans. Continental Tropical Convergence Zone (CTCZ) campaign, focused on the aerosol distribution in the pre-monsoon and monsoon (June–September) seasons was initiated in the year 2008, and covering the continental part of the more common tropical convergence zone over India, including the IGB [92]. During the campaign, Aircraft and ground-based measurements together were carried out over the IGB and Central part of India to quantify the aerosol indirect effect. The details of the campaign and the major findings have been reported in a recent publication in [61].

Even though all these national and international field experiments and campaigns have greatly improved our understanding on aerosol optical, physical as well as chemical properties and have indeed reduced the uncertainty in regional aerosol direct radiative forcing at various parts of India including the IGB region, they are limited to a certain period or location due to their specific goals. In this perspective, long-term experiments with a high spatio-temporal scale can add advantages of understanding aerosol influences on a longer time scale, thereby helping to infer the signs of anthropogenic impacts. This is where satellite data become very useful and can complement the ground-based and/or *in-situ* measurements.

3.2. Space-borne

Satellite retrievals of aerosol properties over land have only been available in recent years and a few studies have been done using these data over the Indian subcontinent, focusing on the IGB region. In reference [12], they were the first to study the spatial distribution of

AOD over India using Multiangle Imaging Spectroradiometer (MISR) during the winter period from 2001 to 2004, where they were able to explain the enormous pollution observed over the IGB based on meteorology, topography and potential aerosol emission sources. Further, subsequent studies using Moderate Resolution Imaging Spectroradiometer (MODIS) data have confirmed this observation [13,15,93] with additional information on the seasonal variability of AOD and fine mode fraction, to some extent. In continuation to that recently in reference [8], they have presented a detailed analysis of a 9 year (2000–2008) seasonal climatology of size- and shape-segregated aerosol properties over the Indian subcontinent derived from the MISR. The spatial heterogeneity of the aerosol parameters are shown in Figure 11 for each season.

The spatial distribution of AOD in the winter season reveals high AOD over the IGB and its outflow to the northern Bay of Bengal because of high anthropogenic emission sources, as previously observed from satellite [12,13,15,93] and ground-based [22,43,45] measurements. It is well known that the IGB region is often enveloped by thick fog/haze during this period, which is typically associated with high aerosol loading over the region [43,49,94-96]. AOD averaged over eight winter seasons is highest (>0.4) over the eastern part of IGB, which was referred to as the 'Bihar pollution pool' in [12]. As pointed out in [12], this observation has strong implications for the large population residing in this area and thus calls for further work. In [97], they have used CO retrieved from MOPITT (version 3 data) and found a corresponding pool of high CO mixing ratios at 850 hPa level in the same area in winter. In continuation of this, in [50], they have further demonstrated the extensive pollution along the eastern parts of the IGB during winter months using the improved version 4 CO data from MOPITT and the new version 3 height resolved aerosol data from CALIPSO as well as the tropospheric column ozone from two different data products. Both the CO and aerosol data from this study confirm the trapping of pollution at low altitudes by subsidence. Aerosols across the IGB was found to be transported from west to east by northwesterly winds, encounter a narrowing valley floor and are trapped efficiently within the atmospheric column in the eastern part of the IGB by subsiding air [8]. Relatively low AE (<0.8) in the eastern IGB than the other parts, suggests high concentration of coarse dust particles emitted possibly by rural activities (e.g., agriculture, etc.) from the densely populated rural population.

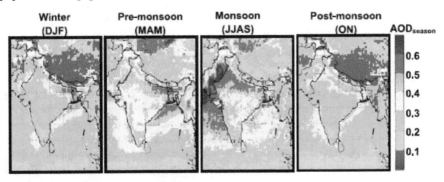

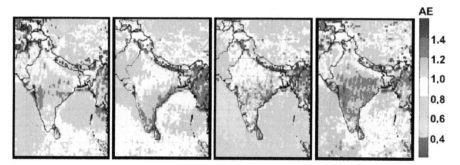

Figure 11. Spatial distribution of climatological mean mid-visible AOD (first panel) and AE (second panel) during the winter, pre-monsoon, monsoon and post-monsoon seasons over the Indian subcontinent (including IGB region) for the period from March 2000 to November 2008 (*Adopted from [8]*).

In the pre-monsoon season, aerosol spectral optical properties change significantly from the preceding winter season because of enhancement in dust loading, particularly over the IGB region [15,31,24]. Large emission of small particles from open biomass burning compensates the relative influence of dust on spectral AOD in the eastern part of the IGB [27] as indicated by an increase of AE during pre-monsoon season compared to the preceding winter season (Figure 11). This also leads to an overall increase in AOD in this region compared to the winter season. Thus, winter to pre-monsoon changes in aerosol properties are not just dominated by an increase in dust, as previously thought, but also by an increase in anthropogenic components, particularly in the regions where biomass combustion is in the common practice during this period. Changing atmospheric aerosol properties caused by anthropogenic activities carries serious implications for climate change and human health [98].

The anthropogenic emissions, particularly BC and sulfate aerosols are present throughout the year in northern India over the IGB [99,100]. Such aerosols form thick layers of haze in winter, termed as Atmospheric Brown Clouds (ABC), which block the solar radiation reaching to the surface [101]. In [102], they have reported in their study over India that the AODs derived from TOMS data from 1979 to 2000 increased by ~11% per decade during the winter with large values over the IGB region, which consequently affect the surface reaching solar radiation, known as "solar dimming" [103-105]. The average solar dimming observed over India is about -0.86 W m^{-2} yr^{-1}, while during winter, pre-monsoon and monsoon seasons the same was observed to be about -0.94, -1.04 and -0.74 Wm^{-2}, respectively [104]. The significant reduction in ground-reaching solar radiation can directly be correlated with the increased aerosol loading in the atmosphere due to enhancement in industrialization, vehicular pollution, biomass burning and dust storm activities over the region [106,107]. Apart from solar dimming effect due to variety of aerosols in terms of haze/fog conditions, our understanding about the role of secondary organic aerosols (particulate organic matters produced by gas-to-particle conversion process), particularly in climate change and its connection to health effects is very limited by numerous uncertainties. In a recent study in [108], they have observed an enhanced production of secondary organic aerosols over

Kanpur in the IGB during the foggy periods of winter, which was hypothesized that the aqueous phase chemistry in fog drops is responsible for increased production of secondary organic aerosols.

During the monsoon, stronger westerly winds were found to be transport greater components of dust from the Arabian Peninsula to the Indian subcontinent [8]. In general, the spatial distribution of AOD (Figure 11) in this season is largely influenced by monsoon precipitation. Suppressed precipitation in the monsoon break phase allows for a rapid buildup of aerosols in the high anthropogenic source regions (e.g. IGB), while particles are being washed out by the precipitation in the active monsoon phase. This also leads to very high intra-seasonal and inter-annual variability in aerosol characteristics. Aerosol regional mean climatology in the post-monsoon season is very similar to that for the winter season (Figure 11), but the spatial distribution differs in several regions. For example, the wintertime high AOD zone in the IGB shows a larger spread and higher inter-annual variability across the basin in this season, owing to a stronger peak in crop waste burning in the western part of IGB than the eastern part [109] and weaker subsidence in the eastern part of IGB compared to the winter season. As a result, IGB is the region with highest aerosol absorption and thus occurred large discrepancy in MISR and MODIS derived AODs [20].

4. Coupling of IGB aerosols to the Himalayan region and their possible impacts

Due to combined effects of IGB topography and the Himalayan orography, aerosols over the IGB region are lifted up quite often and found to be extended up to the Himalayan foothills and also to the other high-altitude regions [23,61,110-113]. Absorbing aerosols in the elevated regions heat the mid-troposphere by absorbing solar radiation, and produce an atmospheric dynamical feedback called as elevated heat pump (EHP) effect. Consequently, this can lead to an increase in the summer monsoon rainfall over India [63] and enhancement in the rate of snow melting in the Himalayan regions [64], which is one of the potential themes for global scientific community and need to be addressed to improve scientific understanding of the regional climate on inter-annual as well as intra-seasonal scales. In particular, the main emphasis of the IGB region coupled with the Himalayan foothills is due to the highest AOD values in this region among the South Asia regions, which are persistent throughout the winter and spring seasons [114].

In a recent study in [115], they have shown a possible influence of desert dust aerosols originated and transported from the Thar Desert region to the high-altitude station at Manora Peak, Nainital in the central Himalayas (Figure 12). The high values of aerosol index (AI) derived from the Ozone Monitoring Instrument (OMI) attest to the presence of absorbing aerosol particles over the region; however, air mass back-trajectory analysis over the station shows different pathways for the transport of air masses from the source region to the experimental site over different time periods (Figure 12). In this study, [115] observed a thick aerosol layer at ~1500 m altitude (Figure 13), above the station level, which was substantiated by the air mass back-trajectory analysis (Figure 14).

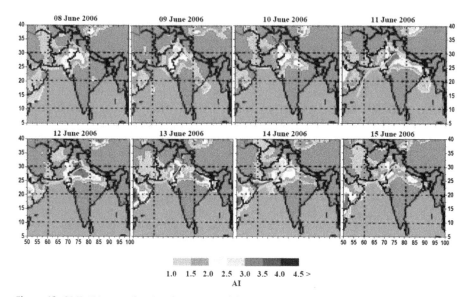

$$1.0 \quad 1.5 \quad 2.0 \quad 2.5 \quad 3.0 \quad 3.5 \quad 4.0 \quad 4.5>$$
AI

Figure 12. OMI AI images showing the source and the progressive movement of the dust aerosols occurred in June 2006. Five-day air mass back-trajectories at Manora Peak for different time periods are superimposed on respective days AI images (*Adopted from [115]*).

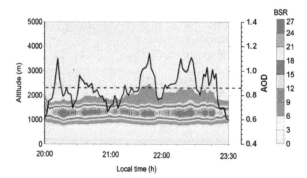

Figure 13. Altitude versus time variations in BSR on a dust day (12 June 2006, with color contour) along with the variations in lidar-derived AOD at 532 nm (with solid line, average AOD = 0.83 ± 0.12) (*Adopted from [115]*).

Apart from dust transport from the Desert regions, recent study in [116], they have also demonstrated significant impact of north Indian biomass burning on aerosols and trace gases and the resultant radiation budget over the central Himalayas during the spring period through air mass back-trajectory analysis coupled with fire counts (Figure 15). The same has also been reported in [117] to be one of the major sources of BC over the same station in the central Himalayas, which was observed to be much lower (in terms of magnitude) as compared to the urban location in the IGB, but was found to have significant contribution to the total aerosol optical depth (~17%) and the resultant atmospheric forcing (~70%) at Manora Peak [62]. Based on BC measured at two different wavelengths at ultra-violet (370nm) and near-infrared (880nm), [117] have distinguished the potential sources of BC at Delhi (one of the densely populated and industrialized urban megacities in Asia and typically represents the plains of Ganga basin) and Manora Peak (one of the high-altitude and sparsely inhabited clean site in the Indian Himalayan foothills situated in the central Himalayas). Based on the analysis, [117] have found the major contribution of BC at Manora Peak is from biomass burning while fossil fuel is found to be the dominating contributor at Delhi.

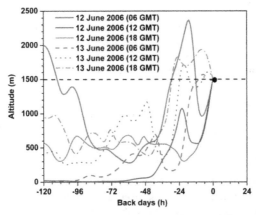

Figure 14. Temporal evolution of air masses at 1500 m altitude for three different time intervals on 12 and 13 June 2006 (*Adopted from [115]*).

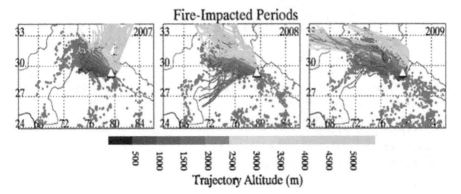

Figure 15. Three day back-air trajectories arriving at Manora Peak during the fire-impacted periods in 2007, 2008 and 2009 (triangle represents the observation site) (*Adopted from [116]*).

5. Summary and future directions

The study over IGB region revealed different aerosol characteristics over the region from western to central and to the eastern parts, which show significant gradient in magnitude of most of the aerosol characteristics. Such gradient can be explained due to the gradual changes in weather parameters and/or emission sources apart from geographical heterogeneity. Such gradient is, ultimately, found to have impact on the Earth-atmosphere system by negative radiative forcing, thus causing cooling, at the surface, and positive aerosol forcing, thus causing heating in the atmosphere. Such gradient in heating rate raises several climatic issues, and is needed to be answered on the basis of longer period investigations at several stations to improve the scientific understanding of the regional climate in inter-annual as well as intra-seasonal scale.

Due to large uncertainty in satellite derived aerosol products over the IGB during pre-monsoon dust periods, long-term ground-based measurements during different seasons can indeed provide useful information of the characteristics of aerosol types over the region on seasonal and inter-annual basis, which are meager and crucial for the regional climate models. Further, the mixing of natural dust with anthropogenically produced aerosol particles, has been hypothesized in [17] over the IGB region, mostly during the pre-monsoon period and corroborated with the AERONET data [36], suggested the complication of the satellite retrieval of aerosol characteristics and quantifying the climatic effects [118]. Hence, it is also one of the important research areas in understanding aerosol characteristics over the IGB region to make realistic assessments of aerosol-hydro-climate interplay.

The issue of black carbon or soot particles and its relationship with climate change has gained enormous scientific and popular interest over the last few years. The knowledge and understanding on aspects such as vertical distribution and mixing of black carbon with other aerosols, effects of cloud cover and monsoon still remains uncertain and incomplete. Few studies have shown that when sulphate or organics is coated over black carbon aerosols, its absorption effects are enhanced by 50% [119]. In case of black carbon mixed with large dust particles, absorption of the composite dust-black carbon aerosol system is enhanced by a factor of two to three compared to sum of black carbon and dust absorption [120]. However, we have no information on the state of mixing of black carbon. The proper assessment of mixing and/or coating of various aerosol species and their impacts on various aerosol characteristics have not been well quantified [121], which makes the investigation a real challenge [122]. IGB, being in proximity to the Thar Desert region, is found to be affected predominantly by the enhanced dust aerosols, mostly during the pre-monsoon period. As a result, the probability of this interaction (i.e. mixing) was suggested to be more over the region during this period [17,36] and is one of the future perspectives. To better understand these crucial issues, National Carbonaceous Aerosol Program (NCAP) was recently launched in India, focusing on the measurement of black carbon; their role in atmospheric stability and the consequent effect on cloud formation, monsoon and retreating of Himalayan glacier [123].

Based on recent observations using aircraft [61] and satellite measurements ([34,46], it has been reported that during pre-monsoon season, IGB region is characterized by the elevated aerosol layers extended up to the altitude from about 3 to 5 km. When the amount of absorbing aerosols such as black carbon and dust, are significant in the atmosphere, the aerosol optical depth and chemical composition are not the only determinants of aerosol radiative effects, but the altitude of the aerosol layer and its altitude relative to clouds (if present) are also essential. Thus, it is also essential to gather information on vertical distribution of aerosols over this region.

Further, fog over the IGB region is observed to be a common feature, occurs mostly during the winter period. The number of foggy days has been increasing in recent years as compared to earlier decades [124], with strong increasing trends of anthropogenic pollution in the IG plains [125]. Fog formation usually begins in the latter half of December and continues till the end of January, thus blanketing some regions for more than a month [126]. The low topography of the IGB, adjacent to the Himalayan range, favors formation of fog and provides high concentration of air pollutants in the plains which serve as additional CCN for nucleation. Fog affects day to day lives of millions of people living in this region, resulting in poor visibility down to less than 100 meters causing frequent flight and train delays and even a significant number of deaths from vehicular accidents in many severe events [127]. Though few studies were done focusing on fog-induced aerosol characteristics over the IGB region and their impacts to the aerosol radiative forcing [49,128], detailed studies of aerosol composition and inter-annual variation of aerosols are required to better understand the interaction of winter haze with the formation of fog over the IGB.

Apart from the measurements for various aerosol characteristics through different ground-based and space-born instrumentations, a 1-D aerosol optical model named as optical properties of aerosols and clouds (OPAC) has been developed by [129], estimating crucial optical properties of aerosols such as AOD and SSA, under the assumption of spherical aerosol particles and external mixing. In [130], they have shown that the optical depth and SSA of aerosol particles have strongest sensitivity on the direct radiative forcing, and these optical properties have found to be large deviation with shape and composition [18]. Further, with the model studies reported in [47,118], the particle composition (i.e. mixing state) and shape (i.e. morphology) attributes to more cooling at both top of the atmosphere and surface, and the combined effect is ~6% more warming than the spherical particles. The significance of consideration of particle shape is more in the regions where black carbon mixes with pure mineral dust, which are the most probable case over the IGB in northern India, because enhancement in the atmospheric warming will be under-estimated if particle morphology is not considered [47]. Thus, there is an urgent need for modeling studies over the IGB region to examine quantitatively the influence of particle morphology along with their mixing states on optical and radiative characteristics of aerosols with their size distribution.

Author details

A.K. Srivastava*
Indian Institute of Tropical Meteorology (Branch), Prof. Ramnath Vij Marg, New Delhi, India

* Corresponding Author

Sagnik Dey
Centre for Atmospheric Sciences, Indian Institute of Technology Delhi, New Delhi, India

S.N. Tripathi
Department of Civil Engineering and Centre for Environmental Science and Engineering, Indian Institute of Technology Kanpur, Kanpur, India

Acknowledgement

Authors are thankful to the various Journals for allowing the use of published materials for this chapter. Figures used in this chapter are considered after taking permission from the respective Journals. SNT acknowledges the support under the program, Changing Water Cycle funded jointly by the Ministry of Earth Sciences, India and Natural Environment Research Council, UK. AKS thanks to Prof. B. N. Goswami, Director, IITM, Pune and Dr. P. C. S. Devara for their encouragement and support.

6. References

[1] Pöschl U (2005) Atmospheric aerosols: composition, transformation, climate and health effects. Angew. Chem. Int. Ed. 44: 7520-7540.

[2] Intergovernmental Panel on Climate Change (IPCC) (2001) Climate Change 2001: The Scientific Basis, edited by J. T. Houghton et al., Cambridge Univ. Press, New York, 881 p.

[3] Intergovernmental Panel on Climate Change (IPCC) (2007) Climate change 2007: The physical science basis: Contribution of Working Group I to the Fourth Assessment Report of the Intergovernmental Panel on Climate Change, Chapter 2, Cambridge Univ. Press, New York, 129 p.

[4] Schwartz SE, et al. (1995) Group Report: Connections between aerosol properties and forcing of climate, John Wiley, Hoboken, N. J, 251–280 p.

[5] Panicker AS, Pandithurai G, Dipu S (2010) Aerosol indirect effect during successive contrasting monsoon seasons over Indian subcontinent using MODIS data. Atmos. Environ. 44: 1937-1943.

[6] Tripathi SN, Pattnaik A, Dey S (2007a) Aerosol indirect effect over Indo-Gangetic plain. Atmos. Environ. 41: 7037–7047.

[7] Ganguly D, Ginoux P, Ramaswamy V, Dubovik O, Welton J, Reid EA, Holben BN (2009) Inferring the composition and concentration of aerosols by combining AERONET and MPLNET data: Comparison with other measurements and utilization to evaluate GCM output. J. Geophys. Res. 114: D16203, doi:10.1029/2009JD011895.

[8] Dey S and Di Girolamo L (2010) A climatology of aerosol optical and microphysical properties over the Indian subcontinent from 9 years (2000–2008) of Multiangle Imaging Spectroradiometer (MISR) data. J. Geophys. Res. 115: D15204, doi:10.1029/2009JD013395.

[9] Moorthy KK, Babu SS, Satheesh SK (2007) Temporal heterogeneity in aerosol characteristics and the resulting radiative impact at a tropical coastal station—Part 1: Microphysical and optical properties. Ann. Geophys. 25: 2293–2308.

[10] Devara PCS, Raj PE, Dani KK, Pandithurai G, Kalapureddy MCR, Sonbawne SM, Rao YJ, Saha SK (2008) Mobile lidar profiling of tropical aerosols and clouds. J. Atmos. Oceanic Technol. 25: 1288–1295, doi:10.1175/2007JTECHA995.1.

[11] Dey S and Tripathi SN (2008) Aerosol direct radiative effects over Kanpur in the Indo-Gangetic basin, northern India: Long-term (2001–2005) observations and implications to regional climate. J. Geophys. Res. 113: D04212, doi:10.1029/2007JD009029.

[12] Di Girolamo L, et al. (2004) Analysis of Multi-angle Imaging Spectro-Radiometer (MISR) aerosol optical depths over greater India during winter 2001-2004. Geophys. Res. Lett. 31: L23115, doi:10.1029/2004GL021273.

[13] Jethva H, Satheesh SK, Srinivasan J (2005) Seasonal variability of aerosols over the Indo-Gangetic basin. J. Geophys. Res. 110: D21204, doi:10.1029/2005JD005938.

[14] Prasad AK and Singh RP (2007) Changes in aerosol parameters during major dust storm events (2001–2005) over the Indo-Gangetic Plains using AERONET and MODIS data. J. Geophys. Res. 112: D09208, doi:10.1029/2006JD007778.

[15] Ramachandran S, Cherian R (2008) Regional and seasonal variations in aerosol optical characteristics and their frequency distributions over India during 2001-2005. J. Geophys. Res. 113: D08207, doi:10.1029/2007JD008560.

[16] El-Metwally M, Alfaro SC, Abdel Wahab MM, Favez O, Mohamed Z, Chatenet B (2011) Aerosol properties and associated radiative effects over Cairo (Egypt). Atmos. Res. 99: 263–276.

[17] Dey S, Tripathi SN, Mishra SK (2008) Probable mixing state of aerosols in the Indo-Gangetic Basin, northern India. Geophys. Res. Lett. 35: L03808, doi:10.1029/2007GL032622.

[18] Mishra SK, Tripathi SN (2008) Modeling optical properties of mineral dust over the Indian Desert. J. Geophys. Res. 113: D23201, doi:10.1029/2008JD010048.

[19] Tripathi SN, Dey S, Chandel A, Srivastva S, Singh RP, Holben B (2005a) Comparison of MODIS and AERONET derived aerosol optical depth over the Ganga basin, India. Ann. Geophys. 23: 1093-1101.

[20] Kahn RA, et al. (2009) MISR aerosol product attributes and statistical comparisons with MODIS. IEEE Transactions on Geoscience and Remote Sensing 47(12): 4095-4114.

[21] Holben BN, et al. (1998) AERONET–A federated instrument network and data archive for aerosol characterization. Remote Sens. Environ. 66: 1–16, doi:10.1016/S0034-4257(98)00031-5.

[22] Singh RP, Dey S, Tripathi SN, Tare V, Holben B (2004) Variability of aerosol parameters over Kanpur, northern India. J. Geophys. Res. 109: D23206, doi:10.1029/2004JD004966.

[23] Gautam R, Hsu NC, Tsay SC, Lau KM, Holben B, Bell S, Smirnov A, Li C, Hansell R, Ji Q, Payra S, Aryal D, Kayastha R, Kim KM (2011) Accumulation of aerosols over the Indo-Gangetic plains and southern slopes of the Himalayas: distribution, properties and radiative effects during the 2009 pre-monsoon season. Atmos. Chem. Phys. 11: 12841–12863.

[24] Srivastava AK, Tiwari S, Devara PCS, Bisht DS, Srivastava MK, Tripathi SN, Goloub P, Holben BN (2011a) Pre-monsoon aerosol characteristics over the Indo-Gangetic Basin: Implications to climatic impact. Ann. Geophys. 29: 789–804.

[25] Srivastava AK, Singh S, Tiwari S, Bisht DS (2012a) Contribution of anthropogenic aerosols in direct radiative forcing and atmospheric heating rate over Delhi in the Indo-Gangatic Basin. Environ. Sci. Pollut. Res. 19: 1144-1158, doi: 10.1007/s11356-011-0633-y.

[26] Ramanathan V, et al. (2001) Indian Ocean Experiment: An integrated analysis of the climate forcing and effects of the great Indo-Asian haze. J. Geophys. Res. 106(D22): 28371–28398, doi:10.1029/2001JD900133.

[27] Srivastava AK, Tripathi SN, Dey S, Kanawade VP, Tiwari S (2012b) Inferring aerosol types over the Indo-Gangetic Basin from ground based sunphotometer measurements. Atmos. Res. 109-110: 64–75.

[28] Guttikunda SK, Carmichael GR, Calori G, Eck C, Woo JH (2003) The contribution of megacities to regional sulfur pollution in Asia. Atmos. Environ. 37: 11–22.

[29] Sharma M, Kiran YNVM, Shandilya KK (2003) Investigations into formation of atmospheric sulfate under high PM10 concentration. Atmos. Environ. 37: 2005–2013.

[30] Monkkonen P, et al. (2004) Relationship and variations of aerosol number and PM10 mass concentrations in a highly polluted urban environment –New Delhi, India. Atmos. Environ. 38: 425– 433.

[31] Dey S, Tripathi SN, Singh RP, Holben B (2004) Influence of dust storm on the aerosol parameters over the Indo-Gangetic basin. J. Geophys. Res. 109: D20211, doi:10.1029/2004JD004924.

[32] Chinnam N, Dey S, Tripathi SN, Sharma M (2006) Dust events in Kanpur, northern India: chemical evidence for source and implications to radiative forcing. Geophys. Res. Lett. 33: L08803, doi:10.1029/2005GL025278.

[33] Gautam R, Liu Z, Singh RP, Hsu NC (2009) Two contrasting dust-dominant periods over India observed from MODIS and CALIPSO data. Geophys. Res. Lett. 36: L06813, doi:10.1029/2008GL036967.

[34] Gautam R, Hsu NC, Lau KM (2010) Pre-monsoon aerosol characterization and radiative effects over the Indo-Gangetic Plains: Implications for regional climate warming. J. Geophys. Res. 115: D17208, doi:10.1029/2010JD013819.

[35] Eck T, et al (2010) Climatological aspects of the optical properties of fine/coarse mode aerosol mixtures. J. Geophy. Res. 115: D19205, doi:10.1029/2010JD014002.

[36] Giles DM, et al. (2011) Aerosol properties over the Indo-Gangetic Plain: A mesoscale perspective from the TIGERZ experiment. J. Geophys. Res. 116: D18203, doi:10.1029/2011JD015809.

[37] Middleton NJ (1986) A geography of dust storms in southwest Asia. Int. J. Climatol. 6: 183–196.

[38] Sikka DR (1997) Desert climate and its dynamics. Curr. Sci. 72(1): 35–46.

[39] Prospero JM, Ginoux P, Torres O, Nicholson SE, Gill TE (2002) Environmental characterization of global sources of atmospheric soil dust identified with the Nimbus 7

Total ozone Mapping Spectrometer (TOMS) absorbing aerosol product. Rev. Geophys. 40(1): 1002, doi:10.1029/2000RG000095.

[40] Washington R, Todd M, Middleton NJ, Goudie AS (2003) Dust storm source areas determined by the Total Ozone Monitoring Spectrometer and surface observations. Ann. Assoc. Am. Geogr. 93: 297–313.

[41] Tripathi JK, Rajamani V (1999) Geochemistry of the loessic sediments on Delhi ridge, eastern Thar Desert, Rajasthan: Implications for exogenic processes. Chem. Geol. 155: 265–278.

[42] Yadav S, Rajamani V (2003) Aerosols of NW India—A potential Cu source. Curr. Sci. 84(3): 278–280.

[43] Tripathi SN, et al. (2006) Measurements of atmospheric parameters during Indian Space Research Organization Geosphere Biosphere Programme Land Campaign II at a typical location in the Ganga basin: 1. Physical and optical properties. J. Geophys. Res. 111: D23209, doi:10.1029/2006JD007278.

[44] Tare V, et al. (2006) Measurements of atmospheric parameters during Indian Space Research Organization Geosphere Biosphere Program Land Campaign II at a typical location in the Ganga Basin: 2. Chemical properties. J. Geophys. Res. 111: D23210, doi:10.1029/2006JD007279.

[45] Nair VS, et al. (2007) Wintertime aerosol characteristics over the Indo-Gangetic Plain (IGP): Impacts of local boundary layer processes and long-range transport. J. Geophys. Res. 112: D13205, doi:10.1029/2006JD008099.

[46] Srivastava AK, Singh S, Tiwari S, Kanawade VP, Bisht DS (2012c) Variation between near-surface and columnar aerosol characteristics during winter and summer at Delhi in the Indo-Gangatic Basin. Journal of Atmospheric and Solar-Terrestrial Physics 77: 57–66.

[47] Mishra SK, Dey S, Tripathi SN (2008) Implications of particle composition and shape to dust radiative effect: A case study from the Great Indian Desert. Geophys. Res. Lett. 35: L23814, doi:10.1029/2008GL036058.

[48] Pandithurai G, Dipu S, Dani KK, Tiwari S, Bisht DS, Devara PCS, Pinker RT (2008) Aerosol radiative forcing during dust events over New Delhi, India. J. Geophys. Res. 113: D13209, doi:10.1029/2008JD009804.

[49] Gautam R, Hsu NC, Kafatos M, Tsay SC (2007) Influences of winter haze on fog/low cloud over the Indo-Gangetic plains. J. Geophys. Res. 112: D05207, doi:10.1029/2005JD007036.

[50] Kar J, Deeter MN, Fishman J, Liu Z, Omar A, Creilson JK, Trepte CR, Vaughan MA, Winker DM (2010) Wintertime pollution over the Eastern Indo-Gangetic Plains as observed from MOPITT, CALIPSO and tropospheric ozone residual data. Atmos. Chem. Phys. 10: 12273–12283.

[51] Habib G, Venkataraman C, Chiapello I, Ramachandran S, Boucher O, Reddy MS (2006) Seasonal and interannual variability in absorbing aerosols over India derived TOMS: Relationship to regional meteorology and emissions. Atmos. Environ. 40: 1909–1921.

[52] Tiwari S, Srivastava AK, Bisht DS, Bano T, Singh S, Behura S, Srivastava MK, Chate DM, Padmanabhamurty B (2009) Black carbon and chemical characteristics of PM_{10} and $PM_{2.5}$ at an urban site of North India. J. Atmos. Chem. 62(3): 193–209.

[53] Ram K, Sarin MM (2010) Spatio-temporal variability in atmospheric abundances of EC, OC and WSOC over Northern India. J. Aerosol Sci. 41: 88–98.

[54] Mani A, Chacko O, Hariharan S (1969) A study of Ångström turbidity parameters from solar radiation measurements in India. Tellus 21: 829–843, doi:10.1111/j.2153-3490.1969.tb00489.x.

[55] Moorthy KK, et al. (1999) Aerosol climatology over India: ISRO GBP MWR network and database. Rep. ISRO GBP SR-03-99, Indian Space Res. Org., Bangalore, India.

[56] Devara PCS, Maheskumar RS, Raj PE, Pandithurai G, Dani KK (2002) Recent trends in aerosol climatology and air pollution as inferred from multi-year lidar observations over a tropical urban station. Int. J. Climatol. 22: 435–449, doi:10.1002/joc.745.

[57] Singh S, Soni K, Bano T, Tanwar RS, Nath S, Arya BC (2010) Clear-sky direct aerosol radiative forcing variations over mega-city Delhi. Ann. Geophys. 28: 1157–1166.

[58] Ram K, Sarin MM, Tripathi SN (2012) Temporal trends in atmospheric $PM_{2.5}$, PM_{10}, EC, OC, WSOC and optical properties: Impact of biomass burning emissions in the Indo-Gangetic Plain. Environ. Sci. and Tech. 46: 686-695.

[59] Venkataraman C, Habib G, Eiguren-Fernandez A, Miguel AH, Friedlander SK (2005) Residential biofuels in South Asia: carbonaceous aerosol emissions and climate. Science 307: 1454–1456.

[60] Arola A, Schuster G, Myhre G, Kazadzis S, Dey S, Tripathi SN (2011) Inferring absorbing organic carbon content from AERONET data. Atmos. Chem. Phys. 11: 215–225.

[61] Devi JJ, Tripathi SN, Gupta T, Singh BN, Gopalakrishnan V, Dey S (2011) Observation-based 3-D view of aerosol radiative properties over Indian Continental Tropical Convergence Zone: implications to regional climate. Tellus 63B: 971-989.

[62] Srivastava AK, Ram K, Pant P, Hegde P, Joshi H (2012d) Black carbon aerosols over central Himalayas: implications to climate forcing. Environ. Res. Lett. 7: 014002, doi:10.1088/1748-9326/7/1/014002.

[63] Lau KM, Kim MK, Kim KM (2006) Asian summer monsoon anomalies induced by aerosol direct forcing: The role of the Tibetan Plateau. Clim. Dyn. 26(7-8): 855-864, doi:10.1007/s00382-006-0114-z.

[64] Lau KM, Kim MK, Kim KM, Lee WS (2010) Enhanced surface warming and accelerated snow melt in the Himalayas and Tibetan Plateau induced by absorbing aerosols. Environ. Res. Lett. 5: 1-10.

[65] Pilewskie P (2007) Climate change: Aerosols heat up. Nature 448: 541-542, doi:10.1038/448541a.

[66] Ramanathan V, Ramana MV, Roberts G, Kim D, Corrigan C, Chung C, Winker D (2007) Warming trends in Asia amplified by brown cloud solar absorption. Nature 448: 575-578.

[67] Srivastava AK, Tiwari S, Bisht DS, Devara PCS, Goloub P, Li Z, Srivastava MK (2011c) Aerosol characteristics during the coolest June month over New Delhi, northern India. Int. J. Remote Sens. 32(23): 8463–8483.

[68] Dey S, Tripathi SN, Singh RP, Holben B (2006) Retrieval of black carbon and specific absorption over Kanpur city, Northern India during 2001-2003 using AERONET data. Atmos. Env. 40(3): 445-456.

[69] Russell PB, Bergstrom RW, Shinozuka Y, Clarke AD, De-Carlo PF, Jimenez JL, Livingston JM, Redemann J, Dubovik O, Strawa A (2010) Absorption Angstrom Exponent in AERONET and related data as an indicator of aerosol composition. Atmos. Chem. Phys. 10: 1155–1169, doi:10.5194/acp-10-1155-2010.

[70] Moorthy KK, Babu SS, Satheesh SK (2005a) Aerosol characteristics and radiative impacts over the Arabian Sea during the inter-monsoon season: Results from ARMEX Field campaign. J. Atmos. Sci. 62: 192–206.

[71] Ganguly D, Jayaraman A, Gadhavi H, Rajesh TA (2005a) Features in wavelength dependence of aerosol absorption observed over central India. Geophys. Res. Lett. 32: L13821, doi:10.1029/2005GL023023.

[72] Ganguly D, Gadhavi H, Jayaraman A, Rajesh TA, Mishra A (2005b) Single scattering albedo of aerosols over the central India: implications for the regional aerosol radiative forcing. Geophys. Res. Lett. 32: L18803, doi:10.1029/2005GL023903.

[73] Moorthy KK, et al. (2005b) Wintertime spatial characteristics of boundary layer aerosols over peninsular India. J. Geophys. Res. 110: D08207, doi:10.1029/2004JD005520.

[74] Singh S, Singh B, Gera BS, Srivastava MK, Dutta HN, Garg SC, Singh R (2006) A study of aerosol optical depth in the central Indian region (17.3–8.6 °N) during ISRO-GBP field campaign. Atmos. Environ. 40: 6494–6503.

[75] Ganguly D, Jayaraman A, Rajesh TA, Gadhavi H (2006) Wintertime aerosol properties during foggy and non-foggy days over urban center Delhi and their implications for shortwave radiative forcing. J. Geophys. Res. 111: D15217, doi:10.1029/2005JD007029.

[76] Niranjan K, Sreekanth V, Madhavan BL, Moorthy KK (2006) Wintertime aerosol characteristics at a north Indian site Kharagpur in the Indo-Gangetic plains located at the outflow region into Bay of Bengal. J. Geophys. Res. 111: D24209, doi:10.1029/2006JD007635.

[77] Pant P, Hegde P, Dumka UC, Sagar R, Satheesh SK, Moorthy KK, Saha A, Srivastava MK (2006) Aerosol characteristics at a high-altitude location in central Himalayas: optical properties and radiative forcing. J. Geophys. Res. 111: D17206, doi:10.1029/2005JD006768.

[78] Ramachandran S, et al. (2006) Aerosol radiative forcing during clear, hazy, and foggy conditions over a continental polluted location in north India. J. Geophys. Res. 111: D20214.

[79] Srivastava MK, Singh S, Saha A, Dumka UC, Hegde P, Singh R, Pant P (2006) Direct solar ultraviolet irradiance over Nainital, India, in the central Himalayas for clear-sky day conditions during December 2004. J. Geophys. Res. 111: D08201. doi:10.1029/2005JD006141.

[80] Rengarajan R, Sarin MM, Sudheer AK (2007) Carbonaceous and inorganic species in atmospheric aerosols during wintertime over urban and high-altitude sites in North India. J. Geophys. Res. 112: D21307.

[81] Ramanathan V, Rehman IH, Ramanathan N (2010) Project Surya Prospectus. University of California, San Diego, USA, 14 p.

[82] Rehman IH, Ahmed T, Praveen PS, Kar A, Ramanathan V (2011) Black carbon emissions from biomass and fossil fuels in rural India. Atmos. Chem. Phys. 11: 7289–7299.

[83] Praveen PS, Ahmed T, Kar A, Rehman IH, Ramanathan V (2012) Link between local scale BC emissions in the Indo-Gangetic Plains and large scale atmospheric solar absorption. Atmos. Chem. Phys. 12: 1173–1187.

[84] Tripathi SN, Dey S, Tare V, Satheesh SK, Lal S, Venkataramni S (2005b) Enhanced layer of black carbon in a north Indian industrial city. Geophys. Res. Lett. 32(12): L12802.

[85] Tripathi SN, Srivastva AK, Dey S, Satheesh SK, Krishnamoorthy K (2007b) The vertical profile of atmospheric heating rate profile due to black carbon at Kanpur (Northern India). Atmos. Env. 41(32): 6909-6915.

[86] Satheesh SK, et al. (2009) Vertical structure and horizontal gradients of aerosol extinction coefficients over coastal India inferred from airborne lidar measurements during the Integrated Campaign for Aerosol, Gases and Radiation Budget (ICARB) field campaign. J. Geophys. Res. 114: D05204, doi:10.1029/2008JD011033.

[87] Moorthy KK, Satheesh SK, Babu SS, Dutt CBS (2008) Integrated Campaign for Aerosols, gases and Radiation Budget (ICARB): An Overview. J. Earth. Sys. Sci. 117: 243-262.

[88] Nair VS, Babu SS, Moorthy KK (2008) Aerosol characteristics in the marine atmospheric boundary layer over the Bay of Bengal and Arabian Sea during ICARB: Spatial distribution and latitudinal and longitudinal gradients. J. Geophys. Res. 113: D15208.

[89] Satheesh SK, Moorthy KK, Babu SS, Vinoj V, Dutt CBS (2008) Climate implications of large warming by elevated aerosol over India. Geophys. Res. Lett. 35: L19809, doi:10.1029/2008GL034944.

[90] Satheesh SK, Vinoj V, Moorthy KK (2010) Radiative effects of aerosols at an urban location in southern India: Observations versus model. Atmos. Environ. 44: 5295-5304.

[91] Babu SS. et al. (2008) Aircraft measurements of aerosol black carbon from a coastal location in the north-east part of peninsular India during ICARB. J. Earth System Science 117 (Sp. Iss. 1): 263-271.

[92] Department of Science and Technology (DST), Continental Tropical Convergence Zone (CTCZ) Programme (2008) Science Plan, Indian Clim. Res. Programme, Gov. of India, New Delhi, 167 p.

[93] Prasad AK, Singh RP, Singh A (2006) Seasonal climatology of aerosol optical depth over the Indian subcontinent: Trend and departures in recent years. Int. J. Remote Sens. 27(12): 2323–2329, doi:10.1080/01431160500043665.

[94] Jenamani RK (2007) Alarming rise in fog and pollution causing a fall in maximum temperature over Delhi. Curr. Sci. 93: 314–322.

[95] Badarinath KVS, Kharol SK, Sharma AR, Roy PS (2009) Fog over Indo-Gangetic Plains—A study using multi-satellite data and ground observations. IEEE J. Selec. Topics Appl. Earth Obs. Remote Sens. 2(3): 185–195, doi:10.1109/JSTARS.2009.2019830.

[96] Eck T, et al (2012) Fog- and cloud-induced aerosol modification observed by the Aerosol Robotic Network (AERONET). J. Geophys. Res. 117: D07206, doi:10.1029/2011JD016839.

[97] Kar J, Jones DBA, Drummond JR, Attie JL, Liu J, Zou J, Nichitiu F, Seymour MD, Edwards DP, Deeter MN, Gille JC, Richter A (2008) Measurement of low altitude CO over the Indian subcontinent by MOPITT. J. Geophys. Res. 113: D16307, doi;10.1029/2007JD009362.

[98] Dey S and Di Girolamo L (2011) A decade of change in aerosol properties over the Indian subcontinent. Geophys. Res. Lett. 38: L14811, doi:10.1029/2011GL048153.

[99] Reddy MS and Venketaraman C (2002a) Inventories of aerosols and sulphur dioxide emissions from India: I. Fossil fuel combustion. Atmos. Environ. 36: 677–697.

[100] Reddy MS and Venketaraman C (2002b) Inventories of aerosols and sulphur dioxide emissions from India: II. Biomass combustion. Atmos. Environ. 36: 699–712.

[101] Ramanathan V, et al. (2005) Atmospheric brown clouds: Impacts on South Asian climate and hydrological cycle, Proc. Natl. Acad. Sci., U.S.A. 102: 5326-5333.

[102] Massie ST, Torris O, Smith SJ (2004) Total Ozone Mapping Spectrometer (TOMS) observations of increases in Asian aerosol in winter from 1979 to 2000. J. Geophys. Res. 109: D18211, doi:10.1029/2004JD004620.

[103] Wild M, Gilgen H, Roesch A, Ohmura A, Long CN, Dutton EG, Forgan B, Kallis A, Russak V, Tsvetkov A (2005) From dimming to brightening: Decadal changes in surface solar radiation. Science 308: 847–850, doi:10.1126/science.1103215.

[104] Kumari BP, Londhe AL, Daniel S, Jadhav DB (2007) Observational evidence of solar dimming: Offsetting surface warming over India. Geophys. Res. Lett. 34: L21810, doi:10.1029/2007GL031133.

[105] Badarinath KVS, Sharma AR, Kaskaoutis DG, Kharol SK, Kambezidis HD (2010) Solar dimming over the tropical urban region of Hyderabad, India: Effect of increased cloudiness and increased anthropogenic aerosols. J. Geophys. Res. 115: D21208, doi:10.1029/2009JD013694.

[106] Streets DG, Wu Y, Chin M (2006) Two-decadal aerosol trends as a likely explanation of the global dimming/brightening transition. Geophys. Res. Lett. 33: L15806, doi:10.1029/2006GL026471.

[107] Porch W, Chyleka P, Dubeya M, Massie S (2007) Trends in aerosol optical depth for cities in India. Atmos. Environ. 41: 7524–7532.

[108] Kaul DS, Gupta T, Tripathi SN, Tare V, Collett Jr JL (2011) Secondary Organic Aerosol: A comparison between foggy and non-foggy days. Environ. Sci. Technol. 45: 7307–7313.

[109] Venkataraman C, Habib G, Kadamba D, Shrivastava M, Leon JF, Crouzille B, Boucher O, Streets DG (2006) Emissions from open biomass burning in India: Integrating the inventory approach with high resolution Moderate Resolution Imaging Spectroradiometer (MODIS) active-fire and land cover data. Global Biogeochem. Cycles 20: GB2013, doi:10.1029/2005GB002547.

[110] Beegum IN, Moorthy KK, Babu SS, Satheesh SK, Vinoj V, Badarinath KVS, Safai PD, Devara PCS, Singh S, Vinod, Dumka UC, Pant P (2009) Spatial distribution of aerosol black carbon over India during pre-monsoon season. Atmos. Environ. 43: 1071–1078.

[111] Bonasoni P, et al. (2010) Atmospheric Brown Clouds in the Himalayas: first two years of continuous observations at the Nepal Climate Observatory-Pyramid (5079 m). Atmos. Chem. Phys. 10: 7515–7531.

[112] Decesari S, et al. (2010) Chemical composition of PM_{10} and PM_1 at the high-altitude Himalayan station Nepal Climate Observatory-Pyramid (NCO-P) (5079m a.s.l.). Atmos. Chem. Phys. 10: 4583–4596.

[113] Gobbi GP, Angelini F, Bonasoni P, Verza GP, Marinoni A, Barnaba F (2010) Sunphotometry of the 2006-2007 aerosol optical/radiative properties at the Himalayan Nepal Climate Observatory-Pyramid (5079ma.s.l.). Atmos. Chem. Phys. 10: 11209–11221, doi:10.5194/acp-10-11209-2010.

[114] Ramanathan V, Ramana MV (2005) Persistent, widespread, and strongly absorbing haze over the Himalayan foothills and the Indo-Gangetic plains. Pure Appl. Geophys. 162: 1609–1626.

[115] Srivastava AK, Pant P, Hegde P, Singh S, Dumka UC, Naja M, Singh N, Bhavanikumar Y (2011d) Influence of south Asian dust storm on aerosol radiative forcing at a high-altitude station in central Himalayas. Int. J. Remote Sens. 32(22): 7827–7845.

[116] Kumar R, et al. (2011) First ground based observations of influences of springtime Northern Indian biomass burning over the central Himalayas. J. Geophys. Res. 116: D19302.

[117] Srivastava AK, Singh S, Pant P, Dumka UC (2012e) Characteristics of black carbon over Delhi and Manora Peak-a comparative study. Atmos. Sci. Let., doi: 10.1002/asl.386 (in press).

[118] Mishra SK, Tripathi SN, Aggarwal A, Arola A (2012) Optical properties of accumulation mode polluted mineral dust: Effects of particle shape, hematite content and semi-external mixing with carbonaceous species. Tellus (accepted).

[119] Bond TC, Streets DG, Yarber KF, Nelson SM, Woo J, Klimont Z (2004) A technology-based global inventory of black and organic carbon emissions from combustion. J. Geophys. Res. 109: D14203, doi:10.1029/2003JD003697.

[120] Chandra S, Satheesh SK, Srinivasan J (2004) Can the mixing state of black carbon aerosols explain the mystery of 'excess' atmospheric absorption?. Geophys. Res. Lett. 31: L19109, doi:10.1029/2004GL020662.

[121] Xue M, Ma J, Yan P, Pan X (2011) Impacts of pollution and dust aerosols on the atmospheric optical properties over a polluted rural area near Beijing city. Atmos. Res. 101: 835–843.

[122] Das SK, Jayaraman A (2011) Role of black carbon in aerosol properties and radiative forcing over western India during pre-monsoon period. Atmos. Res. 102: 320–334.

[123] National Carbonaceous Aerosols Programme (NCEP), Science Plan, Indian Network for Climate Change Assessment (2011) Gov. of India, New Delhi, 44 p.

[124] Singh S, Singh R, Rao VUM (2004) Temporal dynamics of dew and fog events and their impact on wheat productivity in semi-arid region of India, Third International Conference on Fog, Fog Collection and Dew, NetSys Int. (Pty) Ltd., Cape Town, South Africa, 11 – 15 Oct.
(http://www.up.ac.za/academic/geog/meteo/EVENTS/fogdew2003/PAPERS/C65.pdf).

[125] Sarkar S, Chokngamwong R, Cervone G, Singh RP, Kafatos M (2006) Variability of aerosol optical depth and aerosol forcing over India. Adv. Space Res. 37(12): 2153– 2159.

[126] Ali K, Momin GA, Tewari S, Safai PD, Chate DM, Rao PSP (2004) Fog and precipitation chemistry at Delhi, north India. Atmos. Environ. 38: 4215–4222.

[127] Hameed S, Mirza MI, Ghauri BM, Siddiqui ZR, Javed R, Khan AR, Rattigan OV, Qureshi S, Husain L (2000) On the widespread winter fog in Northeastern Pakistan and India. Geophys. Res. Lett. 27: 1891–1894.

[128] Das SK, Jayaraman A, Mishra A (2008) Fog-induced variations in aerosol optical and physical properties over the Indo-Gangetic Basin and impact to aerosol radiative forcing. Ann Geophys. 26: 1345–1354.

[129] Hess M, Koepke P, Schultz I (1998) Optical properties of aerosols and clouds: the software package OPAC. Bull. Am. Meteorol. Soc. 79: 831–844.

[130] McComiskey A, Schwartz SE, Schmid B, Guan H, Lewis ER, Ricchiazzi P, Ogren JA (2008) Direct aerosol forcing: Calculation from observables and sensitivities to inputs. J. Geophys. Res. 113: D09202, doi:10.1029/2007JD009170.

[131] Ganguly D and Jayaraman A (2006) Physical and optical properties of aerosols over an urban location in western India: implications for shortwave radiative forcing. J. Geophys. Res. 111: D24207, doi:10.1029/2006JD007393.

[132] Ganguly D, Jayaraman A, Rajesh TA, Gadhavi H (2006) Wintertime aerosol properties during foggy and non-foggy days over urban center Delhi and their implications for shortwave radiative forcing. J. Geophys. Res. 111: D15217, doi:10.1029/2005JD007029.

[133] Pandithurai G, Pinker RT, Takamura T, Devara PCS (2004) Aerosol radiative forcing over a tropical urban site in India. Geophys. Res. Lett. 31: L12107, doi:10.1029/2004GL019702.

[134] Badarinath KVS, Latha KM (2006) Direct radiative forcing from black carbon aerosols over urban environment. Adv. Space. Res. 37(12): 2183–2188.

[135] Babu SS, Satheesh SK, Moorthy KK (2002) Aerosol radiative forcing due to enhanced black carbon at an urban site in India. Geophys. Res. Lett. 29(18): 1880, doi:10.1029/2002GL015826.

[136] Ramana MV, Ramanathan V, Podgorny IA, Pradhan BB, Shrestha B (2004) The direct observations of large aerosol radiative forcing in the Himalayan region. Geophys. Res. Lett. 31: L05111.

[137] Ramachandran S (2003) Aerosol radiative forcing over Bay of Bengal and Chennai: Comparison with maritime, continental, and urban aerosol models. J. Geophys. Res. 110: D21206, doi:10.1029/2005JD005861.

[138] Satheesh SK (2002) Radiative forcing by aerosols over Bay of Bengal region. Geophys. Res. Lett. 29(22): 2083, doi:10.1029/2002GL015334.

[139] Satheesh SK, Ramanathan V, Holben BN, Moorthy KK, Loeb NG, Maring H, Prospero JM, Savoie D (2002) Chemical, microphysical, and radiative effects of Indian Ocean aerosols. J. Geophys. Res. 107: 4725, doi:10.1029/2002JD002463.

Chapter 3

Temporal Variation of Particle Size Distribution of Polycyclic Aromatic Hydrocarbons at Different Roadside Air Environments in Bangkok, Thailand

Tomomi Hoshiko, Kazuo Yamamoto,
Fumiyuki Nakajima and Tassanee Prueksasit

Additional information is available at the end of the chapter

1. Introduction

Polycyclic aromatic hydrocarbons (PAHs) have been drawing attention as a major hazardous air pollutant due to their potential carcinogenicity and mutagenicity [1-2]. Polycyclic aromatic hydrocarbons are formed during the incomplete combustion of oil, coal, gas, wood and other organic substances. PAHs are initially generated in the gas phase, and they are adsorbed on pre-existing particles undergoing condensation during further cooling of the emission. Thus, most atmospheric PAHs exist in the particulate phase, while some higher volatility PAHs or low molecular weight PAHs remain partly in the gas phase (e.g., [3]). There are basically five major emission source components: domestic, mobile, industrial, agricultural and natural. The relative importance of these sources changes depending on the place or regulatory views; however, in the urban environment with heavy traffic, mobile source, that is vehicle exhaust is the main contributor to the atmospheric PAHs (e.g., [3-6]). Thus, health risk of the dense urban population by the exposure to those PAHs has been of concern both in developed and developing countries.

Currently, PM10 or finer particles are major air pollutants in many urban areas. In the atmosphere, PAHs are partitioned between gaseous and particulate phases as explained earlier. Especially, PAHs of higher molecular weight species, which are often of higher carcinogenic potential, are mostly associated with fine particulate matter (e.g., [2, 7]). However, atmospheric behavior of particulate matter is known to be highly complicated in terms of its chemical compositions, size distributions, physical behavior, reactions, and so on. Accordingly, atmospheric behavior of PAHs associated with particulate matter is subjected to uncertainties and still poorly explained, including their temporal and spatial variations.

Thailand's capital city, Bangkok was selected as the field of this study, where traffic air pollution and its health effects have long been a serious problem due to the heavy traffic and the chronic state of traffic congestion. It was reported that about 88% of PAH emission is attributed to motor vehicles, and a minor contributions are from biomass burning and oil combustion [6]. In Bangkok, road traffic is the main transport. Diesel buses have been the primary public transport, and ownership of passenger cars- both gasoline and diesel- and motorcycles has been increasing. The mass rapid transit network is still insufficient to meet the dramatically increased travel demand, which arose concurrently with the rapid economic development and rise in population in the last several decades. Thus, road traffic is heavily congested. At present, overall air quality in Bangkok has been significantly improved owing to several effective policies taken in the last decade, and the initiation and ongoing extension of railways and reinforcement of vehicle emission controls are quite promising for the further improvement. This is recognized as a successful case of urban air quality improvement in Southeast Asia, where many cities are still suffering from serious air pollution. In spite of the improvement, the present roadside PM10 levels in Bangkok have still been constantly exceeding the standard values; 24 hour standard 120 $\mu g/m^3$ and annual standard 50$\mu g/m^3$ [8]. Given the situation, carcinogenic PAHs associated with particulate matter are suspected to contribute to an increased health risk for the people living in the city.

Previous studies on roadside measurements reported much higher levels of PAHs than those at ambient sites, which has been the case for Bangkok as well [6,9-13]. It is stressed that environmental monitoring of PAHs is needed in more comprehensive ways with higher resolutions of time and space, especially at roadside areas, which are possible hot spots of high levels of exposure. In PAH monitoring, temporal variations of concentrations are an important aspect, for example, seasonal and diurnal changes. As for seasonal variations, monitoring data in developed countries in the temperate regions, such as Western Europe and the USA, are relatively abundant, whereas in developing countries in the tropical regions including Thailand, data are limited. For diurnal variations, there have not been many cases reported because PAH concentrations are usually reported as daily average. However, some previous reports showed remarkable diurnal changes in PAH concentrations, with morning and evening peaks in parallel with traffic rush hours (e.g., [5,12,14-17]). If we further look into the behavior of PAHs, information on diurnal variations of particle size-fractioned PAH concentrations become of interest, because particulate matter of different sizes is known to exert different levels of adverse health effects in the human body and finer particles penetrate into deeper parts of the human body and cause respiratory or cardiovascular disorders. However, studies concerning such information have been quite limited. Therefore, the specific objective of this study is to investigate diurnal variations of particle size distribution of PAHs by conducting field measurements.

2. Methodology

2.1. Study sites and time

Bangkok has a population of more than eight million people. Its climate is classified as tropical savanna with three seasons: hot (March – mid May), wet (mid-May - October) and cool (November - February). In terms of their characteristics, the hot season is hot and dry, the wet season is hot and wet, and the cool season is cool and dry. Field measurements were conducted in March and April, 2006, in the hot season. Concentrations of particulate phase PAHs were measured on the roadside in Bangkok. Diurnal variations of PAH concentrations were investigated by comparing two roadside sites with different road configurations. The measurement sites were Rama6 (R6) and Chockchai4 (CC). The R6 site is located in the area of government offices in the Bangkok city center, where one of the main roads, R6, carries heavy traffic. The R6 road is covered by an elevated highway (Figure 1a), and this configuration, together with large roadside buildings, is likely to cause a stagnant air mass within the road space. By contrast, the CC site has an ordinary open-space configuration along the Ladphrao road, with low-rise small buildings (Figure 1b). The measurement points were approximately three meters distance from the roads at both sites, 1.5 meters height from the ground at R6 (Figure 2a) and three meters height at CC in a Pollution Control Department's (PCD) air monitoring station, where the rooftop space of the station was provided to install measurement equipment for this study (Figure 2b).

(a)   (b)

Figure 1. Figure 1. Study sites. a) Rama6; b) Chockchai4

2.2. Air sampling

Particle mass was collected using a 10-stage Micro Orifice Uniform Deposit Impactor (MOUDI, Model 110, MSP Corporation, U.S.A.) [18] (Figure 3). The principle operation of the MOUDI is the same as any inertial cascade impactor with multiple nozzles. At each stage, jets of particle-laden air impinge upon an impaction plate, and particles larger than the cut-size of that stage cross the air stream lines and are collected upon the impaction plate. The smaller particles with less inertia do not cross the streamline and proceed to the next stage where the nozzles are smaller and where the air velocity through the nozzle is higher, and there, the finer particles are collected. This continues through the cascade impactor until the smallest particles are collected by the after-filter [19]. Figure 4 shows a schematic diagram of one stage of the MOUDI, showing its relation to the above and below stages [19].

(a)   (b)

Figure 2. Measurement locations. a) Rama6; b) Chockchai4

Figure 3. Micro Orifice Uniform Deposit Impactors (MOUDI)

Polytetrafluoroethylene (PTFE) membrane filters of 47-mm diameters (ADVANTEC, Japan) were used as the impaction substrates, and 37-mm glass filters (ADVANTEC, Japan) were used as the after-filter. The aerodynamic diameter size cut points with 50% collection efficiency were 0.18, 0.31, 0.56, 1.0, 1.8, 3.2, 5.6, 10, and 18 μm. The MOUDI operated at 30 L/min, and the particle mass in the filters was determined gravimetrically. Before each weighing, the filters were conditioned in a desiccator with silica gel for about three days to eliminate humidity. Afterward, the filters were wrapped in aluminum foil and stored at 4 °C until the extraction was performed. After ultrasonic extraction, 13 kinds of PAHs with three to six aromatic rings (Table 1 and Figure 4) were determined by Gas Chromatography / Mass Spectrometry (GC/MS) analysis. Among the 13 PAHs, 12 of them, not including Benzo(e)pyrene (BeP), have been included in the priority pollutant list of the Clean Water Act of the United States Environmental Protection Agency (US EPA) since the 1970s.

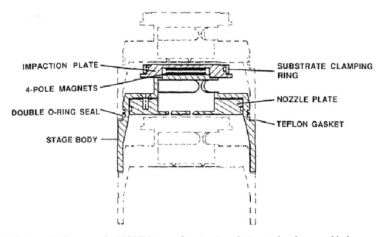

Figure 4. Schematic diagram of a MOUDI stage showing its relation to the above and below stages [19]

Compound	Abbr.	Molecular formula	Molecular weight	No. of aromatic rings	Vapor pressure (Pa at 25 °C)	Solubility in water (μg/L at 25 °C)
Phenanthrene	Phe	$C_{14}H_{10}$	178	3	1.6×10^{-2}	1.3×10^3
Anthracene	Ant	$C_{14}H_{10}$	178	3	8.0×10^{-4}	73
Fluoranthene	Fluo	$C_{16}H_{10}$	202	4	1.2×10^{-3}	260
Pyrene	Pyr	$C_{16}H_{10}$	202	4	6.0×10^{-4}	135
Benzo(a)anthracene	BaA	$C_{18}H_{12}$	228	4	2.8×10^{-5}	5.6
Chrysene	Chr	$C_{18}H_{12}$	228	4	8.4×10^{-5} *	2.0
Benzo(b)fluoranthene	BbF	$C_{20}H_{12}$	252	5	6.7×10^{-5} *	0.80
Benzo(k)fluoranthene	BkF	$C_{20}H_{12}$	252	5	1.3×10^{-8} *	0.76
Benzo(e)pyrene	BeP	$C_{20}H_{12}$	252	5	7.6×10^{-7}	6.3
Benzo(a)pyrene	BaP	$C_{20}H_{12}$	252	5	7.4×10^{-7}	3.8
Indeno(1,2,3-cd)pyrene	IP	$C_{22}H_{12}$	276	6	1.3×10^{-8} *	62
Dibenz(a,h)anthracene	DahA	$C_{22}H_{14}$	278	5	1.3×10^{-8} *	1.0
Benzo(g,h,i)perylene	BghiP	$C_{20}H_{12}$	276	6	1.4×10^{-8}	0.26

*Pa at 20 °C

Table 1. Physico-chemical properties of 13 PAHs measured in this study [2]

To monitor the temporal variations of total concentrations of particulate PAHs, photoelectric aerosol sensors (model PAS2000CE, EcoChem Analytics, Germany) [20] were used for real-time monitoring (Figure 5). Photoelectric aerosol sensors (PAS) work on the basis of photoelectric ionization of PAHs adsorbed onto particles [21]. The measurement techniques of this instrument have been described in detail elsewhere [22]. Briefly, a vacuum pump is used to draw ambient air through a quartz tube around which a UV lamp is mounted. Irradiation with UV light causes particles to emit electrons, which are then captured by surrounding gas molecules. Negatively charged particles are removed from the air stream,

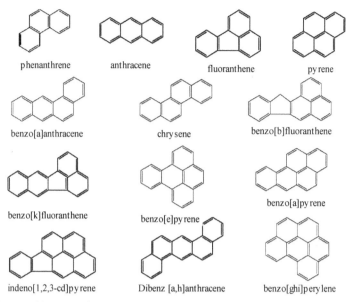

Figure 5. Structural formulas of the 13 PAHs

and the remaining positively charged particles are collected on a particle filter mounted in a Faraday cage. The particle filter converts the ion current to an electrical current, which is then amplified and measured with an electrometer (Figure 6). The electric current establishes signals that are proportional to the concentrations of total PAHs [20]. The target particle size is below 1 μm and the signals are recorded every 2 minutes. In the results of this study, PAS monitoring data are indicated as PAS signals without particular units, because the purpose of use of PAS is to know relative levels of temporal variations of total PAH concentrations, not to know absolute values of the total PAH concentrations, for which actual kinds of PAHs which compose the total concentrations cannot be identified by the PAS.

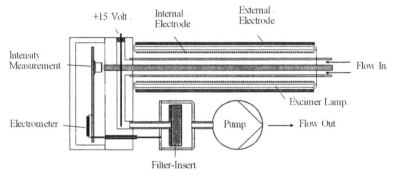

Figure 6. PAS2000CE

Figure 7. Scheme of PAS2000CE [49]

2.3. Particle size distribution

Particle sizes in the atmosphere are known to distribute with certain frequency modes, namely, nuclei mode, accumulation mode and coarse mode [23]. The nuclei mode, or ultrafine mode is mainly primary emission of vehicle exhaust and is carbonaceous particulate matter. The accumulation mode is responsible for formation of secondary organic aerosols especially through photochemical reactions with VOCs including gas phase PAHs and also for coagulation of particles. The coarse mode (>1.8 um) particles are mostly grown particles in the atmosphere and/or re-suspended road dust, which are reported to be subject to condensation of volatile materials including lighter PAHs. From previous studies, concentrations of PAHs are found to be highly dependent upon the size of particles. In view of association mechanisms and atmospheric processes of PAHs to urban aerosols, those particle size modes are applied to this study [7]. Based on previous studies on PAHs measurement using cascade air samplers (e.g., [24-25]) three particle size modes are defined for this study: ultrafine mode (< 0.18 μm), accumulation mode (0.18-1.8 μm) and coarse mode (1.8μm <) according to the particle cut sizes of the MOUDI.

2.4. Traffic and meteorological data

Road traffic was recorded using a video camera for 24 hours or shorter during the air sampling. The traffic volume was counted manually for 10 minutes in every hour, then multiplied by six to estimate hourly average volumes. At the CC site, hourly meteorological data monitored by the PCD were obtained. The meteorological data included temperature, solar radiation, relative humidity, rain, wind speed and wind direction. At the R6 site, wind speed and wind directions were monitored at 10-minute intervals using KADEC wind monitors (Kona Systems, Japan), and temperature, solar radiation, relative humidity and rainfall were monitored at 5-minute intervals using an AutoMet meteorology monitor (MET ONE Instruments, USA).

3. Diurnal variation of particle size distribution of PAHs

3.1. Preliminary measurement

One of the most influential and distinctive factors for diurnal PAH concentration variations was expected to be diurnal variations of vehicle emissions. Thus, it was considered most appropriate to investigate diurnal variations of particle size distributions of PAHs in accordance with morning and evening traffic peak hours and two off-peak hours in between, namely daytime and night time, in total four different time periods of the day for separate measurements. Preliminary PAS real-time monitoring was conducted to see if the PAS signals show morning and evening peaks corresponding to the traffic. The measurements were conducted from March 27th to April 2nd, 2006, at R6 and from April 21st to 23rd, 2006 at CC, right before the MOUDI air sampling, respectively. Figure 7 presents the results from the PAS signals. Although the timing of the PAS signal peaks were somewhat varied on different days, the morning and evening peak hours and the daytime and overnight off-peak hours were confirmed. Based on these observations, the following four time periods were decided for the MOUDI air sampling durations at each site to investigate diurnal variations of particle size distributions of PAHs: 6:00-10:00 (morning (m)), 12:00-16:00 (daytime (d)), 18:00-21:00 (evening (e)) and 22:00-5:00 (overnight (o)) at R6; and 6:00-9:30 (m), 13:00-17:00 (d), 18:30-21:30 (e) and 22:00-5:00 (o) at CC.

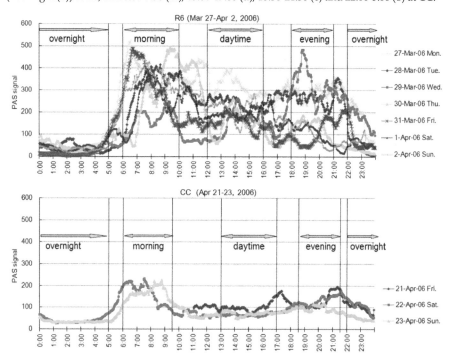

Figure 8. Preliminary real-time monitoring for the selection of four time periods corresponding to peak and off-peak hours of PAS signals for MOUDI air sampling [35]

Temporal Variation of Particle Size Distribution of Polycyclic Aromatic Hydrocarbons at Different
Roadside Air Environments in Bangkok, Thailand

71

3.2. Results and discussion

Air sampling was conducted using the MOUDI during the four selected time periods for three consecutive days. The sampling periods were April 3rd (Mon.)-6th (Thu.), 2006, at R6 and April 24th (Mon.)-27th (Thu.), 2006, at CC. Table 2 shows the 13 particle size-fractioned PAH concentrations (ng/m³) during the four time periods of the day. Particulate matter was collected cumulatively on the same filters in each time period, while the air sampling using

Time period of day	PAH	Particle size fraction (μm)										Total (ng/m³)	Detection limit (ng/m³)
		< 0.18	0.18~0.31	0.31~0.56	0.56~1.0	1.0~1.8	1.8~3.2	3.2~5.6	5.6~10	10~18	18 <		
Morning	Phe	0.064	0.072	0.061	0.026	0.020	0.017	0.0091	0.017	0.012	0.010	0.308	0.0015
	Ant	0.021	0.024	0.018	0.013	0.015	0.013	0.0060	0.012	0.0092	0.018	0.151	0.0015
	Fluo	0.065	0.074	0.054	0.024	0.017	0.014	0.0075	0.014	0.0090	0.0090	0.288	0.0015
	Pyr	0.12	0.13	0.097	0.042	0.030	0.022	0.010	0.020	0.013	0.014	0.497	0.0015
	BaA	0.074	0.094	0.065	0.023	0.014	0.012	0.007	0.014	0.013	0.012	0.329	0.0029
	Chr	0.079	0.11	0.086	0.033	0.021	0.014	0.009	0.019	0.015	0.016	0.400	0.0029
	BbF	0.19	0.17	0.15	0.046	0.017	0.014	0.010	0.017	0.013	0.015	0.642	0.0037
	BkF	0.10	0.11	0.078	0.039	0.018	0.0090	0.0089	0.012	0.0092	0.015	0.398	0.0037
	BeP	0.20	0.19	0.16	0.045	0.017	0.011	0.011	0.012	N.D.	N.D.	0.65	0.0018
	BaP	0.22	0.21	0.18	0.052	0.029	0.017	0.015	0.022	N.D.	N.D.	0.74	0.0037
	IP	0.14	0.11	0.086	0.034	N.D.	N.D.	N.D.	N.D.	N.D.	N.D.	0.37	0.0037
	DahA	0.023	0.010	N.D.	N.D.	N.D.	N.D.	N.D.	N.D.	N.D.	N.D.	0.033	0.015
	BghiP	0.27	0.19	0.18	0.059	N.D.	N.D.	N.D.	N.D.	N.D.	N.D.	0.69	0.0037
	Σ13 PAHs	1.6	1.5	1.2	0.44	0.20	0.14	0.094	0.16	0.094	0.109	5.5	
	Ratio(%)	28.4	26.9	22.1	7.9	3.6	2.6	1.7	2.9	1.7	2.0	100.0	
Daytime	Phe	0.042	0.097	0.11	0.049	0.045	0.039	0.032	0.033	0.039	0.034	0.51	0.0034
	Ant	0.023	0.045	0.053	0.030	0.057	0.037	0.028	0.027	0.035	0.033	0.37	0.0034
	Fluo	0.045	0.081	0.095	0.055	0.048	0.022	0.026	0.023	0.022	0.021	0.44	0.0034
	Pyr	0.075	0.16	0.18	0.091	0.067	0.032	0.034	0.032	0.028	0.027	0.73	0.0034
	BaA	0.048	0.086	0.095	0.059	0.055	0.027	0.047	0.024	0.038	0.032	0.51	0.0067
	Chr	0.073	0.10	0.12	0.058	0.061	0.039	0.059	0.038	0.048	0.038	0.64	0.0067
	BbF	0.11	0.21	0.22	0.079	0.048	0.040	0.045	0.033	0.039	N.D.	0.83	0.0084
	BkF	0.064	0.12	0.14	0.067	0.043	0.047	0.035	0.028	0.036	N.D.	0.58	0.0084
	BeP	0.077	0.19	0.21	0.053	0.023	N.D.	0.027	N.D.	N.D.	N.D.	0.58	0.0042
	BaP	0.099	0.18	0.17	0.049	0.064	N.D.	0.085	N.D.	N.D.	N.D.	0.65	0.0084
	IP	0.19	0.21	0.24	N.D.	N.D.	N.D.	N.D.	N.D.	N.D.	N.D.	0.64	0.0084
	DahA	N.D.	N.D.	N.D.	N.D.	N.D.	N.D.	N.D.	N.D.	N.D.	N.D.	N.D.	0.034
	BghiP	0.23	0.26	0.27	0.035	N.D.	N.D.	N.D.	N.D.	N.D.	N.D.	0.80	0.0084
	Σ13 PAHs	1.1	1.7	1.9	0.63	0.51	0.28	0.42	0.24	0.29	0.19	7.3	
	Ratio(%)	14.8	23.9	26.2	8.6	7.0	3.9	5.8	3.3	3.9	2.6	100.0	
Evening	Phe	0.050	0.096	0.096	0.049	0.036	0.036	0.037	0.038	0.035	0.032	0.50	0.0028
	Ant	0.029	0.053	0.046	0.031	0.028	0.037	0.030	0.031	0.035	0.025	0.35	0.0028
	Fluo	0.058	0.10	0.084	0.054	0.034	0.039	0.035	0.030	0.033	0.027	0.49	0.0028
	Pyr	0.075	0.18	0.15	0.075	0.042	0.059	0.052	0.040	0.034	0.030	0.73	0.0028
	BaA	0.073	0.11	0.12	0.035	0.039	0.032	0.037	0.033	0.036	0.039	0.55	0.0056
	Chr	0.076	0.15	0.11	0.048	0.037	0.034	0.033	0.038	0.033	0.036	0.60	0.0056
	BbF	0.12	0.25	0.23	0.051	0.053	0.029	0.033	0.037	0.041	0.033	0.89	0.0070
	BkF	0.13	0.17	0.13	0.047	0.063	0.033	0.035	0.027	0.027	0.030	0.69	0.0070
	BeP	0.10	0.17	0.22	0.051	N.D.	0.022	0.021	N.D.	N.D.	N.D.	0.58	0.0035
	BaP	0.10	0.25	0.30	0.12	N.D.	0.062	0.050	N.D.	N.D.	N.D.	0.89	0.0070
	IP	0.19	0.20	0.15	0.033	N.D.	N.D.	N.D.	N.D.	N.D.	N.D.	0.57	0.0070
	DahA	0.049	0.080	N.D.	N.D.	N.D.	N.D.	N.D.	N.D.	N.D.	N.D.	0.13	0.028
	BghiP	0.29	0.30	0.26	0.055	N.D.	N.D.	N.D.	N.D.	N.D.	N.D.	0.90	0.0070
	Σ13 PAHs	1.3	2.1	1.9	0.65	0.33	0.38	0.36	0.27	0.27	0.25	7.9	
	Ratio(%)	17.1	26.9	23.9	8.2	4.2	4.9	4.6	3.5	3.5	3.2	100.0	
Overnight	Phe	0.017	0.036	0.031	0.020	0.021	0.014	0.013	0.019	0.011	0.013	0.20	0.0010
	Ant	0.0088	0.017	0.013	0.011	0.011	0.012	0.010	0.022	0.0071	0.010	0.122	0.0010
	Fluo	0.014	0.027	0.026	0.016	0.020	0.012	0.014	0.032	0.0074	0.010	0.177	0.0010
	Pyr	0.028	0.052	0.049	0.031	0.038	0.023	0.018	0.037	0.011	0.013	0.30	0.0010
	BaA	0.024	0.034	0.031	0.014	0.015	0.0083	0.016	0.031	0.0077	0.011	0.190	0.0021
	Chr	0.023	0.043	0.039	0.026	0.024	0.014	0.014	0.031	0.0070	0.010	0.231	0.0021
	BbF	0.022	0.055	0.053	0.012	0.027	0.013	0.018	0.028	0.011	0.012	0.25	0.0026
	BkF	0.032	0.058	0.070	0.031	0.033	0.0094	0.020	0.030	0.011	0.013	0.307	0.0026
	BeP	0.035	0.086	0.082	0.035	0.033	0.012	0.016	0.012	0.0086	0.011	0.329	0.0013
	BaP	0.041	0.112	0.099	0.049	0.027	0.019	0.029	0.023	0.011	0.013	0.42	0.0026
	IP	0.065	0.067	0.083	0.032	0.019	N.D.	N.D.	0.017	N.D.	N.D.	0.28	0.0026
	DahA	0.017	N.D.	N.D.	N.D.	N.D.	N.D.	N.D.	N.D.	N.D.	N.D.	0.017	0.010
	BghiP	0.11	0.11	0.11	0.044	0.030	0.009	0.012	0.027	0.010	0.021	0.48	0.0026
	Σ13 PAHs	0.436	0.69	0.68	0.32	0.30	0.145	0.18	0.31	0.102	0.14	3.3	
	Ratio(%)	13.2	21.0	20.7	9.8	9.0	4.4	5.4	9.3	3.1	4.1	100.0	

(a)

Time period of day	PAH	Particle size fraction (μm)										Total (ng/m³)	Detection limit (ng/m³)
		<0.18	0.18~0.31	0.31~0.56	0.56~1.0	1.0~1.8	1.8~3.2	3.2~5.6	5.6~10	10~18	18<		
Morning	Phe	0.051	0.077	0.070	0.044	0.039	0.037	0.039	0.034	0.036	0.035	0.46	0.0035
	Ant	0.020	0.021	0.019	0.018	0.017	0.016	0.021	0.016	0.019	0.020	0.19	0.0035
	Fluo	0.031	0.040	0.046	0.026	0.023	0.020	0.022	0.016	0.022	0.019	0.27	0.0035
	Pyr	0.058	0.069	0.068	0.040	0.029	0.027	0.025	0.022	0.022	0.023	0.38	0.0035
	BaA	0.031	0.030	0.023	0.019	0.018	0.016	0.024	0.010	0.014	0.025	0.21	0.0070
	Chr	0.057	0.061	0.048	0.025	0.021	0.020	0.023	0.022	0.032	0.034	0.34	0.0070
	BbF	0.094	0.13	0.12	0.11	N.D.	N.D.	N.D.	N.D.	N.D.	N.D.	0.46	0.0087
	BkF	0.093	0.079	0.069	0.038	N.D.	N.D.	N.D.	N.D.	N.D.	N.D.	0.28	0.0087
	BeP	0.10	0.17	0.13	0.039	N.D.	N.D.	N.D.	N.D.	N.D.	N.D.	0.45	0.0044
	BaP	0.10	0.14	0.10	0.036	N.D.	N.D.	N.D.	N.D.	N.D.	N.D.	0.38	0.0087
	IP	0.059	0.11	0.058	N.D.	N.D.	N.D.	N.D.	N.D.	N.D.	N.D.	0.23	0.0087
	DahA	N.D.	N.D.	N.D.	N.D.	N.D.	N.D.	N.D.	N.D.	N.D.	N.D.	N.D.	0.035
	BghiP	0.13	0.20	0.15	N.D.	N.D.	N.D.	N.D.	N.D.	N.D.	N.D.	0.48	0.0087
	Σ13 PAHs	0.84	1.1	0.91	0.39	0.15	0.14	0.15	0.12	0.14	0.16	4.1	
	Ratio(%)	20.2	27.5	22.0	9.5	3.6	3.3	3.7	2.9	3.5	3.8	100.0	
Daytime	Phe	0.050	0.072	0.082	0.042	0.039	0.039	0.038	0.033	0.047	0.044	0.48	0.0032
	Ant	0.024	0.026	0.034	0.024	0.018	0.022	0.022	0.014	0.024	0.022	0.23	0.0032
	Fluo	0.037	0.050	0.051	0.033	0.023	0.024	0.019	0.014	0.026	0.030	0.31	0.0032
	Pyr	0.050	0.081	0.073	0.049	0.032	0.035	0.029	0.016	0.028	0.021	0.42	0.0032
	BaA	0.042	0.045	0.044	0.026	0.020	0.028	0.030	0.015	0.027	0.028	0.31	0.0064
	Chr	0.059	0.054	0.044	0.025	0.014	0.023	0.018	0.014	0.025	0.020	0.30	0.0064
	BbF	0.11	0.12	0.086	0.055	0.041	0.028	N.D.	0.032	0.040	0.046	0.56	0.0081
	BkF	0.057	0.045	0.059	0.028	0.018	0.017	N.D.	0.021	0.023	0.018	0.29	0.0081
	BeP	0.11	0.10	0.097	0.032	N.D.	N.D.	N.D.	N.D.	0.024	0.025	0.40	0.0040
	BaP	0.056	0.055	0.061	0.012	N.D.	N.D.	N.D.	N.D.	0.030	N.D.	0.21	0.0081
	IP	0.052	0.128	0.099	N.D.	N.D.	N.D.	N.D.	N.D.	N.D.	N.D.	0.28	0.0081
	DahA	N.D.	N.D.	N.D.	N.D.	N.D.	N.D.	N.D.	N.D.	N.D.	N.D.	N.D.	0.032
	BghiP	0.070	0.14	0.11	0.030	N.D.	N.D.	N.D.	N.D.	N.D.	N.D.	0.35	0.0081
	Σ13 PAHs	0.72	0.92	0.84	0.36	0.21	0.22	0.16	0.16	0.29	0.25	4.1	
	Ratio(%)	17.6	22.2	20.5	8.6	5.0	5.3	3.8	3.8	7.1	6.2	100.0	
Evening	Phe	0.048	0.070	0.086	0.064	0.053	0.042	0.042	0.033	0.035	0.033	0.51	0.0038
	Ant	0.018	0.025	0.034	0.031	0.022	0.026	0.021	0.022	0.023	0.020	0.24	0.0038
	Fluo	0.030	0.048	0.059	0.040	0.028	0.027	0.024	0.023	0.020	0.018	0.32	0.0038
	Pyr	0.050	0.082	0.080	0.051	0.044	0.041	0.032	0.026	0.030	0.021	0.46	0.0038
	BaA	0.006	0.036	0.033	0.030	0.029	0.020	0.031	0.020	0.023	0.033	0.26	0.0076
	Chr	0.008	0.049	0.048	0.029	0.027	0.020	0.024	0.025	0.020	0.019	0.27	0.0076
	BbF	0.10	0.12	0.11	0.051	0.039	0.031	N.D.	N.D.	N.D.	N.D.	0.45	0.0094
	BkF	0.038	0.050	0.060	0.035	0.023	0.015	N.D.	N.D.	N.D.	N.D.	0.22	0.0094
	BeP	0.096	0.12	0.12	0.044	0.041	N.D.	N.D.	N.D.	N.D.	N.D.	0.41	0.0047
	BaP	0.077	0.075	0.087	0.054	0.057	N.D.	N.D.	N.D.	N.D.	N.D.	0.35	0.0094
	IP	0.044	0.082	0.10	N.D.	N.D.	N.D.	N.D.	N.D.	N.D.	N.D.	0.23	0.0094
	DahA	N.D.	N.D.	N.D.	N.D.	N.D.	N.D.	N.D.	N.D.	N.D.	N.D.	N.D.	0.038
	BghiP	0.065	0.11	0.14	N.D.	N.D.	N.D.	N.D.	N.D.	N.D.	N.D.	0.32	0.0094
	Σ13 PAHs	0.58	0.87	0.95	0.43	0.36	0.22	0.17	0.15	0.15	0.14	4.0	
	Ratio(%)	14.4	21.5	23.6	10.6	9.0	5.5	4.3	3.7	3.7	3.6	100.0	
Overnight	Phe											0.45	0.0026
	Ant											0.28	0.0026
	Fluo											0.33	0.0026
	Pyr											0.41	0.0026
	BaA											0.24	0.0051
	Chr			*data not available*								0.35	0.0051
	BbF											0.42	0.0064
	BkF											0.25	0.0064
	BeP											0.25	0.0032
	BaP											0.24	0.0064
	IP											0.07	0.0064
	DahA											N.D.	0.026
	BghiP											0.22	0.0064
	Σ13 PAHs											3.5	
	Ratio(%)												

(b)

Table 2. Particle size-fractioned 13 PAHs concentrations (ng/m³) in the four time periods of the day a) Rama6; b) Chockchai4

the MOUDI was repeated for three consecutive days. After the sampling on the first and second days, the filters were kept in plastic cases and carried until the sampling on the second and third days. (N.B. Unfortunately, filter samples of CC overnight were mishandled and size-segregated PAH concentration data and particle weight data are not available. However, total atmospheric PAH concentrations are available using the amount of air pumped in by the MOUDI equipment as shown in Table 2.)

Concurrent with the three-day air sampling using the MOUDI, PAS real-time monitoring was also conducted using the PASs, which showed similar trends of diurnal variations to those observed in the preliminary measurements. At R6, the PAH signal values sharply increased from approximately 5 am and reached morning peaks between 9 and 10 am during the three days. Daytime concentrations were lower than in the morning. In the daytime and evening, several small peaks appeared. From around midnight to 5 am, concentrations were remarkably low. At CC, sharp morning peaks were observed at approximately 7 am on April 24th and around 8 am on April 27th. In the evening, broader peaks appeared between 4 and 9 pm, and then the concentration decreased. The sharp increase in the morning was observed at both sites. This observation is consistent with that in previous reports [12,15-16]. The sharp morning peaks can be explained by both the strong atmospheric stability caused by the inversion layer and an increase in emissions from the morning traffic. Although the total traffic volume was smaller at R6 (74,000 vehicle/day on April 5th (Mon.)) than that at CC (92,000 vehicle/day on April 24th (Wed.)), higher concentrations were observed at R6 throughout the day, possibly due to the covered configuration that restricted the atmospheric dilution effect. According to the traffic monitoring, congestion occurred during the daytime at both sites, but daytime PAS signal levels at CC were constantly low compared with the large fluctuations of the daytime levels at R6.

Table 3 shows the average local meteorological data during the three-day monitoring periods. At both sites, the mean wind directions were almost stable at southwest, which situates the both sampling locations at downwind of the road emissions. At R6, the wind speed, which ranged from 0.2-0.6 m/s, was low compared with that at CC. The observation implies the limited dispersion and long residence time of airborne PAHs within the site. On the other hand, the wind speed at CC was much higher, ranging 1.3-2.8 m/s, implying faster dispersion. Solar radiation, which promotes photochemical decomposition of PAHs [26], was more than 40% lower at R6 throughout the day because of the elevated highway and the large buildings along the R6 road. This road configuration may also explain the higher daytime concentrations at R6.

	Temperature (°C)				Solar radiation (W/m^2)				Relative humidity (%)			
	m	d	e	o	m	d	e	o	m	d	e	o
R6	29.5	35.9	31.2	29.7	26.1	284.8	0.2	0	85.3	57.9	76.3	83.8
CC	27.9	29.1	29.6	28.0	129.3	762.5	12.7	0.4	69.5	61.9	55.9	68.2

	Wind speed (m/s)				Wind direction			
	m	d	e	o	m	d	e	o
R6	0.2	0.6	0.3	0.5	SW	WSW	SW	W
CC	1.9	1.3	1.8	2.8	SSW	SSW	SSW	SW

Table 3. Table 3. Local meteorological data; average during the four time periods of the day (April 3-6th, 2006 at R6 and April 24-27th, 2006 at CC).

Diurnal variations of particle size distribution of the 13 PAHs concentrations are shown in Figure 8 with classification of semi-volatile 3-4 ring PAHs and non-volatile 5-6 ring PAHs, respectively. Each graph of the size distribution is shown as Lungren type plots [27]. An overall trend of the size distribution was consistent with that in previous studies that concentrations of PAHs were found to be highly dependent upon the size of particulate matter, with the greatest concentrations being the submicron size range (e.g., [7, 28-30]). The higher concentrations at the submicron range can be explained by the condensation mechanism because larger specific surface areas are associated with such particles (e.g., [31]). On the other hand, there were clear differences of the size distribution between the groups of 3-4 ring and 5-6 ring PAHs. Contribution of coarser, or above 1μm range, was larger for 3-4 ring PAHs, while 5-6 ring PAHs were dominantly distributed in the finer size ranges at both of the sites, regardless of the time periods. The results regarding the relationship of PAH size distribution between ring number of PAHs and associated particle size were consistent with that in many other studies conducted previously, not only at Bangkok but also at various places (e.g., [7,13,32]). The Kelvin effect explains this trend. It determines the relationship between the particle diameter and vapor pressure, where more volatile species, namely lower weight PAHs, are associated with larger diameter particles.

In terms of the temporal variations of PAH size distribution, at R6 in the morning, the distribution of 3-4 ring PAHs had a peak in the 0.18-0.31μm range, while the distribution of 5-6 ring PAHs had a peak in the finest range of below 0.18μm. The ratios of the ultrafine mode were clearly higher than those of other time periods, indicating an elevated burden of primary vehicle exhaust emission in the morning rush. From the morning to daytime, the distribution of both 3-4 ring and 5-6 ring PAHs slightly shifted to a coarser range. It implies atmospheric processes of particles, such as coagulation, condensation and photochemical reactions, which lead to growth of particles under stronger solar radiation and less vehicle emissions in the daytime. In the evening, the distribution peaks shifted back to finer ranges possibly due to traffic increase again in the evening rush. In the nighttime, distribution in coarser range increased apparently, possibly due to increased re-suspension of road dust caused by a faster driving speed of motor vehicles.

At CC, as a whole the coarser range contributed more than that at R6. This may be due to faster wind speed at CC enhancing re-suspension of coarse particles and/or faster dispersion of fine particles because of the open-space configuration of the road. Unlike at R6, any explicable trend of diurnal variations of the size distributions could not be identified at CC, rather some fluctuation of the size distribution was observed. This could be due to shorter residence time of particles in the site by the faster dispersion because of the open-structure configuration and faster wind, so atmospheric processes of particle growth might be limited compared to the R6's case.

To further support the discussion of diurnal variations of size distribution of PAHs, PAH contents in particulate matter (μg/g) were analyzed according to the three particle size modes in the four time periods of day (Figure 9). (N.B. Figure 8 showed PAH concentrations per unit air mass.) The PAH contents varied considerably according to the particle size modes and the different time periods. In general, smaller particle size modes had higher

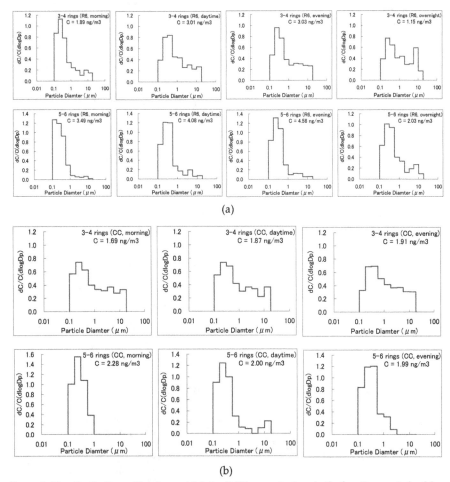

Figure 9. Size distributions of 3-4 ring and 5-6 ring PAH concentrations in the four time periods of day.
a) Rama6; b) Chockchai4

PAH contents. An exceptional case was that low molecular PAHs up to Pyrene in the accumulation mode showed higher contents than those in the ultrafine mode in the daytime at both sites. It might again imply accelerated accumulation processes in the daytime. For example, accumulation mode particles went through photochemical reactions with VOCs accompanying condensation of low molecular weight gaseous phase PAHs in the daytime.

According to previous reports [33-34] Zielinska (2004), elevated concentrations of high molecular weight PAHs, especially BghiP, indicate significant contribution of gasoline exhaust. As shown in the figure, BghiP content in the ultrafine mode were remarkably high in all the four time periods at R6 and in the morning at CC, again implying significant

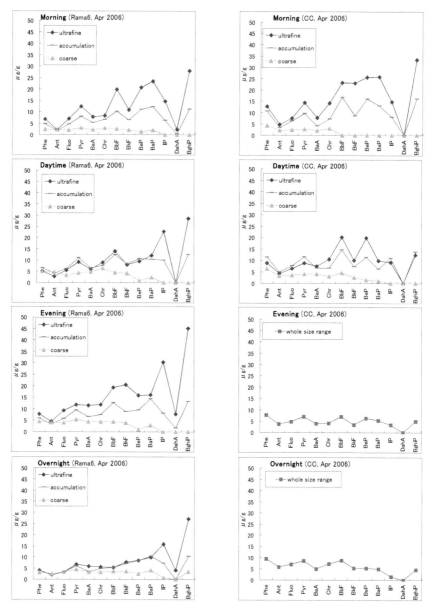

N.B.) The particulate weight data of the ultrafine mode for CC evening, and size distribution data of particle weight for CC overnight were not obtained, thus their mode distribution data are not available in the figure.

Figure 10. 13 PAH contents in the three particle modes in the four time periods of day (ultrafine: < 0.18μm, accumulation: 0.18-1.8μm, coarse: 1.8μm <)

contribution of gasoline vehicles to the roadside atmospheric PAHs. Especially the BghiP content in the ultrafine mode in the evening at R6 had the highest value. In fact at this time period, the ratio of the gasoline vehicles to the total traffic volume was the highest (63%) among the four time periods at the two sites.

4. Conclusion

This study clearly showed diurnal variations of particle size distribution of PAHs in four different time periods of day by field measurements on the two different types of roads in Bangkok, Thailand. The data indicated possible influences on the temporal variations of size distribution of PAHs by diurnal changes in traffic emissions, road configurations in relation to dispersion efficiency, meteorological conditions, atmospheric processes of particles such as coagulation, condensation and photochemical reactions under daytime sunlight, also atmospheric processes of PAHs in the accumulation particle mode. In view of people's exposure to vehicular emissions in an urban area with heavy road traffic, it is desired that urban air quality monitoring should be conducted in more comprehensive ways with higher resolutions of time and space, and actual behavior of pollutants be elucidated by various advanced approaches.

Author details

Tomomi Hoshiko*
Department of Urban Engineering, Graduate School of Engineering, The University of Tokyo, Tokyo, Japan

Kazuo Yamamoto and Fumiyuki Nakajima
Environmental Science Center, The University of Tokyo, Tokyo, Japan

Tassanee Prueksasit
Department of Environmental Science, Faculty of Science, Chulalongkorn University, Bangkok, Thailand

5. References

[1] IARC (2011) IARC monographs on the evaluation of carcinogenic risks to humans. http://monographs.iarc.fr/ENG/Classification/index.php. Accessed 13 Apr 2012.

[2] IPCS (1998) Selected non-heterocyclic polycyclic aromatic hydrocarbons. Environmetal Health Criteria 202. WHO, Geneva.

[3] Beak SO, Field RA, Goldstone ME, Kirk PW, Lester JN, Perry R (1991) A review of atmospheric polycyclic aromatic hydrocarbons: Sources, fate and behavior. Water Air and Soil Pollution 60: 279-300.

[4] European Commission (2001) Ambient air pollution by Polycyclic Aromatic Hydrocarbons (PAH). Position Paper.

* Corresponding Author

[5]   Nielsen T (1996) Traffic contribution of polycyclic aromatic hydrocarbons in the center of a large city. Atmospheric Environment 30: 3481-3490.

[6]   Panther B, Hooper M, Limpaseni W, Hooper B (1996) Polycyclic aromatic hydrocarbon as environmental contaminants: Some results from Bangkok. The third international symposium of ETERNET-APR: Conservation of the hydrospheric environment. Bangkok. December 1996.

[7]   Venkataraman C, Thomas S, Kulkarni P (1999) Size distributions of polycylic aromatic hydrocarbons – Gas/particle partitioning to urban aerosols. Journal of Aerosol Science 30: 759-770.

[8]   Pollution Control Department, Ministry of Natural Resources and Environment, Thailand (2010) Thailand State of Pollution Report 2010.

[9]   European Environment Agency (2004) Air pollution in Europe 1990-2000. Topic report.

[10] Panther B, Hooper MA, Tapper NJ (1999) A comparison of air particulate matter and associated polycyclic aromatic hydrocarbons in some tropical and temperate urban environments. Atmospheric Environment 33: 4087-4099.

[11] Garivait H, Polprasert C, Yoshizumi K, Reutergardh LB (2001) Airborne polycyclic aromatic hydrocarbons (PAH) in Bangkok urban air: Part II. Level and distribution. Polycyclic Aromatic Compounds 18: 325-350.

[12] Chetwittayachan T, Shimazaki D, Yamamoto K (2002) A comparison of temporal variation of particle-bound polycyclic aromatic hydrocarbons (pPAHs) concentration in different urban environments: Tokyo, Japan, and Bangkok, Thailand. Atmospheric Environment 36: 2027-2037.

[13] Boonyatumanond R, Murakami M, Wattayakorn G, Togo A, Takada H (2007) Sources of polycyclic aromatic hydrocarbons (PAHs) in street dust in a tropical Asian megacity, Bang-kok, Thailand. The Science of the Total Environment 384: 420-432.

[14] Nielsen T, Jorgensen HE, Larsen JC, Poulen M (1996) City air pollution of polycyclic aromatic hydrocarbons and other mutagens: occurrence, sources and health effects. The Science of the Total Environment 189/190: 41-49

[15] Chetwittayachan T (2002) Temporal variation of particle-bound polycyclic aromatic hydro-carbons (pPAHs) concentration and risk assessment of their possible human exposure in urban air environments. PhD thesis submitted to department of urban engineering, graduate school of engineering, The University of Tokyo

[16] Chetwittayachan T, Kido R, Shimazaki D, Yamamoto K (2002) Diurnal profiles of particle-bound polycyclic aromatic hydrocarbon (pPAH) concentration in urban environment in To-kyo metropolitan area. Water, Air, and Soil Pollution: Focus 2: 203-221

[17] Fine PM, Chakrabarti B, Krudysz M, Schauer JJ, Sioutas C (2004) Diurnal Variations of Individual Organic Compound Constituents of Ultrafine and Accumulation Mode Particulate Matter in the Los Angeles Basin. Environmental Science and Technology 38: 1296-1304.

[18] MSP Corporation (1998) Micro-Orifice Uniform Deposit Impactor Instruction Manual, Model 100/ Model 110

[19] Marple VA, Rubow KL, Behm, SM (1991) A microorifice uniform deposit impactor (MOUDI): Description, calibration and use. Aerosol Science and Technology 14: 434-446.

[20] EcoChem Analytics (1999) User's guide: Realtime PAH monitor PAS2000CE.

[21] Burtscher H, Schmidt-Ott A (1986) In situ measurement of adsorption and condensation of a polycyclic aromatic hydrocarbon on ultrafine c particles by means of photoemission. Journal of Aerosol Science 17: 699-703.

[22] Burtscher H (1992) Measurement and characteristics of combustion aerosols with special consideration of photoelectric charging and charging by flame ions. Journal of Aerosol Science 23: 549-595.

[23] Kittelson DB (1998) Engines and Nanoparticles: A Review, J. Aerosol. Sci. 29:575-588.

[24] Keshtkar H, Ashbaugh LL (2007) Size distribution of polycyclic aromatic hydrocarbon particulate emission factors from agricultural burning. Atmospheric Environment 41: 2729-2739.

[25] Phuleria HC, Sheesley RJ, Schauer JJ, Fine PM, Sioutas C (2007) Roadside measurements of size-segregated particulate organic compounds near gasoline and diesel-dominated freeways in Los Angeles, CA. Atmospheric Environment 41: 4653-4671.

[26] Kamens RM, Guo Z, Fulcher JN, Bell DA (1988) Influence of humidity, sunlight, and temperature on the daytime decay of polycyclic aromatic hydrocarbons on atmospheric soot particles. Environmental Science and Technology 22: 103-108.

[27] Lungren DA, Paulus HJ (1975) The mass distribution of large atmospheric particles. Journal of the Air Pollution Control Association 25: 1227-1231.

[28] Pierce RC, Katz M (1975) Dependency of polynuclear aromatic hydrocarbon content on size distribution of atmospheric aerosols. Environmental Science and Technology 9: 343-353.

[29] Venkataraman C, Lyons JM, Friendlander SK (1994) Size distributions of polycyclic aromatic hydrocarbons and elemental carbon. 1. Sampling, measurement methods, and source characterization. Environmental Science and Technology 28: 555-562.

[30] Venkataraman C, Friedlander SK (1994) Size distribution of polycyclic aromatic hydrocarbons and elemental carbon. 2. Ambient measurements and effects of atmospheric processes. Environmental Science and Technology 28: 563-572.

[31] Wiest F, Fiorentina HD (1975) Suggestions for a realistic definition of an air quality index relative to hydrocareonaceous matter associated with airborne particles. Atmospheric Environment 9: 951-954.

[32] Garivait H, Polprasert C, Yoshizumi K, Reutergardh LB (1999) Airborne polycyclic aromatic hydrocarbons (PAH) in Bangkok urban air I. Characterization and quantification. Polycyclic Aromatic Compounds 13: 313-327.

[33] Miguel AH, Kirchstetter TW, Harley RA (1998) On-road emissions of particulate polycyclic aromatic hydrocarbons and black carbon from gasoline and diesel vehicles. Environmental Science and Technology 32: 450-455.

[34] Zielinska B, Sagebiel J, McDonald JD, Whitney K, Lawson DR (2004) Emission rates and comparative chemical composition from selected in-use diesel and gasoline fueled vehicles; J. Air & Waste Manage. Assoc. 54(9):1138-50.

[35] Hoshiko T, Nakajima F, Prueksasit T, Yamamoto K (2012): 4. Health risk of exposure to vehicular emissions in wind-stagnant street canyons, In: "Ventilating cities -Air-flow criteria for healthy and comfortable urban living-" (Eds.: Kato S and Hiyama K), pp.59-95, Springer

Natural vs Anthropogenic Background Aerosol Contribution to the Radiation Budget over Indian Thar Desert

Sanat Kumar Das

Additional information is available at the end of the chapter

1. Introduction

In recent times, atmospheric aerosols are receiving increasing attention as they directly affect the Earth's radiation balance by altering incoming shortwave solar radiation that can cause positive (heating) or negative (cooling) radiative forcing depending on their scattering and absorption properties, the reflectivity of the underlying surface [10, 24] and the position of aerosols with respect to the global cloud coverage [8, 88]. Aerosols also affect the outgoing longwave radiation by absorption, emission and scattering. Presently, effects of radiative forcing of atmospheric aerosols on climate is a subject of great concern to atmospheric researchers. An accurate quantification of the aerosol direct radiative forcing is critical for the interpretation of existing climate records and also for the projection of future climate change [11, 47]. Significant amount of atmospheric radiative forcing causes high atmospheric heating due to strong absorption of solar radiation which can change the regional atmospheric stability and may alter the large scale circulation and the hydrological cycle, enough so, apparently, to account for observed temperature and precipitation changes in China and India [1, 46, 62, 70]. Therefore, the effect of aerosols on the radiation budget in terms of radiative forcing calculations is challenging and demanding, especially on the regional scale for the exclusive understanding of climate change.

The uncertainties involved in the climate models are mainly due to optical properties of aerosols on the regional scale, specially underestimated absorption of solar radiation by aerosols, both, naturally and anthropogenically produced [34], their residence time [57, 58], etc, which arise mostly due to lack of observations. Black carbon (BC) or soot and dust aerosols are playing the leading role in aerosol interaction with the solar radiation due to their strong absorption properties. BC comes into the atmosphere during combustion of fossil fuels, principally, diesel and coal, and from biomass burning. BC demands large attention due to its strong absorption of incoming solar radiation and produces positive radiative forcing which is sometimes comparable to the forcing of the green-house gas methane [31].

Therefore, underestimation of BC can introduce large uncertainty in the climate models. On the other hand, dust, mainly coming from arid regions, is generally known for scattering of solar radiation. However, dust also has a strong absorption in the UV and infrared regimes and therefore, can influence radiative forcing not only in the shortwave region but also in the longwave region. Hence, the study of dust aerosols is equally important. In addition, long-range transported dust aerosols can enhance the atmospheric radiative forcing in the presence of soot aerosols [14, 54].

South-East Asia, with its fast growing urbanization and industrialization, is one of the major hot-spot regions on the global aerosol map. A study of historical records from different locations on the globe reported an increasing trend of BC emissions in South and Central Asia [6]. In addition, dust aerosols are also transported from the Middle-East region to over South-East Asia. A mixture of locally produced anthropogenic aerosols with natural aerosols like mineral dust and seasalt, reported over this hot spot region [42, 60–62] aids in the warming of the atmosphere. There were several campaigns of ship-, land- and air-borne measurements over Indian subcontinent and surrounding marine regions to investigate the regional effects of anthropogenic aerosols [32, 48, 75, etc.]. In-situ measurements during the Indian Ocean Experiment (INDOEX) and several campaigns under Indian Space Research Organisation – Geosphere Biosphere Programme (ISRO–GBP) found that the sources of the anthropogenic aerosols are biomass burning and fossil-fuel combustions [33, 61]. The second phase of the ISRO-GBP land campaign during winter conducted in the Indo-Gangetic Plain (IGP), a hot-spot region over India, reported significant anthropogenic aerosol loading in the atmosphere coming from industries and vehicular emissions [15, 18, 50]. Satellite-based observations suggested that significant amount of dust is also transported over IGP from Thar Desert located in western India during premonsoon (March to May) [16, 17, 54]. This transported dust helps to sustain the hot-spot over IGP maintaining the large background aerosol loading. Majority of the earlier research works focused on aerosol contribution, either locally produced anthropogenic aerosols or transported natural dust, to regional climate change over this hot-spot region. However, uncertainties in those results are found to be relatively large, especially in studies on transported dust, as the dust becomes aged by externally and internally mixing with locally produced pollutants.

This chapter investigates and quantifies the natural and anthropogenic contribution of background aerosols over western India where both the source regions, Thar Desert, source of natural dust, and IGP, hot spot region of anthropogenic aerosols, are present. The contributions of both types of aerosols are estimated for the years 2006 and 2007 from ground-based and satellite measurements of aerosol optical and physical properties. Ground-based observations have been conducted at Mt. Abu (24.65°N, 72.78°E, 1.7 km asl), the highest location in Aravalli mountains in western India. The main advantage of the location is its proximity to both, Thar Desert and IGP. Also, due to the high altitude, the observation site is less affected by the boundary layer aerosols. The hill-top background aerosols are significantly influenced by wind that carries aerosols from either Thar Desert or IGP and show strong seasonal variation. Therefore, the site becomes a unique location for the investigation of both, natural and anthropogenic aerosols. The present study investigates the seasonal variation of aerosol properties at Mt. Abu and estimates the contribution of both aerosols on the radiation budget during the four seasons – winter (Dec-Feb), premonsoon (Mar-May), monsoon (Jun-Aug), and postmonsoon (Sep-Nov).

2. Datasets

2.1. Ground-based instruments

2.1.1. Microtops

Aerosol Optical Depth (AOD) was measured using a hand-held Microtops II (Solar Light Co., Inc., USA) [49] at every five minutes interval during daytime from 0730 to 1600 hours. This instrument can measure AOD at five different wavelengths centered at 0.380, 0.440, 0.500, 0.675, 0.870 μm simultaneously. Another Mictotops II was used to measure AOD at 1.020 μm associated with ozone and columnar water vapor . Both Microtops were regularly calibrated, once in a month, and all calibrated constants were obtained from Langley's plot analysis [30]. There is only 1% variation in the calibration constant since 2002. The absolute uncertainty of measured AOD is not more than 0.03 at all wavelengths [33].

2.1.2. QCM

Aerosol Mass Concentration was measured using a 10-stage Quartz Crystal Microbalance (QCM) cascade impactor (model PC-2, California Measurements Inc., USA) and the aerosol size distribution at the ground level was determined. Aerosols were collected in 10 stages of the impactor with cut-off radii at 12.5, 6.25, 3.2, 1.6, 0.8, 0.4, 0.2, 0.1, 0.05 and 0.025 μm from stage 1 to 10 respectively. The air flow rate through the impactor was kept at 240 ml/min and the typical sampling period was 300 sec for each measurement. The QCM was operated from the terrace of the observatory building at a height of about 6m. The air inlet was installed vertically to minimize the loss of aerosol particles within the inlet tube. The relative temperature change of the crystals during each sampling period of 5 minute is too small and can be neglected. Uncertainties involved in the QCM measurements are mainly due to variations in RH [15, 33, 59]. In an earlier study, the QCM was operated simultaneously with an Anderson impactor to investigate the measurement accuracy of each stage and it was observed that measurement error is always less than 15% [33].

2.1.3. Aethalomater

Absorbing aerosol mass concentrations at seven different wavelengths (centered at 0.37, 0.47, 0.52, 0.59, 0.66, 0.88 and 0.95 μm) were obtained using a multichannel Aethalometer (model AE-42) manufactured by Magee Scientific, USA [21]. The flow rate of ambient air was maintained at 3.0 l.min^{-1} and the data was stored in the memory disk at a time interval of two minutes during the measurement period. BC mass concentration is estimated by detecting the light transmitted through the particle deposited sample spot and particle free reference spot on the filter as follows [5, 89]

$$BC = -\frac{A\,100\ln\left(\frac{I_2}{I_1}\right)}{kQ\Delta t} \tag{1}$$

where I_1 and I_2 are the ratios of light intensities of the sample beam to the reference beam before and after particle sampling at time interval Δt, Q is the volume flow rate of the ambient air through the filter, A is the area of the sample spot and k is the absorption coefficient. The real-time BC mass concentration is considered at 0.88 μm wavelength channel because the spectral response of elemental carbon particles has a peak near this wavelength [5]. The manufacturer quoted the overall uncertainty in aethalometer data to be about 10%, which is calculated by comparing the data of the aethalometer to other instruments that measure BC using different techniques [2]. However, Weingartner et al. [89] reported that BC measurements using filter techniques have significant uncertainty due to "shadowing effect" after investigating several types of carbon aerosols. This effect causes underestimation of BC measurement due to its high loading on the filter. This effect is very pronounced for pure BC while almost negligible for aged atmospheric aerosols. This uncertainty is found to be less than 10% [14].

2.2. Space-borne measurements

2.2.1. MODIS

AOD over Mt. Abu is also obtained from observations of the Moderate Resolution Imaging Spectroradiometer (MODIS) sensors on-board Terra and Aqua satellites. Terra and Aqua spacecrafts pass over the equator at 10:30 and 13:30 Local Solar Time, respectively [43]. Global images of the full disc are produced due to larger swath widths and instrument-scanning angle of 110° [44]. MODIS has 36 channels spanning the spectral range from 0.41 to 14.4 μm at three spatial resolutions: 250 m (2 channels), 500 m (5 channels) and 1 km (29 channels). MODIS aerosol algorithm consists of three independent algorithms to retrieve the aerosol characteristics, two over land and one over oceans, and makes use of eight of these channels (0.47-2.13 μm) [29, 35, 67]. The measurements at other wavelengths provide information to identify clouds and river sediments [20, 45]. MODIS provides an accurate retrieval of spectral AOD and the parameters characterizing aerosol size [79, 80]. The retrieved data used in this study include both Terra and Aqua MODIS aerosol products; such estimations are made over cloud-free regions only [67]. Long-term analysis of MODIS aerosol retrievals collocated with AERONET measurements confirm that MODIS retrieved AOD agrees with AERONET observations to within 0.10 over land and to within 0.035 over oceanic island sites. There are several studies demonstrating that MODIS AOD has a strong correlation with AERONET AOD [41, 83, etc.] and thereby provide enough confidence to use the MODIS AOD over western India, the region of interest in the current study.

2.2.2. OMI

Aerosol index (AI) is obtained from observations in the UV region (UV-1, 0.270 to 0.314 μm; UV-2, 0.306 to 0.380 μm) of the Ozone Monitoring Instrument (OMI) on-board Aura satellite [82]. AI is defined as the difference between satellite measured (including aerosol effects) spectral contrast at 0.360 and 0.331 μm radiances and the contrast theoretically calculated from radiative transfer model for pure molecular (Rayleigh) atmosphere [9, 25, 28]. The Aura satellite launched in July 2004, flies eight minutes after the Aqua satellite as a part of NASA A-train constellation. OMI has been designed for the replacement of Total Ozone Mapping Spectrometer (TOMS) to continue recording the total ozone and other atmospheric parameters related to ozone and climate study. OMI is sensitive to aerosol absorption even when aerosols are present above the cloud. Therefore, AI can be successfully derived for clear as well as cloudy conditions. OMI has a spatial resolution of 13×24 km at nadir and uses the same algorithm that is used for TOMS observations. AI provides a quantitative measurement of UV-absorbing aerosols over all the terrestrial surfaces including deserts and ice sheets. AI is positive for absorbing aerosols and negative for non-absorbing aerosols. Zero AI indicates cloud presence. High OMI-AI values with high MODIS-AOD and low Ångström exponent represent dust dominated regions and such high AI values are mainly observed over arid regions [78]. OMI-AI Level 3 global-gridded product with spatial resolution of $0.25° \times 0.25°$ is obtained over western India for identifying the dust dominating periods in the present study.

2.2.3. CALIOP

The Cloud-Aerosol Lidar and Infrared Pathfinder Satellite Observations (CALIPSO) provides a new insight into the role of atmospheric aerosols and clouds in regulating the study of Earth's climate change and air quality. It is a part of the A-train satellite constellation that includes Aqua, CloudSat, and Aura satellites. CALIPSO is in a sun-synchronous orbit at 705 km at an inclination of 98° and provides the vertical distribution of aerosols and clouds. It consists of three sensors: a Cloud-Aerosol Lidar with Orthogonal Polarisation (CALIOP), an Imaging Infrared Radiometer (IIR), and a moderate spatial resolution Wide Field-of-view Camera (WFC). CALIPSO passes over the equator at 13:31 local hours, one minute behind Aqua. The primary instrument, CALIOP, transmits linearly polarized laser light of 0.532 μm and 1.064 μm at a pulse rate of 20.16 Hz. Its receiver measures the backscattered intensity at 0.532 μm and 1.064 μm with the former divided into two orthogonally polarized components which help to calibrate the optically thick clouds and aerosols. CALIOP observes both clouds and aerosols at high spatial resolution, but must be spatially averaged to increase signal to noise ratio. From the surface to 8 km, the vertical resolution is 30 m and the nominal horizontal resolution is 1/3 km. CALIPSO data products provide the aerosol vertical distribution along with aerosol layer height and AOD [7, 85]. CALIPSO LEVEL 2 Vertical Feature Mask (VFM) products provide vertical mapping of the locations of aerosols and clouds together with information about the types of each layer and the discrimination between aerosols and clouds is expected to be good in these products [4, 52, 87, etc.].

2.3. Models

2.3.1. OPAC

OPAC (Optical Properties of Aerosols and Clouds) model [26] is used to derive aerosol optical depth from the measured atmospheric aerosol chemical compositions obtained from literature [39, 40] at Mt. Abu. OPAC model contains two major parts: (1) a dataset of microphysical properties and the resulting optical properties of cloud and aerosol components at different wavelengths and for different humidity conditions, (2) a FORTRAN program that allows the user to extract data from this dataset, to calculate additional optical properties, and to calculate optical properties of mixtures of the stored clouds and aerosol components. In the present study, OPAC model has been used for obtaining the aerosol optical properties in shortwave region (0.25-4 μm) from the known chemical compositions. OPAC, based on Mie theory, can compute aerosol optical properties at 61 wavelengths starting from 0.25 μm to 40 μm. It mainly has 10 aerosol components which are as follows - insolubles (mostly soil particles), water soluble aerosols (mainly sulfate and nitrate aerosols of anthropogenic origin), soot (of anthropogenic origin), sea salts (naturally produced on the oceanic surface by wind and also available in the atmosphere of coastal regions) in accumulation and coarse mode, mineral dust (generally coming into atmosphere from the arid surface by wind) in three modes, transported mineral dust and sulfate droplets (mainly found at stratospheric altitudes). This model is used to derive the AOD spectrum using a combination of aerosol components and in the present study the sulfate droplets are not considered. Some of the aerosol components which are hygroscopic in nature, may change their optical properties, and hence OPAC outputs are available for eight different relative humidity (0%, 50%, 70%, 80%, 90%, 95%, 98% and 99%) conditions. Optical properties for different aerosols are different. Single scattering albedo (SSA) is one of the important optical parameters for aerosol radiative effect calculations. OPAC derived SSA is the weighted average of SSA of all aerosol components. Water soluble (SSA $\approx$ 0.9 at 0.5 μm) aerosols which contain mainly sulfate, nitrate, etc. and seasalt (SSA $\approx$ 0.99 at 0.5 μm) do not absorb significantly in the visible range but they do absorb significantly in the infrared region (SSA $\leq$ 0.4 at 10.0 μm). Major aerosol components are scattering type in the shortwave range (0.25-4.0 μm), whereas, in the longwave range (4.0-40.0 μm) they can be totally absorbing. The SSA of soot in the shortwave is 0.22 (at 0.5 μm), whereas, in the longwave range it is totally absorbing. Dust (SSA $\approx$ 0.98 at 0.5 μm) is mainly scattering in nature in the shortwave range but exhibits strong absorption in UV region and also in the longwave range. On one hand, in the longwave region absorption decreases the outgoing radiation, while on the other hand, the energy re-emitted consequent to this absorption increases the surface reaching infrared radiation. The net SSA over a particular location is the weighted average of SSA of all aerosol components.

2.3.2. SBDART

Atmospheric radiative transfer code, named Santa Barbara DISORT Atmospheric Radiative Transfer (SBDART) [68] developed at the University of California, Santa Barbara, is used to estimate aerosol radiative forcing over the study area. SBDART is a well established code for estimation of radiation flux in the shortwave (0.25-4.0 μm) as well as longwave (4.0-40.0 μm) range. It is a radiative transfer code that computes plane-parallel radiative transfer in clear and cloudy conditions within the Earth's atmosphere and at the surface. In the present study only clear sky conditions are considered. All the important processes that affect the ultraviolet, visible, and infrared radiation, are included in this code. For molecular absorption SBDART uses the low-resolution band models of LOWTRAN-7 atmospheric transmission code [56]. LOWTRAN-7 codes can take into account the effects of all radiatively active molecular species found in the Earth's atmosphere with wavelength resolution of about 5 nm in the visible and about 200 nm in the thermal infrared. In SBDART, the radiative transfer equations are numerically integrated with DISORT (Discreet Ordinate Radiative Transfer) code [74]. This discrete ordinate method provides a numerically stable algorithm to solve the equations of plane-parallel radiative transfer in a vertically inhomogeneous atmosphere. The intensity of both scattered and thermally emitted radiation can be computed at different heights and directions. Presently, SBDART is configured to allow up to 65 atmospheric layers and 20 radiation streams (20 zenith angles and 20 azimuthal modes).

The ground surface cover is an important determinant of the overall radiation environment because spectral albedo of the surface which defines the ratio of upwelling to downwelling spectral irradiance at the surface determines upwelling irradiance from the surface. In SBDART there are five basic surface types, namely (1) ocean water [76], (2) lake water [36], (3) vegetation [65], (4) snow [91], and (5) sand [73]. The spectral albedo describing a given surface is often well approximated by combinations of these basic surface types. Input parameters in SBDART allow the user to specify a mixed surface consisting of weighted combinations of water, snow, vegetation and sand. SBDART can compute the radiative effects of several lower and upper atmosphere aerosol types. In the lower atmosphere, typical rural, urban,

or maritime conditions can be simulated using the standard aerosol models of Shettle &
Fenn [72]. SBDART gives the opportunity to specify up to five aerosol layers (i.e., at five
different altitudes), with radiative characteristics that model fresh or aged volcanic, meteoric,
and upper-tropospheric background aerosols.

The major inputs required to estimate the aerosol radiative effects for DISORT module in
SBDART include spectral values of solar radiation incident on the atmosphere, spectral
values of columnar AOD, SSA and angular phase function of the scattered radiation or
asymmetry factor. The asymmetry factor is used to generate a scattering phase function
through the Henyey-Greenstein approximation. The Henyey-Greenstein parameterization
provides good accuracy when applied to radiative flux calculations [22, 84]. It can also
compute radiation fluxes with less uncertainty from the aerosol optical properties at 0.55
micron wavelength obtained from satellite observations. Spectral values of AOD, SSA and
asymmetry parameter are also obtained from OPAC using the chemical properties of the
atmospheric aerosols. OPAC model derived aerosol optical parameters are obtained by
varying the number concentration of individual components in small steps until the model
derived parameters satisfactorily match the observed values. Another important input
parameter that is required for accurate computation of the aerosol radiative effects over land is
the surface reflectance [71, 90]. Radiative forcing is determined from the difference of the solar
radiation with and without aerosols during clear-sky conditions in the short wave (0.25-4.0
μm) by running SBDART for every one hour interval in a day using the profiles for tropical
atmosphere. The present work gives more realistic results considering the aerosol vertical
profiles from CALIPSO and MODIS surface reflectance over Mt. Abu. The seasonal forcing
is estimated from the diurnally averaged forcing which represents the mean of the hourly
forcing as derived from SBDART for 24 h/day.

3. Site location and meteorology

Major aerosol parameters have been monitored during 2006 and 2007 inside the campus
of Physical Research Laboratory situated at Gurushikhar, Mt. Abu – the highest peak (1.7
Km asl) of Aravalli range in India. Topography of the Indian Peninsula, Himalayas and the
Tibetan plateau are shown in Figure 1a. Arid (dashed line) and semi-arid (solid line) regions
of Thar Desert in western India are shown in Figure 1b. More details on physical features
of Thar Desert are described in literature [92]. Mt. Abu is situated within the semi-arid
region of Thar Desert. A picture of the campus is shown in Figure 1c, which is better known
for the astronomy observatory. Aravalli mountains are located in between Thar Desert and
IGP. Major part of these mountains on the western side is in the semi-arid region of Thar
Desert while the north-east region of the mountains is in IGP. The highest location, Mt. Abu
is situated in the south-west of the mountain range. The observatory being a prohibited
hilltop area makes the measurement site anthropogenic free and hence, is a suitable place
for background aerosol measurements in western India. The observatory is built on rocky
mountainous terrain surrounded by forest and therefore, there is significantly less soil dust
coming from the surface of the nearby mountain region. Being very close ($\sim$300 Km) to Thar
Desert, measurement site gives an opportunity to study desert dust. Freshly generated desert
dust aerosols are transported within few hours to Mt. Abu and thereby are exposed to local
pollutants minimally. Also, due to the high altitude, these aerosols are less influenced by the
boundary layer aerosols that consist mostly of locally produced anthropogenic aerosols.

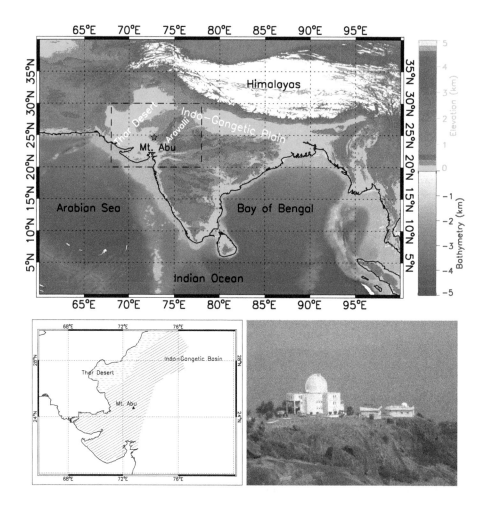

Figure 1. (a) Topography of the Indian subcontinent. The box is showing western India including Thar Desert and Indo-Gangetic Plain. Star shows the location of Mt. Abu, highest location in the Aravalli Mountains. (b) Arid (dashed lines) and semi-arid (solid lines) region of Thar Desert. Mt. Abu is situated in the semi-arid region. (c) A photograph of the measurement site - PRL observatory at Mt. Abu.

Diurnal variations of surface temperature and relative humidity (RH) at Mt. Abu during different seasons are shown in the top row of Figure 2. The vertical bars represent the $\pm 1\sigma$ variation about the hourly mean values. Temperature is found to be minimum at $15.5 \pm 3.0°$C during winter and maximum at $23.0 \pm 3.3°$C during premonsoon, followed by monsoon ($19.3 \pm 2.0°$C) and postmonsoon ($18.3 \pm 1.4°$C). In case of RH, minimum is observed at $24.6 \pm 6.4\%$ during premonsoon and maximum at $88.3 \pm 10.5\%$ during monsoon, followed by $54.7 \pm 22.8\%$ during postmonsoon and $30.7 \pm 8.3\%$ during winter. There is a strong diurnal variation in the hourly averaged surface temperature at Mt. Abu during all seasons, whereas, there is no significant variation present in RH. During monsoon and postmonsoon RH shows large variations about the means. During monsoon all the measurements of aerosol parameters were carried out only in June (mean RH = 63.9±12.6%). Observations were very few in July and August due to heavy rain and high RH. The seasonal variation of wind pattern over India subcontinent, obtained from National Center for Environmental Prediction (NCEP) reanalysis data is shown in the bottom row of Figure 2. Wind speed over study region was minimum and mainly coming from IGP during winter,. During premonsoon, wind over western India was westerly and coming from desert areas. During monsoon and postmonsoon, wind became stronger coming from coastal region of Arabian Sea.

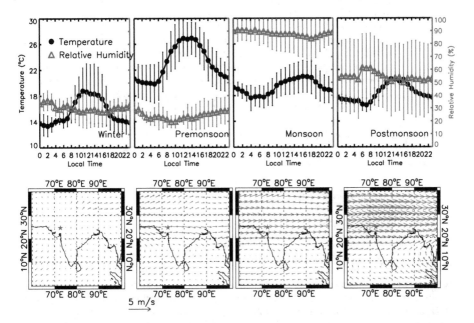

Figure 2. (Top) Diurnal variation of temperature and relative humidity during different seasons. The vertical bars represents ±1σ variation about the mean. (Bottom) Seasonal variation of wind speed and direction, obtained from NCEP reanalysis data.

3.1. Land surface properties

Underlying surface plays an important role in the aerosol radiative effects towards climate change [24, 27, etc.]. Aerosols over high surface reflectance (bright surface) can produce relatively higher positive radiative forcing than those over low surface reflectance (dark surface). Space-borne observations suggest that there is a strong seasonal variation of surface over western India. Figures 3a and 3b show images of land surface over western India during premonsoon and postmonsoon seasons, respectively, captured by MODIS-Terra satellite. The surface is very bright during premonsoon due to open bare land, while it is relatively dark during postmonsoon due to green vegetation born during monsoon rain. As a result, surface reflectance is maximum during premonsoon and minimum during postmonsoon.

In the present study, MODIS derived surface reflectance data over Mt. Abu is used in the estimations of radiative forcing. It is obtained from Nadir BRDF-Adjusted Reflectance 16-Day L3 Global 0.5 km SIN Grid product which is derived at the mean solar zenith angle of Terra overpasses for every successive 16-day period, calculating surface reflectance as if every pixel in the grid was viewed from nadir direction. Surface reflectance data available in seven wavelength bands of MODIS centered around 0.47, 0.56, 0.65, 0.86, 1.24, 1.64 and 2.13 μm are used to reproduce the spectral dependence of surface reflectances for the entire SW range using a combination of three different surface types, namely, vegetation, sand and water. The monthly variation of surface reflectance at 1.64 μm during 2007 is shown in Figure 3c. Vertical lines in this Figure represent $\pm 1\sigma$ variation about the monthly mean values. Average surface reflectance is found to be high at about 0.35 during premonsoon (Apr-May) and low at about 0.20 during postmonsoon (Sep-Nov) and winter (Dec-Feb). Space-borne observations show that the land over western India increases its brightness by about 75% during premonsoon season. This could be due to bare surface and deposited dust that is transported from arid region. Model simulations to fit the surface reflectance combining the three surfaces suggest that during premonsoon sand surface contributes a maximum of about 70% and during postmonsoon and winter it contributes a minimum of about 20% while vegetation surface contributes 15% and 60%, respectively. These varying land properties are also considered in the radiative forcing calculations.

4. Background aerosol optical and physical properties over Thar desert

4.1. Aerosol optical depth

The seasonal variation of AOD spectrum at the hilltop station over western India is shown in Figure 4. Vertical lines represent $\pm 1\sigma$ variation about the mean AOD. The solid lines are the OPAC model fitted AOD spectrum and the shaded regions indicate the variation of simulated AOD within that season. At all wavelengths, AODs are maximum during premonsoon followed by postmonsoon and monsoon and are minimum during winter. At 0.5 μm, the AODs are 0.20±0.08, 0.18±0.04, 0.10±0.02, and 0.08±0.03 during premonsoon, postmonsoon, monsoon and winter, respectively. The reasons could be as follows. Firstly, there is a large variation of boundary layer height. In an earlier research work carried out over a tropical Indian station, Gadanki (13.5^{o}N, 79.2^{o}E), Krishnan & Kunhikrishnan [37] studied the annual boundary layer height variation and observed a minimum during winter and maximum during premonsoon. In the present study, during winter the boundary layer height is lower than the observation altitude and hence the observation site is in the free troposphere region. As a result, AOD is minimum during this season. During premonsoon the observation site is within the boundary layer and hence AOD increases. In addition, there is significant amount of dust transported from arid region which results in maximum AOD. On the other hand, during monsoon though the boundary layer height is significantly high, AOD is low due to wash out of aerosols from the atmosphere by the heavy monsoonal rain events. Monsoon rain has a major role to wash out the aerosol loading from the atmosphere causing significant decrease of AOD. A case study over tropical Indian station reported about 64% decrease of AOD due to heavy rain [69]. In the present study, there is no significant decrease of AOD during monsoon. This is because of the presence of a very stable aerosol layer of about 1.5

Figure 3. Picture of land surface over western India during premonsoon (top) and potmonsoon (middle) seasons, obtained from MODIS-Terra satellite. Dark black, gray and green colors represent oceanic surface, arid bare land, forest regions, respectively. (Bottom) Monthly variation of surface reflectance at 1.64 μm wavelength during 2007 and the vertical bars represents $\pm 1\sigma$ deviation about the monthly mean. Note that the surface reflectance increases by 70% over study area during premonsoon.

km thickness over the inversion layer during monsoon in the western India, as reported by Ganguly et al. [19].

The spectral dependence of AOD is parameterized through Ångström exponent (α) which is the slope of the logarithm of AOD versus the logarithm of wavelength (in micron) and provides the basic information about the columnar particle size distribution [66]. α is higher for relatively higher number of smaller particles and as the number of bigger particles increases α decreases. It can even reduce to ~ 0 for very large number of coarse-mode soil particles [51, 77, 81, etc]. In the present study, α is obtained from Microtops measured AOD for the entire wavelength (0.380 - 1.020 μm) and is given in Figure 4 along with the variation in the parenthesis. It varies from 0.2 to 0.6. During monsoon α is minimum at 0.2 ± 0.15 indicating dominance of bigger aerosols. It is due to the presence of bigger water soluble aerosols which increase in size due to accumulation and coagulation processes in high relative humidity conditions. During premonsoon also, when RH is low and the atmosphere is dry and warm, α is low at about 0.3 ± 0.25 indicating the dominance of bigger aerosols. These are the soil born dust aerosols produced by the frequently occurring dust storms in the Thar desert and transported to other parts of India [17, 55] including Mt. Abu. α is found to be maximum at about 0.6 ± 0.01 during postmonsoon indicating dominance of smaller aerosols. This is probably due to dominance of fine seasalt aerosols transported from Arabian Sea [19, 64] . During winter also α is found to be high showing dominance of smaller aerosols which could be due to anthropogenic aerosols coming from burning tree branches and dry leaves by the poor villagers living in the surrounding hill areas to keep themselves warm during cold mornings and evenings.

4.2. Aerosol mass concentration

Aerosol mass concentration measured separately in ten different sizes by Quartz Crystal Microbalance (QCM) cascade impactor has been classified into three different categories, viz., nucleation (radius<0.1 μm), accumulation ($0.1\mu m \leq radius \geq 1.0$ μm) and coarse (radius>1.0 μm) mode particles. Nucleation mode aerosols represent total aerosols collected in stages 9-10, accumulation mode aerosols are the total aerosols in stages 5-8 and coarse mode aerosols are the total aerosols collected in stages 2-4. Aerosols collected in stage 1 are not considered in the calculations because all aerosols whose radius is greater than 12.5 μm are collected in this stage and thus, there is no definite aerosol radius representing this stage.

Figure 5 shows the seasonal variation of aerosol mass concentration (μg.m^{-3}) of all three modes, viz., nucleation, accumulation and coarse modes at the hill top region, Mt. Abu from January 2006 to December 2007. Total aerosol mass concentration observed was minimum at 16.5 ± 1.5 μg.m^{-3} during winter and maximum at 25.8 ± 2.7 μg.m^{-3} during postmonsoon followed by premonsoon (19.9 ± 5.6 μg.m^{-3}) and monsoon (16.7 ± 6.0 μg.m^{-3}). This variation is similar to that observed in columnar AOD at Mt. Abu. The accumulation aerosol mass was contributing maximum to the total aerosol mass during all seasons and the coarse mode aerosol mass was contributing equivalently only during premonsoon. This is due to large transportation of dust aerosols from Thar desert during this season that enhanced the coarse mode aerosol mass.

In general, nucleation aerosols contribute least to the total aerosol mass concentration. This contribution was maximum during postmonsoon when the wind speed was almost calm and RH was relatively high. This atmospheric condition helps in gas-to-particle conversion and

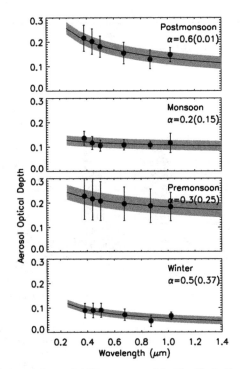

Figure 4. Seasonal variation of observed AOD spectrum at Mt. Abu. Vertical bars represent ±1σ variation about the mean. Solid line is the OPAC model derived AOD spectrum and the shaded regions are the variation in the simulated AOD. Ångström exponent (α) is also given along with the standard deviations in parenthesis.

enhances the nucleation mode aerosols which explains the maximum mass of $6.4 \pm 1.1 \ \mu g.m^{-3}$ observed during this season. During winter the nucleation mass concentration decreased to $3.2 \pm 0.1 \ \mu g.m^{-3}$ as the boundary layer height decreased and the measurement site was in free troposphere. The nucleation aerosol mass was 2.9 ± 1.8 and $3.6 \pm 0.7 \ \mu g.m^{-3}$ during premonsoon and monsoon, respectively. During premonsoon, the boundary layer height was maximum which gives more room for these fine aerosols to dilute and high temperature with low RH are not favorable for gas-to-particle conversion processes. In addition, strong wind also helps in removing the aerosols from the measurement site during this season and makes mass of the nucleation mode aerosols minimum. Monsoon also experiences high boundary layer height and strong wind condition, however, nucleation mode aerosols are significant compared to premonsoon. This could be due to the transport of seasalt coming from Arabian Sea.

The seasonal variation observed in the accumulation aerosols is similar to the nucleation aerosols. The accumulation aerosol mass concentration was minimum at $8.4 \pm 2.8 \ \mu g.m^{-3}$ during premonsoon and maximum at $12.6 \pm 0.6 \ \mu g.m^{-3}$ during postmonsoon followed by monsoon ($10.0 \pm 1.0 \ \mu g.m^{-3}$) and winter ($9.6 \pm 1.6 \ \mu g.m^{-3}$). Accumulation aerosols are mainly produced by the condensation growth and coagulation of nucleation aerosols. During

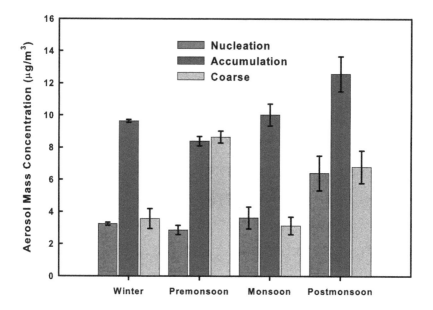

Figure 5. Seasonal variation of aerosol mass concentrations in nucleation (radius < 0.1 micron), accumulation (0.1≥radius≤1.0 micron) and coarse (radius >1.0 micron) modes. The vertical bars represent ±1σ variation about the mean.

premonsoon these processes are slowed down due to low RH and hence the low accumulation aerosol mass. During monsoon seasalt aerosols are transported from the Arabian sea and high RH maintains the accumulation mode aerosols, increasing the mass concentration. During postmonsoon, minimum wind speed results in further increase resulting in the observed maximum. And during winter it is minimum as the the measurement site is in the free troposphere.

The coarse mode aerosols show a slightly different seasonal behaviour at Mt. Abu. During premonsoon, they mainly consist of dust aerosols transported from Thar desert and the mass concentration is maximum at 8.6±0.4 μg.m^{-3}. It is minimum at 3.1±0.5 μg.m^{-3} during monsoon due to wash out of the dust aerosols by heavy rains. During postmonsoon, the coarse aerosol mass concentration was slightly enhanced to 6.8±1.0 μg.m^{-3} as the accumulation aerosols, which mainly consist of seasalt particles, swell up by absorbing water vapor at high RH conditions and become coarse mode particles. During winter, their mass concentration becomes 3.6±0.6 μg.m^{-3} when low boundary layer height helps to keep them low at the hill-top region.

4.3. Aerosol number concentration

Aerosol number concentration is also obtained from the observed aerosol mass concentration from QCM observations for the hilltop area using appropriate mass density valid for semi-arid background atmosphere and prevailing relative humidity conditions [13, 26]. Figure 6 shows the typical aerosol size distributions for the four seasons. The vertical bars represent $\pm 1\sigma$ variation about the monthly mean number concentration of different sizes of aerosols. In all seasons, the size distribution showed tri-modal distribution and each mode could be fitted using using three lognormal modes of the following form.

$$\frac{dn(r)}{dr} = \frac{N}{\sqrt{2\pi} \log \sigma_m} \exp \left[-\frac{\log^2 \left(\frac{r}{r_m} \right)}{2 \left(\log \sigma_m \right)^2} \right] \tag{2}$$

where N is the number concentration (cm^{-3}), σ_m is the width of the distribution and r_m is the mode radius for a particular mode. The three modal parameters for all the seasons are given in Table 1. At Mt. Abu the number concentrations (N) of nucleation and accumulation modes are lower by an order of magnitude than that at other urban region in western India, Ahmedabad while for coarse mode it is comparable [19]. Since Mt. Abu is far from anthropogenic activity, the anthropogenically influenced modes (nucleation and accumulation) have smaller number concentrations. However, the proximity to Thar desert and similarity of the surface conditions of Mt. Abu make the coarse mode number concentrations comparable. The radii of nucleation mode lie in the range 0.018-0.020 μm and number concentration for this mode is found to be maximum during postmonsoon and minimum during premonsoon and monsoon. Similarly the radii for corresponding accumulation and coarse modes lie in the range of 0.12-0.19 μm and 1.1-2.1 μm, respectively.

Season	Nucleation			Accumulation			Coarse		
	N cm^{-3}	r_m μm	σ μm	N cm^{-3}	r_m μm	σ μm	N cm^{-3}	r_m μm	σ μm
Winter	12000	0.019	2.0	18	0.14	2.0	0.02	1.4	1.9
Premonsoon	10000	0.018	1.9	22	0.19	1.8	0.01	2.2	1.8
Monsoon	15000	0.018	1.9	50	0.13	1.9	0.01	1.7	1.8
Postmonsoon	17000	0.020	1.9	60	0.12	1.9	0.08	1.1	1.8

Table 1. Average values of size distribution parameters obtained by fitting lognormal curves to the measured aerosol number distribution over Mt. Abu

Accumulation aerosols are mainly produced by the condensation growth and coagulation of nucleation aerosols. During winter accumulation mode aerosols number concentration (N) was minimum at 18 cm^{-3}. During premonsoon, the anthropogenic activities were maximum at Mt. Abu which increased N of accumulation mode to 22 cm^{-3}. During monsoon, it further increased to 50 cm^{-3}. It is due to the wind coming from Arabian sea (Figure 2) that carried large amount of sea salt and enriched the sea salt aerosols at the hill top region [64]. During high RH conditions these sea salt aerosols belong to the accumulation mode. Later during postmonsoon, wind was south-easterly and the transported sea salt reduced. However, burning of biomass like garbage and fallen leaves increased and hence, BC particle concentration was enhanced. Therefore, high production, shallow boundary layer height and low wind speed made the accumulation mode aerosol number concentration reach a maximum at 60 cm^{-3} during this season.

During premonsoon, there is large transportation of mineral dust aerosols from Thar Desert which enhanced the abundance of coarse mode aerosols at the hill top area. The coarse mode radius was maximum at 2.2 μm. During monsoon, rain washes out these dust aerosols from the atmosphere and reduces their number and mode radius. However, the abundance

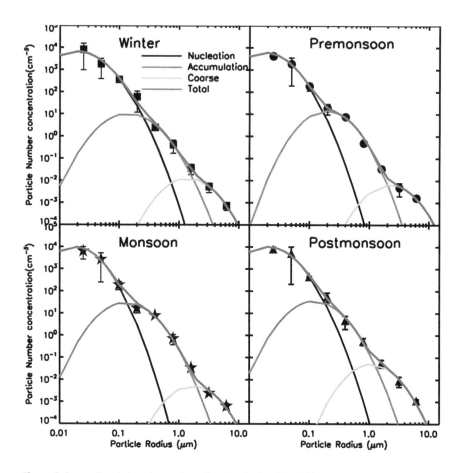

Figure 6. Seasonal variation of aerosol number distribution. Vertical bars represent $\pm 1\sigma$ variation about the mean. The solid lines are the best-fitted curves representing nucleation, accumulation and coarse modes.

of coarse aerosols is found to be maximum with minimum mode radius of about 1.1 μm during postmonsoon which indicates the transfer of aerosols from accumulation mode due to hygroscopic and coagulation growth of particles at high RH conditions.

4.4. Black carbon mass concentration

Black carbon (BC) produced due to incomplete combustion of carbon-based fuels [3, 31, 53, 86, etc] is the most efficient light absorbing aerosol component in the atmosphere. BC has major contribution to alter the radiative balance by absorbing the solar radiation in the visible spectrum. As a result, it cools the surface and warms the atmosphere [24, 38]. A recent study of BC contribution to radiative forcing by Jacobson [31] showed that BC has a great contribution towards global warming and is the second most important component of global warming after CO_2 and has a larger impact on direct radiative forcing than that of methane. As a result, in populated countries like China and India, the large production of BC aerosols has a large impact on the hydrological cycle and precipitation pattern [46, 61, 71]. In India the fraction of BC production from fossil fuel burning, open burning and biofuel combustion to the global emission is significantly large and hence, it is necessary to estimate radiative impact of different kinds of BC not only on global scale but also in the regional scale.

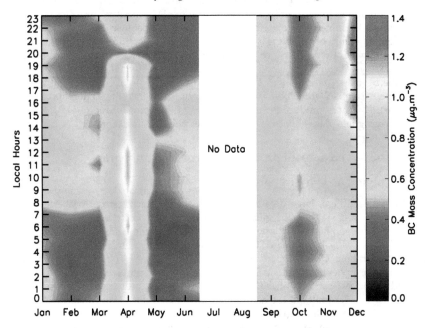

Figure 7. Diurnal variation of BC mass concentration during each month in 2007. White region indicates no data during Jul-Aug due to heavy monsoonal rain.

In recent years, global climate has received considerable attention due to increase in the percentage contribution of anthropogenic aerosols on the Earth's radiation budget [23]. BC particles exist mainly in the accumulation mode and can be transported over long distances [12] from source regions to far off pristine environment and perturb the climate of the latter, like that of Mt. Abu. The diurnal variation of BC mass concentration during different months over Mt. Abu is shown in Figure 7. Observations were not possible in July and August due to heavy rain. Minimum BC concentration was observed during monsoon (0.428±0.128 μg.m^{-3}) and maximum was observed during premonsoon (0.665±0.478 μg.m^{-3}) followed by winter (0.608±0.246 μg.m^{-3}) and postmonsoon (0.620±0.158 μg.m^{-3}). The annual mean BC mass concentration was 0.580±0.104 μg.m^{-3}. At Mt. Abu the BC concentration is an order of magnitude less than that at any other urban region in India. BC during April is found to be as high as 1.00±0.170 μg.m^{-3} which is a factor of 2 higher than the previous month. Backtrajectory analysis indicates wind coming from IGP which increases the BC concentration. In another study at a high altitude station, Nainital (29.4°N, 79.5°E, 1950m asl), in central Himalayas, mean BC was observed to be 1.36 ± 0.99 μg.m^{-3} during December 2004 [55]. This shows that Mt. Abu is less affected by anthropogenic activities.

The diurnal variation of BC mass concentration does not show any significant morning and nocturnal peaks like other urban regions. However, increased BC was observed during the noon hours except during November and December. The reason for such an increase is during the day time the thermal convection becomes stronger and as a result, the pollutants at the foothill area rise up to the hilltop region and enhance the BC concentration. This day time enhancement was prominent during winter and postmonsoon because during these seasons there is a large difference between the day and night time temperatures. During November and December the night time BC concentration was larger by a factor of two. During these months the nearby villagers burn wood and fallen leaves to keep themselves warm thereby increasing the BC mass concentration. During January this nocturnal enhancement was not observed. The reason is that the boundary layer height is less than the station altitude and the night time BC that is produced cannot reach the hill top region due to weak thermal convection. During this period hill top region becomes pollutant free region.

5. Satellite observed aerosol properties over Mt. Abu region

5.1. Aerosol optical and physical properties

In the current satellite era, large databases are available to study aerosol properties from space, both in the regional and the global scale, that are essentially demanding. For the present study, Terra and a series of satellite sensors flying on the A-train platform provide the required data. MODIS on board Terra and Aqua provide aerosol parameters in the morning and afternoon. OMI on board Aura satellite provides AI. The joint information of AOD, Ångström exponent (α) and small mode fraction (SMF) retrieved from MODIS and AI retrieved from OMI can be utilized to estimate the optical properties of aerosols with their size and type. In addition, the aerosol vertical distribution obtained from CALIPSO fulfills the requirement for the regional climate change study. In the present study, AOD, α, SMF and AI obtained from above multi-satellite observations are considered to distinguish the dominant natural and anthropogenic aerosols during different seasons. Figure 8 shows the multi-satellite observed AOD, α, SMF and AI over the study area during 2006–2007. Open circles represent the parameters obtained from MODIS-Aqua and filled circles are MODIS-Terra observations. AI gives information about the dust aerosols while SMF provides information about the anthropogenic and natural aerosols. Low SMF with low α indicates the presence of natural aerosols and the reverse represents the dominance of anthropogenic aerosols. AI has large values during Mar-Jul, whereas, α and SMF have low values. These combined observations suggest the abundance of coarse dust aerosols during these periods. OMI captures many dust storms over Thar Desert during premonsoon season in the AI images and enhancement of AI is due to transport of the dust plume from the desert region. During Aug-Feb, SMF is found to be high and AI is found to be very low indicating less abundance of dust aerosols in the atmosphere. In addition, α is also found to be very low during Aug-Sep but AOD is significantly high. Earlier studies from chemical composition of aerosol samples collected at Mt. Abu reported the enhancement of seasalt aerosols transported from Arabian sea during these periods and ground-based lidar observations at Ahmedabad, located 300 km to the south of Mt. Abu, reported the existence of a layer of seasalt aerosol in between 2–4 km [19, 64]. It can thus be inferred that the increase of monsoonal AOD is due to transport of seasalt aerosols from Arabian Sea. During Oct-Feb, high values of α are found indicating enhancement of anthropogenic aerosols. Ground-based observations show high abundance of BC on the hill-top region during winter. All these observations suggest that dust is dominating during premonsoon, anthropogenic aerosols during winter and natural seasalt are present in the atmosphere during monsoon season.

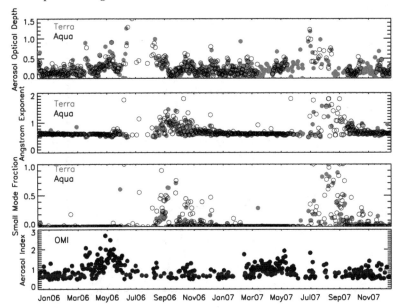

Figure 8. Space-borne daily observations of AOD, Ångström exponent, small mode fraction (SMF) obtained from MODIS onboard Terra and Aqua satellites and aerosol index obtained from OMI onboard Aura satellite during 2006 and 2007.

5.2. Aerosol vertical profile

Seasonal variation of aerosol vertical profiles over the study region is obtained from CALIPSO observations. Figure 9 shows the seasonal variation of aerosol extinction coefficient (km^{-1}). The horizontal dotted line at 1.7 km represents the height of Mt. Abu. The extinction coefficient is directly proportional to the total aerosol loading. It is clearly seen from the figure that aerosol loading over Mt. Abu is minimum during winter and higher during other seasons. There is a peak found near 2.2 km altitude during monsoon which becomes weak during postmonsoon. Ganguly et al. [19] reported that this peak is due to seasalt aerosols transported from Arabian sea and chemical analysis also supports this result showing significantly high amount of seasalt present over Mt. Abu during monsoon [64]. During premonsoon, there is a peak at 4.2 km which is due to the transported dust layer. MODIS and OMI observations also indicate significant amount of dust present in the atmosphere.

Near surface region also shows high extinction coefficient values. This could be due to locally produced anthropogenic aerosols. In the present study, the properties of aerosols at the

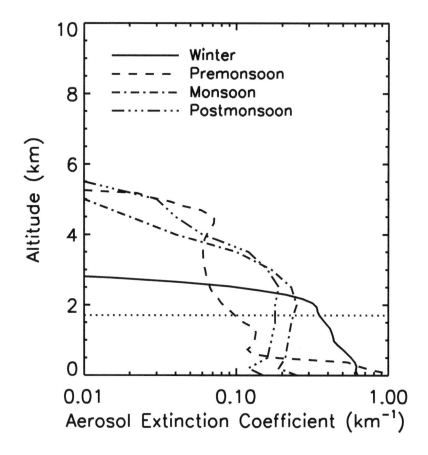

Figure 9. Seasonal variation of vertical distribution of aerosol extinction coefficient obtained from CALIPSO observations. The horizontal dashed line at 1.7 km is the altitude of Mt. Abu.

hill-top region are considered and defined as the 'background aerosols'. The vertical profile of aerosols indicate that these background aerosols are less influenced from these locally produced anthropogenic aerosols. Therefore, the aerosol properties observed over Mt. Abu are assumed to represent those of the background aerosols over semi-arid region of western India.

6. Natural and anthropogenic background aerosol properties

6.1. Estimation of natural and anthropogenic aerosols

The estimation of natural and anthropogenic aerosols over this background site is a challenging task because many aerosol compositions have both origins. For example, sulphates are mainly considered as anthropogenic components over urban regions as they are coming from factories and vehicular emissions. However, there is significant contribution from marine sources as di-methyl sulphate. On the other hand, BC is mainly anthropogenic, but it becomes natural when produced during natural forest fires. In the present study, dust and seasalt are considered as natural aerosols and BC, sulphate and nitrates as anthropogenic aerosols. BC is obtained from ground-based measurements using Aethalometer. Other aerosols like dust, sulphate and nitrates are obtained from the chemical analysis of aerosols samples collected over this hill-top region [39, 40]. These chemical compositions are used as input to the OPAC model to obtained aerosol optical properties and compared with measured values. OPAC model is also used to distinguish the natural and anthropogenic aerosols by separating the natural and anthropogenic components. A scatter plot of monthly averaged AODs obtained from Microtops observations and OPAC model is shown in Figure 10. The solid line represents the 1:1 line. Model derived and observed AODs are linearly varying with a slope of 0.90 and very close to the 1:1 line which indicates that the model derived AOD are very close to the observed values. However, the model is underestimating the AOD by about 10%. This is due to the cut-off radius of aerosols at 7.5 micron considered by the model, but in reality, aerosols are larger, especially over semi-arid regions, though their residence period is only for a few hours and their contribution towards optical depth is small.

6.2. Source identification of natural and anthropogenic aerosols

Seven days air parcel back trajectories are considered to identify the possible source regions of the natural and anthropogenic aerosols at Mt. Abu. The back-trajectories during premonsoon and winter are shown in Figure 11(a) and (b), respectively. Air parcels are mainly coming from IGP during winter and the heights of the trajectories are within 2 km. Ground based observations show that BC values at Mt. Abu are higher during winter and it is also clearly seen that there is long-range transportation of anthropogenic aerosols like BC from IGP within the boundary layer height. On the other side, air parcels are direction during premonsoon season. The heights are also greater than 3 km. Earlier chemical analyses report that dust concentration during this season is maximum of about 80% (in mass) of the total aerosols [39]. Therefore, one can easily conclude by these trajectories that the source of these dust aerosols is the nearby desert region. The back-trajectory analysis indicates that there is significant contribution of IGP during winter enhancing anthropogenic aerosols and that by nearby arid region during premonsoon increasing natural dust aerosols.

7. Natural vs anthropogenic background aerosol radiative forcing

7.1. Seasonal variation of aerosol radiative forcing

Aerosol radiative forcing is estmated using SBDART model considering aerosol optical properties obtained from OPAC, aerosol vertical profile from CALIPSO and MODIS surface reflectance. Aerosol radiative forcings in different seasons are given in table 2. Aerosol radiative forcing is found to vary from -3.2 to +0.2 Wm^{-2} at TOA and from 6.1 to 23.6 Wm^{-2} within the atmosphere. Aerosol radiative forcing at TOA is found to be maximum of about $0.2\pm2.5\,Wm^{-2}$ during premonsoon, followed by -1.3$\pm$0.5, -2.7$\pm$1.6, and -3.1$\pm$1.3 Wm^{-2} during monsoon, winter and postmonsoon, respectively. Forcing within the atmosphere is maximum of about 23.6$\pm$5.5 Wm^{-2} during premonsoon, followed by 12.5$\pm$3.9, 7.4$\pm$1.8, and 6.1$\pm$1.8 Wm^{-2} during monsoon, postmonsoon and winter, respectively. Annual mean aerosol forcing at Mt. Abu is found to be 8.7$\pm$3.4 Wm^{-2} which is lower than other urban

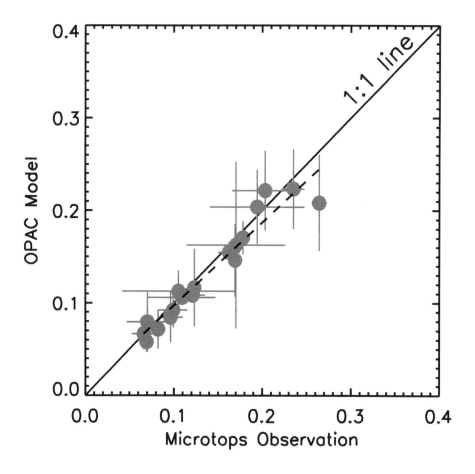

Figure 10. Scatter plot of monthly averaged AOD obtained from Microtops observations and OPAC model simulations. The solid line is the 1:1 line. The dashed line is the best-fitted line with a slope of 0.90, indicating that the OPAC underestimates AOD by 10%.

regions (mean forcing, 50 Wm^{-2}) and hill-top regions (mean forcing, 31 Wm^{-2}) in the Indian subcontinent [14]. For example, aerosol forcing over other hill-top regions like Pune, Kathmandu, Dibrugarh are about 33, 25 and 35.7 Wm^{-2}, respectively. These hilly areas are mainly influenced by anthropogenic aerosols. However, maximum forcing over Mt. Abu is found of about 23.6 Wm^{-2} during premonsoon which is lower but comparable with their forcing values. This is due to the maximum natural dust loading in the atmosphere at Mt. Abu.

Aerosol radiative forcing mainly depends on the amount of aerosol loading and underlying surface. Also, the sign of forcing at TOA is influenced by the aerosol type. An increase of absorbing aerosol loading causes positive forcing at TOA. In addition, bright surface which

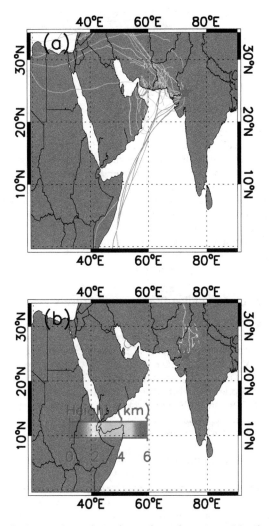

Figure 11. Seven days back trajectory analysis of aerosol parcels coming to Mt. Abu (location indicated by star) during (a) premonsoon and (b) winter. The color bar represents the height of the air parcels during their travel from source regions to the measurement site. Aerosols are mainly coming from desert region at higher altitude during premonsoon and from IGP within the boundary layer height (2 km) during winter.

reflects more solar radiation back to the space can cause positive TOA forcing. Radiative forcing at TOA changes its sign from negative to positive during premonsoon. This could be due to the combined effects of the relatively brighter land surface over western India and high dust loading in the atmosphere by frequently occurring dust storms over Thar Desert. TOA forcing becomes minimum during postmonsoon as land surface becomes darker by

Season	Aerosol Radiative		Forcing
	TOA	Surface	Atmosphere
Winter	-2.7	-8.8	6.1
Premonsoon	0.2	-23.4	23.6
Monsoon	-1.3	-13.8	12.5
Postmonsoon	-3.2	-10.6	7.4

Table 2. Seasonal variation of aerosol radiative forcing over Mt. Abu

the growing forest area over western India after monsoonal rain. Atmospheric forcing is proportionally varying with amount of aerosol loading. Maximum atmospheric forcing is found during premonsoon due to the maximum dust loading in this season while minimum forcing is observed during winter since the boundary layer height becomes lower than observational site which makes the site a free tropospheric station over western India with minimum aerosol loading in the atmosphere. During monsoon, heavy rains wash out the aerosols from the atmosphere, though atmospheric forcing is observed to be significantly high. This is due to the existence of aerosol layer, as found in the CALIPSO observations, that consist of large abundance of seasalt aerosols transported from Arabian Sea. This layer reflects the solar radiation significantly to the space which also causes relatively positive TOA forcing than that during winter and postmonsoon.

7.2. Contribution of natural and anthropogenic aerosols

Mt. Abu experiences large variation in aerosol properties and hence in the radiation forcing. During premonsoon there is large transportation of natural dust aerosols from surrounding arid region by the strong westerly wind and during monsoon large amount of seasalt is transported from the Arabian sea by the southwesterly wind. Figure 12 shows the seasonal variation of contributions of natural and anthropogenic forcings to the total aerosol radiative forcing within the atmosphere over Mt. Abu. The contributions of anthropogenic radiative forcing are 52%, 40%, 33%, and 56% and those of natural forcing are 48%, 60%, 67%, and 44% during winter, premonsoon, monsoon, and postmonsoon, respectively. Natural forcing is dominating at Mt. Abu during premonsoon and monsoon, whereas, the contributions of anthropogenic and natural forcing during winter and postmonsoon are almost equal. It is to be noted that natural and anthropogenic aerosol radiative forcings are calculated on the basis of their optical properties derived from OPAC model and OPAC model considers 7.5 μm as the upper limit of aerosol radius. In the present study dust is considered as natural aerosols, which in reality can be larger than this cut off limit over arid region, especially during premonsoon. Therefore, the contribution of natural forcing could be underestimated due to these large dust aerosols even though their AOD is very low. The comparison between OPAC and Microtops AOD indicates that this underestimation is not more that 10% (Figure 10).

Due to the proximity of Mt. Abu to the Thar desert dust aerosols are transported to this hill-top region during premonsoon and hence natural forcing is higher. During monsoon also, natural forcing is higher due to the large amount of seasalt coming from over the Arabian sea and simultaneously, dust and boundary layer anthropogenic aerosols are washed out by the heavy rains. Chemical analysis also shows that during monsoon, anthropogenic compositions like non-seasalt potassium, ammonium and nitrate are relatively less and the natural compositions like seasalt are enhanced over Mt. Abu [40, 63]. During postmonsoon, there is less transportation of seasalt aerosols to the measurement site due to low wind speed

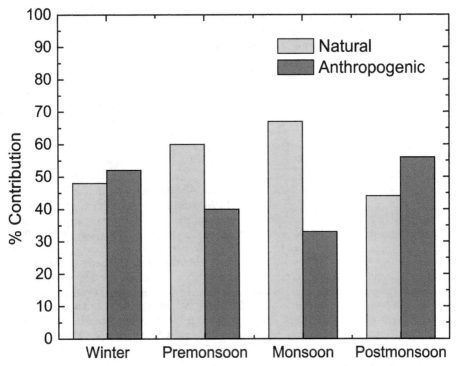

Figure 12. Seasonal variation of the contribution of natural and anthropogenic forcing to the total atmospheric radiative forcing over Mt. Abu.

and hence the natural forcing reduces and anthropogenic forcing increases. During winter, total aerosol loading is minimum as the measurement site becomes a free tropospheric station and thereby, both natural and anthropogenic forcings contribute equally.

Annual mean contributions of natural and anthropogenic forcing are about 55% and 45%, respectively. This indicates that anthropogenic aerosols are also significantly contributing to total radiative forcing within the atmosphere. This could be due to the close proximity of IGP which is a potential source of anthropogenic aerosols over semi-arid region. Therefore, it is concluded that western India is influenced by natural as well as anthropogenic aerosols significantly.

8. Conclusions

Western India is known for the presence of Thar Desert, which is a potential source of dust aerosols in the Indian subcontinent. Therefore, it is commonly believed that the atmosphere over western India is largely influenced by natural dust aerosols. With this motivation, the present study investigates the natural and anthropogenic contribution to the background aerosols and their radiative effects over western India. The optical and physical properties of aerosols over Mt. Abu, highest peak of the Aravalli mountains in western India are obtained from a variety of ground-based and satellite-borne instruments. Mt. Abu is situated in the semi-arid region of Thar Desert and is less influenced by the local anthropogenic aerosols. It is therefore, a unique site for the observation of background aerosols over semi-arid region. Also, the Aravalli mountains are located in between Thar Desert and IGP which has large abundance of anthropogenic aerosols. Therefore, there is a significant variation of aerosol properties over Mt. Abu during different seasons, namely, winter (DJF), premonsoon (MAM), monsoon (JJA) and postmonsoon (SON). Ground-based observations show that AOD is maximum during premonsoon due to the large dust loading in the atmosphere by frequently occurring dust storms over Thar desert and minimum during winter due to low boundary layer height. Space-borne observations suggest that natural dust aerosols are dominating during premonsoon while anthropogenic aerosols are dominating during winter over western India. An interesting observation of CALIPSO is a layer of transported seasalt aerosols during monsoon over western India coming from Arabian Sea. These aerosols increase the contribution of natural forcing to the total atmospheric radiative forcing. Atmospheric radiative forcing is found to be maximum of about 23.6 ± 5.5 Wm^{-2} during premonsoon and minimum of about 6.1 ± 1.8 Wm^{-2} during winter. Another interesting result is TOA forcing is positive due to the bright land surface over western India during premonsoon, while it is negative during other seasons. The contribution of natural aerosols is found to be higher during premonsoon and monsoon and that of anthropogenic aerosols is higher during postmonsoon. During winter, they contribute equally. The annual average of natural and anthropogenic contribution is about 55% and 45%, respectively, indicating that the anthropogenic effects are also very significant. Thus the background aerosols over western India are not only influenced by desert dust aerosols but also by seasalt coming from Arabian Sea and anthropogenic aerosols transported from IGP.

Acknowledgment

Author would like to thank his Ph. D. thesis supervisor, Prof. A. Jayaraman, for his guidance, constant help, inspiration and support during this research work initiated at PRL. Author would also like to acknowledge NOAA National Center for Environmental Prediction and Air Resources Laboratory for providing reanalysis data and online HYSPLIT model output for back-trajectories analysis. Author acknowledges Terra, Aqua, Aura and CALIPSO mission scientists and associated NASA personnel for the production of the data used in this research effort. Special thanks to Prof. J. P. Chen and Dr. U. Das for the valuable scientific discussions. This work is partially supported by NSC grant 100-2119-M-002 -023 -MY5, Taiwan.

Author details

Sanat Kumar Das
Department of Atmospheric Sciences, National Taiwan University, Taiwan

9. References

[1] Ackerman, A. S., Toon, O. B., Stevens, D. E., Heymsfield, A. J., Ramanathan, V. & Welton, E. J. [2000]. Reduction of tropical cloudiness by soot, *Science* 288: 1042–1047.

[2] Allen, A. G., Joy, L. & Petros, K. [1998]. Field validation of a semi-continuous method for aerosol black carbon (Aethalometer) and temporal patterns of summertime hourly black carbon measurements in southwestern PA, *Harvard School of Public Health*.

[3] Andreae, M. O. [1995]. *Climate effect of changing atmospheric aerosol levels, in Future Climates of the World, vol. 16 of World Survey of Climatology*, A. Henderson-Sellers, Ed. (Elsevier, Amsterdam).

[4] Badarinath, K. V. S., Kharol, S. K., Kaskaoutis, D. G., Sharma, A. R., Ramaswamy, V. & Kambezidis, H. D. [2010]. Long-range transport of dust aerosols over the Arabian Sea and Indian region – A case study using satellite data and ground-based measurements, *Global and Planetary Change* 72(3): 164–181.

[5] Bodhaine, B. A. [1995]. Aerosol absorption measurements at Barrow and Mauna Loa and the South Pole, *J. Geophy. Res.* 100: 8967– 8975.

[6] Bond, T. C., Bhardwaj, E., Dong, R., Jogani, R., Jung, S., Roden, C., Streets, D. G. & Trautmann, N. M. [2007]. Historical emissions of black and organic carbon aerosol from energy-related combustion, 1850–2000, *Global Biogeochemi. Cy.* 21(2): GB2018.

[7] Chand, D., Anderson, T. L., Wood, R., Charlson, R. J., Hu, Y., Liu, Z. & Vaughan, M. [2008]. Quantifying above-cloud aerosol using space borne lidar for improved understanding of cloudy-sky direct climate forcing, *J. Geophys. Res.* 113: 1–12.

[8] Chand, D., Wood, R., Anderson, T. L., Satheesh, S. K. & Charlson, R. J. [2009]. Satellite derived direct radiative effect of aerosols dependent on cloud cover, *Nature Geosc.* .

[9] Chiapello, I., Goloub, P., Tanre, D., Marchand, M., Hermann, J. & Torres, O. [2000]. Aerosol detection by TOMS and POLDER over oceanic region, *J. Geophy. Res.* 105: 7133–7142.

[10] Chýlek, P. & Coakley, J. A. [1974]. Aerosol and climate, *Science* 183: 75–77.

[11] Chýlek, P., Lohmann, U., Dubey, M., Mishchenko, M., Kahn, R. & Ohmura, A. [2007]. Limits on climate sensitivity derived from recent satellite and surface observations, *J. Geophys. Res.* 112.

[12] Chýlek, P., Ramaswamy, V. & Srivastava, V. [1984]. Graphitic carbon content of aerosols, clouds and snow and its climatic implications, *Sci. Total Environ.* 36: 117–120.

[13] d'Almeida, G. A., Koepke, P. & Shettle, E. P. [1991]. *Atmospheric Aerosols: Global Climatology and Radiative Characteristics*, A. Deepak, Hampton, Va.

[14] Das, S. K. & Jayaraman, A. [2011]. Role of black carbon in aerosol properties and radiative forcing over western india during premonsoon period, *Atmos. Res.* 102(3): 320–334.

[15] Das, S. K., Jayaraman, A. & Misra, A. [2008]. Fog-induced variations in aerosol optical and physical properties over the indo-gangetic basin and impact to aerosol radiative forcing, *Ann. Geophys.* 26: 1345–1354.

[16] Deepshikha, S., Satheesh, S. K. & Srinivasan, J. [2005]. Regional distribution of absorbing efficiency of dust aerosols over India and adjacent continents inferred using satellite remote sensing, *Geophys. Res. Lett.* 32.

[17] Dey, S., Tripathi, S. N., Singh, R. P. & Holben, B. N. [2004]. Influence of dust storms on the aerosol optical properties over the Indo-Gangetic basin, *J. Geophy. Res.* 109.

[18] Ganguly, D. & Jayaraman, A. [2006]. Physical and optical properties of aerosols over an urban location in western India: Implications for shortwave radiative forcing, *J. Geophy. Res.* 111: 1–13.

[19] Ganguly, D., Jayaraman, A. & Gadhavi, H. [2006]. Physical and optical properties of aerosols over an urban location in western India: Seasonal variabilities, *J. Geophy. Res.* 111.

[20] Gao, B.-C., Kaufman, Y. J., Tanre, D. & Li, R.-R. [2002]. Distinguishing tropospheric aerosols from thin cirrus clouds for improved aerosol retrievals using the ratio of 1.38-μm and 1.24-μm channels, Geophys. Res. Lett. 29(18): 1890.

[21] Hansen, A. D. A., Rosen, H. & Novakov, T. [1982]. Real-time measurement of the absorption coefficient of aerosol particles, Appl. Opt. 21: 3060–3062.

[22] Hansen, J. E. [1969]. Exact and approximate solutions for multiple scattering by cloudy and hazy planetary atmospheres, J. Atmos. Sci. 26: 478–487.

[23] Haywood, J. M. & Ramaswamy, V. [1998]. Global sensitivity studies of the direct radiative forcing due to anthropogenic sulfate and black carbon aerosols, J. Geophy. Res. 103 (D3): 6043–6058.

[24] Haywood, J. M. & Shine, K. P. [1995]. The effect of anthropogenic sulfate and soot aerosol on the clear sky planetary radiation budget, Geophys. Res. Lett. 22: 603–606.

[25] Herman, J. R., Bhartia, P. K., Torres, O., Hsu, N. C., Seftor, C. J. & Celarier, E. [1997]. Global distribution of uv-absorbing aerosols from Nimbus-7/TOMS data, J. Geophy. Res. 102: 16,911–16,922.

[26] Hess, M., Keopke, P. & Schult, I. [1998]. Optical properties of aerosols and clouds: The software package OPAC, Bull. Am. Meteorol. Soc. 79: 831–844.

[27] Houghton, J. T., Ding, Y., Griggs, D. J., Noguer, M., van der Linden, P. J. & Xiaosu, S. [2001]. Climate Change 2001: The Scientific Basis, Cambridge Univ. Press, New York.

[28] Hsu, N. C., Herman, J. R. & Weaver, C. [2000]. Determination of radiative forcing of Saharan dust using combined TOMS and ERBE data, J. Geophy. Res. 105: 20,649–20,661.

[29] Hsu, N. C., Tsay, S.-C., King, M. D. & Herman, J. R. [2004]. Aerosol properties over bright-reflecting source regions, IEEE Trans, Geosci. Remote Sens. 42(3): 557–569.

[30] Ichoku, C., Levy, R., Kaufman, Y. J., Remer, L. A., Li, R.-R., Martins, V. J., Holben, B. N., Abuhassan, N., Slutsker, I., Eck, T. F. & Pietras, C. [2002]. Analysis of the performance characteristics of the five-channel Microtops II sun photometer for measuring aerosol optical thickness and precipitable water vapor, J. Geophy. Res. 107 (D13): 417.

[31] Jacobson, M. Z. [2001]. Strong radiative heating due to the mixing state of black carbon in atmospheric aerosols, Nature 409: 695–697.

[32] Jayaraman, A., Gadhavi, H., Ganguly, D., Misra, A., Ramachandran, S. & Rajesh, T. A. [2006]. Spatial variations in aerosol characteristics and regional radiative forcing over India: Measurements and modeling of 2004 road campaign experiment, Atmos. Environ. 40: 6504–6515.

[33] Jayaraman, A., Lubin, D., Ramachandran, S., Ramanathan, V., Woodbridge, E., Collins, W. D. & Zalpuri, K. S. [1998]. Direct observation of aerosol radiative forcing over the tropical Indian Ocean during the January-February 1996 pre-INDOEX cruise., J. Geophy. Res. 103 (D12): 13827–13836.

[34] Kaufman, Y. J., Tanre, D., Holben, B. N., Mattoo, S., Remer, L. A., Eck, T. F., Vaughan, J. & Chatene, B. [2002]. Aerosol radiative impact on spectral solar flux at the surface, derived from principal-plane sky measurements, J. Atmos. Sci. 59: 635–646.

[35] Kaufman, Y. J., Tanré, D., Remer, L. A., Vermote, E. F., Chu, A. & Holben, B. N. [1997]. Operational remote sensing of tropospheric aerosol over land from EOS moderate resolution imaging spectroradiometer, J. Geophy. Res. 102(D14): 17,051–17,067.

[36] Kondratyev, K. Y. [1969]. Radiation in the Atmosphere, Academic Press.

[37] Krishnan, P. & Kunhikrishnan, P. K. [2004]. Temporal variations of ventilation coefficient at a tropical indian station using uhf wind profiler, Curr. Sci. 86(3): 477–451.

[38] Krishnan, R. & Ramanathan, V. [2002]. Evidence of surface cooling from absorbing aerosols, *Geophys. Res. Lett.* 29(9): 1340.

[39] Kumar, A. & Sarin, M. M. [2009]. Mineral aerosols from western india: Temporal variability of coarse and fine atmospheric dust and elemental characteristics, *Atmos. Environ.* 43(26): 4005–4013.

[40] Kumar, A. & Sarin, M. M. [2010]. Atmospheric water-soluble constituents in fine and coarse mode aerosols from high-altitude site in western india: Long-range transport and seasonal variability, *Atmos. Environ.* 44(10): 1245–1254.

[41] Kumar, N., Chu, A. & Foster, A. [2007]. An empirical relationship between PM2.5 and aerosol optical depth in Delhi Metropolitan, *Atmos. Environ.* 41(21): 4492–4503.

[42] Lelieveld, J., Crutzen, P. J., Ramanathan, V., Andreae, M. O., Brenninkmeijer, C. A. M., Campos, T., Dickerson, G. R. C. R. R., de Gouw, H. F. J. A., Hansel, A., Jefferson, A., Kley, D., de Laat, A. T. J., Lal, S., Lawrence, M. G., Lobert, J. M., Mayol-Bracero, O. L., Mitra, A. P., Novakov, T., Oltmans, S. J., Prather, K. A., Reiner, T., Rodhe, H., Scheeren, H. A., Sikka, D. & Williams, J. [2002]. The Indian Ocean Experiment: Widespread Air Pollution from South and Southeast Asia, *Science* 291: 1031–1036.

[43] Levy, R. C., Remer, L. A., Mattoo, S., Vermote, E. F. & Kaufman, Y. J. [2007]. Second-generation operational algorithm: Retrieval of aerosol properties over land from inversion of moderate resolution imaging spectroradiometer spectral reflectance, *J. Geophys. Res.* 112(D13): D13211.

[44] Levy, R. C., Remer, L. A., Tanré, D., Kaufman, Y. J., Ichoku, C., Holben, B. N., Livingston, J. M., Russell, P. B. & Maring, H. [2003]. Evaluation of the Moderate-Resolution imaging spectroradiometer (MODIS) retrievals of dust aerosol over the ocean during PRIDE, *J. Geophy. Res.* 108(D19): 8594.

[45] Martins, J. V., Tanré, D., Remer, L., Kaufman, Y., Mattoo, S. & Levy, R. [2002]. MODIS cloud screening for remote sensing of aerosols over oceans using spatial variability, *Geophys. Res. Lett.* 29(12): 8009.

[46] Menon, S., Hansen, J., Nazarenko, L. & Luo, Y. [2002]. Climate effects of black carbon aerosols in China and India, *Science* 297: 2250–2253.

[47] Mishchenko, M. I., Geogdzhayev, I. V., Rossow, W. B., Cairns, B., Carlson, B. E., Lacis, A. A., Liu, L. & Travis, L. D. [2007]. Long-term satellite record reveals likely recent aerosol trend, *Sscience* 315: 1543.

[48] Moorthy, K. K., Satheesh, S. K., Babu, S. S. & Dutt, C. B. S. [2008]. Integrated Campaign for Aerosols, gases and Radiation Budget (ICARB): An overview, *J. of Earth System Sci.* 117: 243–262.

[49] Morys, M., Mims, F. M., Hagerup, S., Anderson, S. E., Baker, A., Kia, J. & Walkup, T. [2001]. Design and calibration and performance of MICROTOPS II handheld ozone monitor and Sun photometer, *J. Geophy. Res.* 106 (D3): 14,573–14,582.

[50] Nair, V. S., Moorthy, K. K., Denny, A. P., Kunhikrishnan, P. K., George, S., Nair, P. R., Babu, S. S., Abish, B., Satheesh, S. K., Tripathi, S. N., Niranjan, K., Madhavan, B. L., Srikant, V., Dutt, C. B. S., Badarinath, K. V. S. & Reddy, R. R. [2007]. Wintertime aerosol characteristics over the Indo-Gangetic Plain (IGP): Impacts of local boundary layer processes and long-range transport, *J. Geophy. Res.* 112.

[51] Nakajima, T., Tanaka, M., Yamano, M., Shiobara, M., Arao, K. & Nakanishi, Y. [1989]. Aerosol optical characteristics in the yellow sand events observed in May, 1982 at Nagasaki, part 2, Models, *J. Meteorol. Soc. Jpn* 67: 279–291.

[52] Niranjan, K., Devi, T. A., Spandana, B., Sreekanth, V. & Madhavan, B. L. [2012]. Evidence for control of black carbon and sulfate relative mass concentrations on composite aerosol radiative forcing: Case of a coastal urban area, *J. Geophy. Res.* 117(D5).

[53] Novakov, T., Andreae, M. O., Gabriel, R., Kirchstetter, T. W., Mayol-Bracero, O. L. & Ramanathan, V. [2000]. Origin of carbonaceous aerosols over the tropical Indian Ocean: Biomass burning or fossil fue, *Geophys. Res. Lett.* 27(4): 4061–4064.

[54] Pandithurai, G., Dipu, S., Dani, K. K., Tiwari, S., Bisht, D. S., Devara, P. C. S. & Pinker, R. T. [2008]. Aerosol radiative forcing during dust events over New Delhi, India, *J. Geophy. Res.* 113.

[55] Pant, P., Hegde, P., Dumka, U. C., Sagar, R., Satheesh, S. K., Moorthy, K. K., Saha, A. & Srivastava, M. K. [2006]. Aerosol characteristics at a high-altitude location in central Himalayas: Optical properties and radiative forcing, *J. Geophy. Res.* 111.

[56] Pierluissi, J. H. & Peng, G. S. [1985]. New molecular transmission band models for LOWTRAN, *Opt. Eng.* 24(3): 541–547.

[57] Quinn, P. K. & Bates, T. S. [2005]. Regional aerosol properties: Comparisons of boundary layer measurements from ACE 1, ACE 2, Aerosols99, INDOEX, ACE Asia, TARFOX, and NEAQS, *J. Geophys. Res.* 110.

[58] Quinn, P. K., Bates, T. S., Coffman, D. J., Miller, T. L., Johnson, J. E., Covert, D. S., Putaud, J. P., Neususs, C. & Novakov, T. [2000]. A comparison of aerosol chemical and optical properties from the 1st and 2nd Aerosol Characterization Experiments, *Tellus, Ser. B* 52: 239–257.

[59] Ramachandran, S., Rengarajan, R., Jayaraman, A., Sarin, M. M. & Das, S. K. [2006]. Aerosol radiative forcing during clear, hazy, and foggy conditions over a continental polluted location in north India, *J. Geophy. Res.* 111.

[60] Ramanathan, V., Chung, C., Kim, D., Bettge, T., Buja, L., Kiehl, J. T., Washington, W. M., Fu, Q., Sikka, D. R. & Wild, M. [2005]. Atmospheric brown clouds: Impacts on South Asian climate and hydrological cycle, *PNAS* 102 (4): 5326–5333.

[61] Ramanathan, V., Crutzen, P. J., Lelieveld, J., Mitra, A. P., Althausen, D., Anderson, J., Andreae, M. O., Cantrell, W., Cass, G. R., Chung, C. E., Clarke, A. D., Coakley, J. A., Collins, W. D., Conant, W. C., Dulac, F., Heintzenberg, J., Heymsfield, A. J., Holben, B., Howell, S., Hudson, J., Jayaraman, A., Kiehl, J. T., Krishnamurti, T. N., Lubin, D., McFarquhar, G., Novakov, T., Ogren, J. A., Podgorny, I. A., Prather, K., Priestley, K., Prospero, J. M., Quinn, P. K., Rajeev, K., Rasch, P., Rupert, S., Sadourny, R., Satheesh, S. K., Shaw, G. E., Sheridan, P. & Valero, F. P. J. [2001]. Indian ocean experiment: An integrated analysis of the climate forcing and effects of the great Indo-Asian haze, *J. Geophy. Res.* 106: 28,371–28,398.

[62] Ramanathan, V., Ramana, M. V., Roberts, G., Kim, D., Corrigan, C., Chung, C. & Winker, D. [2007]. Warming trends in Asia amplified by brown cloud solar absorption, *Nature* 448: 575–578.

[63] Rastogi, N. & Sarin, M. M. [2005a]. Chemical characteristics of individual rain events from a semi-arid region in India: Three-year study, *Atmos. Environ.* 39: 3313–3323.

[64] Rastogi, N. & Sarin, M. M. [2005b]. Long-term characterization of ionic species in aerosols from urban and high-altitude sites in western India: Role of mineral dust and anthropogenic sources, *Atmos. Environ.* 39: 5541–5554.

[65] Reeves, R. G., Anson, A. & Eds, D. L. [1975]. *Manual of Remote Sensing*, Amer. Soc. Photogrammetry.

[66] Reid, J. S., Eck, T. F., Christopher, S. A., Hobbs, P. V. & Holben, B. N. [1999]. Use of the Ångström exponent to estimate the variability of optical and physical properties of aging smoke particles in brazil and, *J. Geophy. Res.* 104: 473–489.

[67] Remer, L. A., Kaufman, Y. J., Tanre, D., Mattoo, S., Chu, D. A., Martins, J. V., Li, R.-R., Choku, C., Levy, R. C., Kleidman, R. G., Eck, T. F., Vermote, E. & Holben, B. N. [2005]. The MODIS aerosol algorithm, products and validation, *J. Atmos. Sci.* 62: 947–973.

[68] Ricchiazzi, P., Yang, S., Gautier, C. & Sowle, D. [1998]. SBDART, A research and teaching tool for plane-parallel radiative transfer in the Earth's atmosphere, *Bull. Am. Meteorol. Soc.* 79: 2101–2114.

[69] Saha, A. & Krishna Moorthy, K. [2004]. Impact of precipitation on aerosol spectral optical depth and retrieved size distributions: A case study, *J. Appl. Meteor.* 43(6): 902–914.

[70] Satheesh, S. K. & Ramanathan, V. [2000]. Large differences in tropical aerosol forcing at the top of the atmosphere and Earth's surface, *Nature* 405: 60–63.

[71] Satheesh, S. K. & Srinivasan, J. [2002]. Enhanced aerosol loading over Arabian Sea during the pre-monsoon season: Natural or anthropogenic?, *Geophys. Res. Lett.* 29(18): 1874.

[72] Shettle, E. P. & Fenn, R. W. [1975]. Models of the atmospheric aerosols and their optical properties., *AGARD Conf. Proc., Optical Propagation in the Atmosphere, Lyngby, Denmark, NATO Advisory Group for Aerospace Research* pp. 2.1–2.16.

[73] Staetter, R. & Schroeder, M. [1978]. Spectral characteristics of natural surfaces, *Proc. Int. Conf. on Earth Observation from Space and Management of Planetary Resources, Toulouse, France, Council of Europe, Commission of the European Communities, and European Association of Remote Sensing Laboratories,* p. 661.

[74] Stamnes, K., Tsay, S., Wiscombe, W. & Jayaweera, K. [1988]. Numerically stable algorithm for discrete-ordinate-method radiative transfer in multiple scattering and emitting layered media, *Appl. Opt.* 27: 2502–2509.

[75] Subbaraya, B. H., Jayaraman, A., Krishnamoorthy, K. & Mohan, M. [2000]. Atmospheric aerosol studies under isroâÁŹs geosphere biosphere programs, *J. Ind. Geophys. Union* 4: 77–90.

[76] Tanre, D., Deroo, C., Duhaut, P., Herman, M., Morcrette, J. J., Perbos, J. & Deschamps, P. Y. [1990]. Description of a computer code to simulate the satellite signal in the solar spectrum: The 5S code, *Int. J. Remote Sens.* 11(4): 659–668.

[77] Tanré, D., Deschamps, P. & Herman, C. D. M. [1988]. Estimation of saharan aerosol optical thickness from blurring effects in thematic mapper data, *J. Geophy. Res.* 93(D12): 15955–15964.

[78] Tanré, D., Haywood, J., Pelon, J., FLeón, J., Chatenet, B., Formenti, P., Francis, P., Goloub, P., Highwood, E. J. & Myhre, G. [2003]. Measurement and modeling of the saharan dust radiative impact: Overview of the Saharan Dust Experiment (SHADE), *J. Geophy. Res.* 108(D18): 8574.

[79] Tanré, D., Herman, M. & Kaufman, Y. J. [1996]. Information on aerosol size distribution contained in solar reflected spectral radiances, *J. Geophy. Res.* 101(D14): 19,042–19,060.

[80] Tanré, D., Kaufman, Y. J., Herman, M. & Mattoo, S. [1997]. Remote sensing of aerosol properties over oceans using the MODIS/EOS spectral radiances, *J. Geophys. Res.* 102(D14): 16,971–16,988.

[81] Tomasi, C., Caroli, E. & Titale, V. [1983]. Study of the relationship between Ångström's wavelength exponent and junge particle size distribution exponent, *J. Clim. Appl. Meteorol.* 22: 1707–1716.

[82] Torres, O., Bhartia, P. K., Herman, J. R., Ahmad, Z. & Gleason, J. [1998]. Derivation of aerosol properties from satellite measurements of backscattered ultraviolet radiation, Theoretical basis, *J. Geophy. Res.* 103: 17,099 –17,110.

[83] Tsai, T.-C., Jeng, Y.-J., Chu, D. A., Chen, J.-P. & Chang, S.-C. [2011]. Analysis of the relationship between MODIS aerosol optical depth and particulate matter from 2006 to 2008, *Atmos. Environ.* 45(27): 4777–4788.

[84] van de Hulst, H. C. [1968]. Asymptotic fitting, a method for solving anisotropic transfer problems in thick layers, *J. Comput. Phys.* 3: 291–306.

[85] Vaughan, M. A., Winker, D. M. & Powell, K. A. [2005]. CALIOP algorithm theoretical basis document, part 2: Feature detection and layer properties algorithms,Rep.PC-SCI-202, *NASA Langley Res. Cent., Hampton, Va* pp. 1–87.

[86] Venkataraman, C., Habib, G., Eiguren-Fernandez, A., Miguel, A. H. & Friedlander, S. K. [2005]. Residential biofuels in South Asia:carbonaceous aerosol emissions and climate impacts, *Science* 307: 1454–1456.

[87] Vernier, J. P., Pommereau, J. P., Garnier, A., Pelon, J., Larsen, N., Nielsen, J., Christensen, T., Cairo, F., Thomason, L. W., Leblanc, T. & McDermid, I. S. [2009]. Tropical stratospheric aerosol layer from calipso lidar observations, *J. Geophys. Res.* 114.

[88] Weare, B., Temkin, R. & Snell, F. [1974]. Aerosols and Climate: Some further considerations, *Science* 186: 827–828.

[89] Weingartner, E., Saathoff, H., Schnaiter, M., Streit, N., Bitnar, B. & Baltensperger, U. [2003]. Absorption of light by soot particles: Determination of the absorption coefficient by means of aethalometers, *J. Aerosol Sci.* 34: 1445–1463.

[90] Wielicki, B. A., Wong, T., Allan, R. P., Slingo, A., Kiehl, J. T., Soden, B. J., Gordon, C. T., Miller, A. J., Yang, S.-K., Randall, D. A., Robertson, F., Susskind, J. & Jacobowitz, H. [2002]. Evidence for large decadal variability in the tropical mean radiative energy budget, *Science* 295: 841–844.

[91] Wiscombe, W. J. & Warren, S. G. [1980]. A model for the spectral albedo of snow. Part I: Pure snow, *J. Atmos. Sci.* 37: 2712–2733.

[92] Yadav, S. & Rajamani, V. [2004]. Geochemistry of aerosols of northwestern part of India adjoining the Thar Desert, *Geochim. Cosmochim. Ac.* 68: 1975–1988.

Chapter 5

Distribution of Particulates in the Tropical UTLS over the Asian Summer Monsoon Region and Its Association with Atmospheric Dynamics

S.V. Sunil Kumar, K. Parameswaran and Bijoy V. Thampi

Additional information is available at the end of the chapter

1. Introduction

Particulates in the Upper Troposphere and Lower Stratosphere (UTLS) gained considerable interest due to their role in the dehydration of tropospheric air entering the stratosphere [1,2] as well as their potential to influence the radiation budget of the Earth-atmosphere system [3]. The upper troposphere region, which is also conducive for the formation of cirrus, plays a major role in the transport of water vapour and other chemical constituents into the stratosphere. The physical processes responsible for maintaining the observed aerosol distribution in the tropical UTLS, the process with which it interacts with cirrus clouds and the effect of these particulates on the radiation budget are not fully understood [4]. Studies have shown that the microphysical (such as particle shape, size, and size distribution) as well as the chemical properties of particles in the UTLS region [5-7] are mainly governed by the strength of tropospheric convection and the prevailing dynamics of the underlying troposphere. The formation and persistence of cirrus clouds in the upper troposphere is mainly governed by the concentration of available condensation nuclei in this region and their physical and chemical properties. These clouds are believed to be a significant contributor to atmospheric greenhouse effect [8,9] as well as hypothesized to play a major role in the dehydration of the lower stratosphere [1,2] and thus becomes an important factor governing global climate, through their positive feedback.

The last four decades of the 20th century have been marked by relatively intense volcanic activity [10] and hence long-term measurements of aerosols during this period mostly characterize the volcanically perturbed aerosols system rather than 'background' conditions [11-13].During this period the increase in aerosol loading in the stratosphere could accelerate the heterogeneous chemistry of sulfate aerosols leading to a decrease in ozone amount [14-17], altering the NO_2 concentration [18-20] and hence modifying the earth's radiation budget [21]. However, long-term studies on stratospheric aerosols show that on a global scale the stratospheric aerosol loading has returned to the pre-eruption levels (prevailed in the late 1970s) after the eruption of Mt. Pinatubo in 1991 [13,22] for the first time in 1998 and continued to remain almost at the same level for a couple of years. Only very few studies are carried out on the characteristics of these background stratospheric aerosols using observational data [7,13,23]. These studies, however, have shown that the global distribution of stratospheric aerosols will be significantly influenced by the atmospheric dynamics which includes periodic variations, such as the Quasi Biennial Oscillation (QBO), seasonal cycles and long-term secular changes in addition to small perturbations due to the feeble volcanic eruptions and also due to degassing from the Earth's crust. Identifying these secular trends in the background stratospheric aerosol system is crucial to predict future aerosol levels [24]. While *Deshler et al.* [13,25] observed no discernable long-term trends in the non-volcanic component of stratospheric aerosols over an extended period (1970-2004), *Hofmann et al.* [26] could observed a significant enhancement in the lower stratospheric aerosol load for the past several years (2000-2008) which they attributed to the increase in anthropogenic sulfur emission. Moreover, eruptions of a few minor volcanoes such as Manam, Ruang, Revantador and Soufriere also might have disturbed the background stratospheric aerosol level to a smaller extent. Although the volcanic degassing during the quiescent and the small eruptive volcanic periods contributes only 14% to total global SO_2 emissions, its efficiency on total atmospheric SO_2 burden is found to be much higher (factor of 5) than that of anthropogenic emissions. Model calculations [27] shows that even though the source strength of volcanic emissions is less than 20 % of the anthropogenic component, the flux of sulfur gases from volcanoes during this period leads to a sulfate burden in the free troposphere which is comparable to that from anthropogenic emissions. This is caused mainly by the altitude-latitude distribution of volcanic emissions, and is most pronounced in tropical latitudes [27]

The quasi-biennial oscillation in stratospheric zonal wind (QBO_U) is found to influence significantly the distribution of volcanic stratospheric aerosols mainly over the tropics [28-30]. Even though, recent observational studies [26,31] revealed significant seasonal and inter-annual variations in the stratospheric aerosol load during the volcanically quiescent period, studies on the influence of these types of periodic oscillations on the stratospheric aerosol distribution over the tropics during volcanically quiescent periods are very rare [11]. This study involves an attempt to study the features of particulates in the UTLS region during the relatively quiescent volcanic period (1998-2005) using global aerosol data from Stratospheric Aerosol and Gas Experiment (SAGE-II) archive and lidar data from Gadanki [13.5°N, 79.2°E]. Formation of semitransparent cirrus (STC) is very common in the upper troposphere. The characteristics of these STCs and their contribution to particulate scattering in UTLS region are investigated.

2. Extinction/Backscatter data from LIDAR and SAGE-II

The biaxial, monostatic dual polarization Lidar at the National Atmospheric Research Laboratory (NARL), Gadanki, is used to study the scattering properties of atmospheric particulates in the UTLS region. This Lidar [7] is equipped with Nd:YAG laser (Model: PL8020, Continuum, USA) emitting linearly polarized pulses with 7 ns width and 20 Hz repetition rate at its second harmonic wavelength of 532 nm with a pulse energy of 550 mJ. The basic beam emerging from the laser source with a divergence of 0.45 m rad is expanded using a 10X beam expander to reduce the divergence to <0.1 m rad before transmitting vertically into the atmosphere. The time series of backscattered photons from different altitudes corresponding to each transmitter pulse are received using a 350 mm diameter Schmidt-Cassegrain telescope having a field of view of ~1 m rad. Both the transmitted beam and vertically looking receiving telescope are configured with a fixed horizontal separation of ~3m. For this lidar configuration as the lowest altitude at which the full overlap of the transmitter beam with the receiver field-of-view (beam-filled condition) is encountered around 7 km, the data from altitudes above 8 km only are used for retrieving the aerosol properties. A polarized beam splitter in the receiver beam path splits the beam into co-polarized and cross-polarized components which are detected independently using two photomultiplier tubes operated in photon counting mode and acquired with a bin width of 2µs corresponding to an altitude resolution of 300 m. These photon-number profiles corresponding to each transmitted pulse are summed over 250s to achieve a good signal to noise ratio up to altitude above about 40 km.

The SAGE-II onboard the Earth Radiation Budget Satellite (ERBS) employs solar occultation technique to measure the attenuation of solar radiation at the Earth's limb between the satellite and the Sun due to scattering and absorption by different atmospheric species [32]. These measurements provide the altitude profile of the volume extinction coefficients of atmospheric particulates which includes particles of thin sub-visual cirrus clouds and aerosols at four different wavelengths in the visible and near-IR range (1020, 525, 453 and 385 nm) with a horizontal resolution of about 200 km and a vertical resolution of 0.5 km [33]. This sensor takes 30 occultation observations on a single day, which are equally spaced in longitude round the globe but vary in latitude by a few degrees giving a near global coverage over a period of 25–40 days. Details regarding the SAGE instrumentation and algorithms are discussed in earlier publications [32,34]. The upper limit of the particulate extinction measurable by SAGE sensor at 1020 nm is ~2 x 10^{-2} km^{-1}, which is much larger than that in the UTLS region (~2 x 10^{-4} km^{-1}) under volcanically quiescent period. Cirrus cloud with extinction greater than this value, generally referred to as 'opaque clouds' [35,36], are not measurable by the SAGE-II sensor. Presence of such clouds, limits the SAGE measurements below tropopause.

3. Estimation of particulate extinction from lidar data

The lidar data (backscattered signal) on different nights from Gadanki are used to derive the altitude profiles of particulate backscatter coefficient (β_P) and volume depolarization ratio (δ) [37-39]. In the lidar system, received backscattered signal (at 532 nm wavelength) is separated into co-polarized and cross-polarized components ($\perp$ and $\parallel$ channels, respectively) and recorded separately in two channels. The data in these two channels are analyzed separately employing the Fernald's algorithm [40] to estimate the total backscattering coefficients ($\beta_\perp$ and $\beta_\|$, respectively) taking 30 km as the reference altitude where the aerosol contribution is assumed to be negligible. For this inversion a value of 40 Sr^{-1} is assigned for the lidar ratio (S_P) and its variation with altitude depending on δ is also accounted appropriately. With this correction the value of S_P reduces to ~26 Sr^{-1} within the STC [39] in the upper troposphere. Further incorporating the correction for multiple scattering the value of S_P within the STC reduces to 20 Sr^{-1} (which is used to study the properties of STCs). The molecular backscatter coefficients for the two polarized components are estimated from the mean molecular number density profile taking a molecular depolarization factor (δ_m) of 0.028 [41]. The molecular backscatter coefficient of the co-polarized component ($\beta_{m\perp}$) is related to that of the cross-polarized component ($\beta_{m\|}$) as $\beta_{m\perp} = \delta_m \beta_{m\perp}$. Subtracting $\beta_{m\perp}$ and $\beta_{m\|}$ from the altitude profiles of $\beta_\perp$ and $\beta_\|$, respectively obtained from lidar data employing the Fernald's algorithm, the altitude profiles of particulate backscatter coefficient, $\beta_{P\perp}$ and $\beta_{P\|}$, are estimated. The respective backscatter ratios for the co-polarized ($R_\perp$) and cross-polarized ($R_\|$) components are estimated [37] as $R_\perp = \beta_\perp / \beta_{m\perp}$ and $R_\| = \beta_\| / \beta_{m\|}$. As far the net atmospheric backscattering is concerned, the "unbiased" or effective backscatter ratio (R) is to be defined , to quantify the gross property of the medium, which on mathematical simplification can be written as $R(h) = [R_\perp(h) + \delta_m R_\|(h)]/(1+\delta_m)$. The volume depolarization ratio is obtained from the ratios of $R_\perp$ and $R_\|$ as $\delta(h) = [\delta_m R_\|(h)] / R_\perp(h)$. This ratio is a good indicator for distinguishing the cirrus based on 'particle habit'. While for small spherical particles, the values of δ will be relatively small, its value increases significantly as they become large and non- spherical. Using this property of cloud particles, structure and altitude extent of cirrus can be estimated from each lidar profile, which will be used to study the temporal variation of cirrus properties during the entire period of lidar observation. Based on a detailed scrutiny of a number of profiles at different cloud conditions a threshold value of $\delta \geq 0.04$ is assigned for discriminating the STC [42]. If the value of δ exceeds this threshold value it is classified as STC.

4. Altitude structure of backscatter/extinction, scatting ratio, and depolarization from lidar and comparison with SAGE measurements

Figure 1 shows the altitude profiles of mean particulate backscatter coefficient (β_P), effective backscatter ratio (R) and volume depolarization ratio (δ) for a few nights at Gadanki obtained from Lidar data during the year 1999 as typical samples. In general, β_P and R show a general decrease with increase in altitude (eg. 05 and 12 April 1999) in the troposphere and stratosphere. But on a few nights a significant enhancement is observable over a small region between 9 and 17 km. This sharp increase in β_P is due to strong scattering from ice particles of thin STC layers, formed at these altitudes either by *in situ* condensation of water vapour or originated from the outflow of convective anvils [43-45]. On a few occasions this layer of enhanced β_P extends down up to 5–6 km, depicting typical case of dense cirrus (9 June 1999), occurring predominantly during the monsoon period. Note that, these STCs are so thin that the lidar beam could penetrate the cloud and provide measurable signal even from higher altitudes. The opacity of STCs are quantified using the cloud optical depth (τ_c)

Distribution of Particulates in the Tropical UTLS over the Asian Summer Monsoon Region and Its
Association with Atmospheric Dynamics

117

which is the height integrated particulate extinction coefficient (α_P), obtained by multiplying β_P with the S_P, from the cloud base (h_{cb}) up to the cloud top (h_{ct}). In case if the cirrus is too dense (with τ_c exceeding 1.5) the lidar beam will not be able to penetrate the cirrus layer impeding useful lidar observations. In association with the enhancement in β_P and R, a significant increase in δ also can be observed at these altitudes. This suggests that the scatters in within the STCs are relatively large and significantly non-spherical in nature. Depending on τ_c, STCs are further classified [46] in three classes viz., sub-visual cirrus (SVC) with $\tau_c<0.03$, thin cirrus (TC) with $0.03 < \tau_c <0.3$ and dense cirrus (DC) with $\tau_c >0.3$. General features of STCs from this tropical station [37] showed that while the occurrence of SVC is larger during winter, TC and DC occur more frequently during the monsoon period. The upper and lower boundaries of STCs are identified from altitude profiles of R and δ using a threshold condition [37,42] for R to exceed 2 in either of the two lidar channels ($\perp$ or $\parallel$ channels) along with the value of δ exceeding 0.04 in the altitude region where the STCs are usually observed (8 to 20 km).

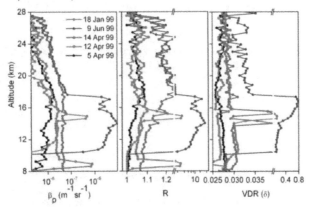

Figure 1. Altitude profiles of mean particulate backscatter coefficient (β_P), effective backscatter ratio (R) and volume depolarization ratio (δ) for few nights during the year 1999

A detailed error analysis [42] showed that the estimated values of β_P is less sensitive to the variability in S_P. For a given uncertainty of 25% in S_P, the maximum uncertainty in β_P is 10% in the absence of clouds, ~15% for thin cirrus and ~30% for thick cirrus. For the same uncertainty in S_P, the maximum uncertainty in the retrieved backscatter coefficient and effective backscatter ratio are around 0.6%, 2% and 10%, respectively, for clear atmosphere, atmosphere with thin cirrus and atmosphere with thick cirrus. Including the possible errors in the lidar signal inversion associated with the uncertainty in the molecular backscatter coefficient, the resultant error in the derived optical depth would be ~20%. As the signal-to-noise ratio is >2 up to ~45 km, for altitudes <30 km the system induced errors will be significantly small (<1%) compared to that from other sources. The error due to the influence of background and system noise is negligible (<□0.001%) compared to that due to the uncertainty in S_P.

Altitude profiles of α_P at 525 and 1020 nm from Global SAGE-II aerosol data archive (version 6.2) during the period 1998-2005 are obtained through the NASA website *http://wwwsage2.larc.nasa.gov/data/v6_data*. Typical estimated error in SAGE-II measured α_P at 525 and 1020 nm are in the range 10 to 15% [34]. Figure 2 show a comparison of α_P obtained from the SAGE-II at 525 and 1020 nm along with that of the lidar at 532 nm in the altitude region 10–30 km for a few sunset occultation events during the period 1998–2003. These comparisons are made when SAGE-II had an occultation pass within a grid size of ± 5° in latitude and ±10° in longitude centered at Gadanki within a time-duration of 1 day with respect to the lidar observation. As the difference between the SAGE-II wavelength of 525nm and and lidar wavelength of 532nm is less than 1.5%, the expected absolute differences in α_P for these two wavelengths would be almost insignificant [47]. The latitude and longitude of line-of-sight tangent point of SAGE-II for each occultation event are shown in the respective frames of Figure 2. The radial distance'd' between the tangent point and the lidar location estimated from the latitudinal and longitudinal differences between the two is also marked in this figure. In general, the shape of the SAGE-II and lidar-derived extinction profiles show a good agreement especially in the stratosphere. The mean percentage difference for extinction measured by the two instruments in the altitude range18–25km is <40%, which is comparable in magnitude with those obtained in other similar inter-comparisons [48-50]. For altitudes<17 and >25 km, both mean differences and standard deviations are relatively large (< 40%). This could partly be due to the temporal variations during the course of the two measurements as well as the spatial heterogeneity (between the locations of the two measurements). The observed increase in deviation below 17 km is partly due to the influence of STCs in the UT region as well as their spatial heterogeneity.

5. Semitransparent cirrus in the upper troposphere

Space-borne lidars, *in situ* measurements and ground-based experiments indicates frequent manifestation of cirrus clouds in the upper troposphere [36,51-53]. The frequency of occurrence of STC (F_{STC}) over a wide geographical region can be derived from VHRR data from remote sensing satellites. The geostationary meteorological satellite, KALPANA-1, positioned at 74°E over the equator for continuous measurements of clouds and convective systems, provides the required information over the Indian region. This satellite observes the earth in three wavelength bands: Visible (0.55–0.75 mm), Water vapor band (WV: 5.7–7.1 mm), and the atmospheric window of thermal infrared (TIR: 10.5–12.5 mm). In TIR and WV bands, the data is recorded at a pixel resolution of 8 km (nadir) with a digital resolution of 10 bits. Unless the cloud is optically thick, the radiance observed in TIR band does not correspond only to the cloud top, but is also weighted by the radiation emitted from the altitudes below. The brightness temperatures measured in these two channels are used to detect STC following the bi-spectral approach of Roca et al. [54]. In this method all the cloudy pixels having WV brightness temperature < 246 K and TIR brightness temperature > 270 K are treated as STC. In addition to the above, those cloudy pixels having a brightness temperature difference > 20 K between the two channels and having WV brightness

Distribution of Particulates in the Tropical UTLS over the Asian Summer Monsoon Region and Its
Association with Atmospheric Dynamics

119

temperature < 246 K, a condition imposed mainly to detect STC above other low level clouds, also are treated as STC [55]. Figure 3 shows the frequency of occurrence of STC (F_{STC}) over the Indian region in different months for the year 2005 derived from KALPANA-1 data. Any thin cirrus cloud above a deep convective cloud cannot be detected by the present bi-spectral algorithm, unless the difference in brightness temperature in the two channels exceeds 20 K (indicating significant altitudinal separation between the top of the STC and the optically dense high-altitude cloud), which may not be the case if STC forms just above the high altitude cloud. Because of this inherent limitation, the estimated F_{STC} will always be an underestimate over the region where high-altitude clouds are present. Hence the regions where the monthly mean frequency of occurrence of high-altitude clouds larger than 20% are masked (dark brown) in Figure 3.

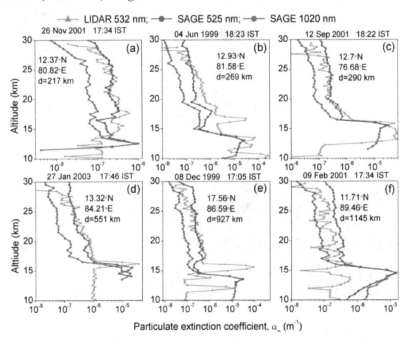

Figure 2. A comparison of the altitude profiles of α_P on a few days derived from lidar data (532nm) at Gadanki, along with that of SAGE-II (525 and 1020 nm) sunset occultation events near this region.

For studying the role of deep convection on the genesis of STC, the monthly mean spatial distribution of the frequency of occurrence of deep clouds (F_D) with TIR brightness temperature <235 K derived from Kalpana-1 during January-December 2005 are presented in Figure 4. A comparison of Figures 3 and 4 suggests that that the longitudinally extended band of high F_{STC} (~40-60%) in the region 10°S-20°S over the western tropical Indian Ocean and around the equator over the eastern Indian Ocean in January (Figure 3) is closely associated with the deep convection associated with the Inter tropical convergence Zone.

However, the values of F_{STC} far exceed F_D over all these regions. During this month, the highest values of F_{STC} (~60%) are observed at the north of Madagascar in the western Indian Ocean and over Sumatra/Indonesia in the eastern Indian Ocean, where the frequency of occurrence of very large deep convection is quite large. A region of less cloudiness (with F_{STC}<15%) is observed in the central Arabian Sea and Indian Peninsula centered around 5°N-15°N, which runs parallel to the equatorial band of high STC occurrence and is well separated from the deep convective regions. A similar STC-free zone is also observed in the southeast Indian Ocean centered on 20°S. Clearly, these regions with low occurrence of STC are caused by the large subsidence in the upper troposphere associated with the descending limb of the Hadley circulation cell.

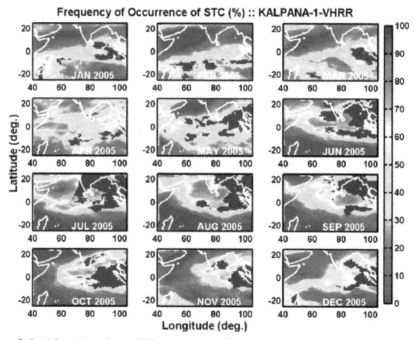

Figure 3. Spatial variation of mean STC occurrence in different months during the year 2005 (brown color shows the region where the STC retrieval was not possible due to the presence of high clouds) derived from the KALPANA-1 VHRR data

6. General features of STC over the Indian region

Lidar studies from Gadanki indicate that the occurrence of STC in this geographical region is the largest from May-October, associated with the formation of intense convection and the subsequent onset of Asian summer monsoon (ASM). Figure 5 shows the mean feature of STC occurrence and its altitude extend averaged for the period 1998-2001. While the STCs occurring during the period May- October are generally thick and optically denser [37],

those occurring during the rest of the year (dry months) are relatively thin (both
geometrically and optically). As the Gadanki region is almost free from deep convection
(Figure 4) during winter, the STCs forming during this period could be of in situ origin.
While the dense STCs observed during the May-October period are associated with deep
convection over the Indian land mass and Bay of Bengal.

Frequency of Occurrence of Deep Cloud (%) :: KALPANA-1-VHRR

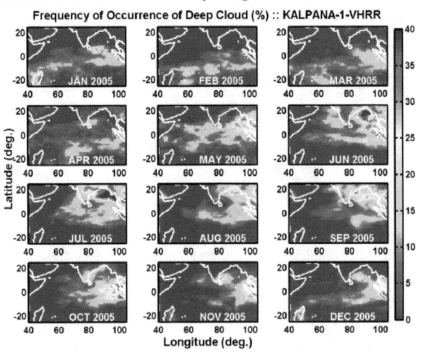

Figure 4. Spatial variation of deep convective clouds in different months during the year 2005 derived
from the KALPANA-1VHRR data

On examining the frequency of occurrence of STC during the period 1998-2001(Figure 6a) it
can be seen that the frequency of occurrence of SVC is much larger than TC and DC. In
general, STCs are observed in the altitude region 10 to 18 km with a preferred altitude
(frequent occurrence) region between 14 and 16 km (Figure 6b) Thin clouds occur more
frequently than thick clouds (Figure 6c). Though the vertical extent of STC generally vary
from 0.4 to ~ 4.0 km in majority of the cases it is less than 1.7km. Though the volume
depolarization,δ, in these clouds varies in the range 0.03 to 0.6 its distribution peaks (Figure
6d)in the lowest value. The value of δ for SVC and TC are generally very small compared to
DC. The particulate depolarization (δ_p) of STC (Figure 6e) generally varies from zero to
unity. The distribution of δ_p peaks around a value of 0.15. The properties of STC vary
significantly with cloud temperature (or altitude). Figure 7 shows the variation of thickness,
depolarization and optical depth of STC with cloud temperature. As can be seen from

Figure 7a, the thickness of STC is a maximum for temperatures in the range −55° to −75°C. Above and below this temperature range the cloud thickness decreases. Similarly the depolarization is maximum around −75°C and decreases steadily with increase in temperature. The value of δ also shows a sharp decrease when the cloud temperature decreases below −75°C. The decrease in δ with increase in temperature can be attributed to the melting and evaporation of ice crystals and subsequent blunting of their edges [56]. Decrease in temperature leads formation of particles with sharp edges. But when the temperature decreases below a threshold value, the particle size [56] becomes small (needle type). This can lead to a decrease in δ. Figure 7c shows that on an average τc increases with increase in temperature. The cloud becomes more opaque at higher temperatures. This has important implication in the radiative effects of STCs [38] in the context of global change.

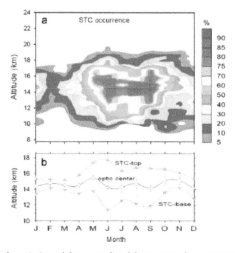

Figure 5. Month-to-month variation of the normalized frequency of encountering STC at different attitudes in the upper troposphere (a) along with the month-to-month variation of the mean top, base and optic center of STC with vertical bars indicating the standard error (b) from lidar data at Gadanki for the period 1998–2003.

7. Mean annual variation of particulate scattering in the UTLS region

Figure 8a shows a contour plot of the logarithm of mean β_P as a function of month and altitude for the period 1998–2003. The values of β_P are relatively large during the May to September period and small during the winter months. This is due to the influence of STC. Relatively high values of β_P in the UT region during the monsoon period are due to the presence of relatively dense cirrus and low values during winter are due to the presence of SVC. In contrast to the UT region, β_P in the LS is generally large (as high as $6 - 10^{-8}$ m^{-1} sr^{-1}) during the winter (November to January) and pre-monsoon (April–May) months and low (as low as 10^{-9} m^{-1} sr^{-1}) during the summer (July and August) months. Prominent peaks are observed during May–June and November–January periods with low values in July–August and February. As the SAGE-II provides the altitude profile of particulate extinction coefficient (α_P) for making a direct comparison with lidar data the altitude profiles of α_P is derived from lidar by multiplying the derived values β_P with the lidar ratio. However, in this estimation the altitude variation of S_P is taken in to account [39]. The mean annual pattern of annual pattern of α_P at different altitudes from the lidar derived β_P (Figure8a) is presented in Figure 8b. Similar grading scheme is used in both these plots to make an easy direct visual comparison of the pattern. The major features of the annual variation of β_P and α_P in different altitudes are very similar in these two plots, even though the lidar ratio is assumed to be variable with altitude depending on aerosol properties.

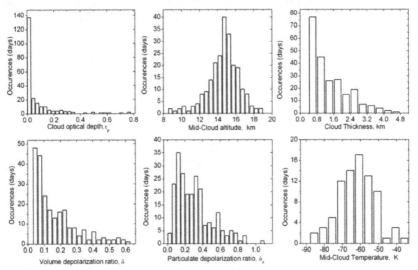

Figure 6. Frequency distribution of (a) cloud optical depth, (b) Cloud mean (mid-cloud) altitude, (c) Geometrical thickness, (e) Volume Depolarization Ratio, (f) Particle depolarization, and (g) Mid-cloud temperature of STCs observed at Gadanki for the period 1998-2003.

Microphysical properties like the size and shape of particles in the UTLS region can be delineated from the depolarization of backscattered radiation [57,58]. Figure 8c shows a contour plot of monthly mean δ in the altitude region 8–28 km. The value of δ varies in the range 0.03 to 0.6 in UT region and from 0.03– 0.04 in the LS region. High values of δ are generally confined to a narrow altitude region (14–16 km) during winter while these extend to a wider altitude region (12– 18 km) during summer monsoon period. High values of δ (>0.2) are observed in the altitude region 14– 16 km during April–October period. Values of δ exceeding 0.04 observed in the UT (above 10 km) are mainly due to presence of highly non-spherical ice particles associated with STC [37]. The overall low values of δ suggests that particles in the LS region are very small and tend to become more spherical in nature.

Figure 9 shows a contour plot of α_P similar to Figure 8b but generated from SAGE-II derived mean particulate extinction at 525 nm and 1020 nm over a small geographical grid size of 10-

16°N and 73-86°E centered around Gadanki. As thin STCs in the UT region significantly attenuates the SAGE-II wavelengths, (especially that at 525 nm) there will be a large data gap at the lower altitudes. The region bound between X-axis and the rectangular vertical bars (shown white) are the data gaps. As the attenuation for 1020 nm is less than that for 525 nm this wavelength can penetrate to lower altitudes to yield useful data. The data gap is relatively less for 1020 nm. However, in generating the contours, the data gap is appropriately interpolated. General similarity of the pattern in Figure 9a and 9b suggests that the interpolation did not influence the major features of Figure 9b. Except for an overall decrease in the values of α_P derived from SAGE-II (at 525nm) compared to those derived from lidar data, the major spatio-temporal features in Figure 9 also matches well with those of Figure 8b. Thus, the inferences derived from lidar data is reconfirmed by SAGE-II observations during the same period.

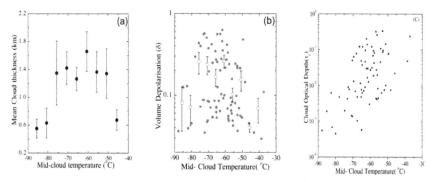

Figure 7. Mean variation of Cloud thickness, Volume Depolarization Ratio and cloud optical depth with mid-cloud temperature for STCs observed at Gadanki during the period 1998-2003.

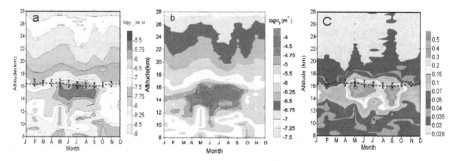

Figure 8. Contour plot of mean particulate backscatter coefficient (β_P), extinction coefficient (α_P) and volume depolarization (δ) in logarithmic scale as a function of month and altitude for the period 1998–2003 derived from lidar data. Mean cold point tropopause altitude is superposed over the contour along with its standard deviation.

To make the features more concise, the month to month variation of altitude weighted δ (altitude integral of δ normalized to the slab thickness) for different altitude regions with a

Distribution of Particulates in the Tropical UTLS over the Asian Summer Monsoon Region and Its
Association with Atmospheric Dynamics

125

slab thickness of 2 km are examined in the UT and LS region. Though a sharp definition of
UT and LS region is rather difficult, for the present analysis we use the term UT for the
altitude region from 10 to 16 km and LS from 18 to 24 km. Though the cold point tropopause
shows a small variation with time of the day and day of the year, it lies always in the range
16–18 km (a transition region), such that the region defined as UT is always below cold
point and region defined as LS is above. Figure 10a shows the mean annual variation of the
cold point tropopause altitude and tropopause temperature.

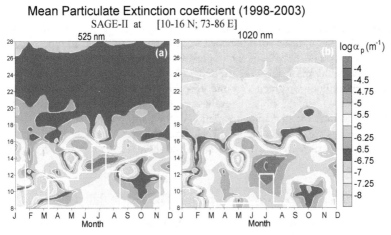

Figure 9. Contour plots of mean particulate extinction coefficient (α_P), in logarithmic scale, at 525nm
and 1020 nm as a function of month and altitude for the period 1998–2003 from SAGE-II observations

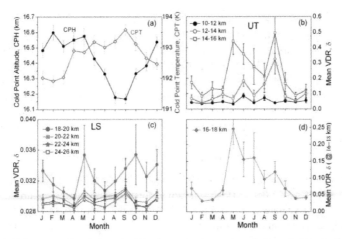

Figure 10. (a) Mean annual variation of cold point tropopause temperature and altitude along with the
mean annual variation of integrated VDR normalized to the slab thickness of 2 km at different altitudes
(b) in the upper troposphere, (c) in the lower stratosphere and (d) in the tropopause region.

The annual variation of δ for different altitudes in UT is shown in Figure 10b and that in LS in Figure 10c. Though, in general, on an average, the value of δ in LS region is < 0.04, it shows a pronounced oscillation with prominent peaks in May and September–January period. Above 20 km, the value of mean δ (~0.03) is very close to the molecular depolarization, it is relatively large (in the range 0.03 – 0.04) in the altitude region 18–20 km just above the cold point tropical tropopause. Thus, in the LS region, the particle size and non-sphericity decreases with increase in altitude. Note that, the value of δ encountered in this region is less than the threshold value (0.04) used for identifying the cirrus particles. The value of δ is largest in the altitude region 14–16 km, just below the mean level of tropical tropopause. The general similarity in the annual variation of δ in the to altitude regions 14–16 km and 10-12 km indicates that the particle microphysics in these altitudes are strongly coupled. Even though the cold point tropopause altitude shows a small variation from month-to-month, the mean level lies around 16.5 km Figure 10d shows the mean annual variation of the altitude weighted δ around the cold point tropopause for a slab thickness of 2 km. The observed general similarity of the annual variation of δ in the entire region from 10 to 20 km indicates that the annual variation of the particle habit (size and shape) in the UT and LS regions are strongly coupled. The observed general decrease in δ from UT to LS suggests that the particles tend to become small and more regular in shape with increase in altitude.

8. Effect of tropospheric convection on UTLS particulates

During the ASM period, the upper troposphere is significantly influenced by the tropospheric convection. This in turn could influence the microphysical properties of UTLS particulates. The outgoing long wave radiation (OLR), which is directly influenced by cloud cover and in turn by convection, could be effectively used as a proxy to the strength of tropospheric convection [59] and the seasonal variation of deep convection. The value of mean OLR is generally high for clear sky condition (~460 W m^{-2}) and decreases with increase in cloud cover and cloud vertical extent. A threshold value [60] of 200 Wm^{-2} for OLR which when declines below could be treated as in index for deep convection. The percentage of occurrence of daily mean OLR with its value < 200 Wm^{-2}, within a grid size of $2.5° \times 2.5°$ over a geographical region $12.5° –15°$ N in latitude and $77.5° –80°$ E in longitude obtained from NCEP/NCAR reanalysis provided by the Climate Diagnostic Center through their Web site at http://www.cdc.noaa.gov/ averaged in different months for the period 1998–2003, which is almost negligible during the December to April period, starts increasing from May and shows a broad peak during the period June to October (Figure 11a).

The strength of tropospheric convection also can be assessed from the convective available potential energy (CAPE), defined as the altitude integrated buoyancy of lifted air parcel from the level of free convection to the level of neutral buoyancy [61], is estimated from the altitude profile of potential temperature as CAPE = $\int [g(q_v - q_T)/q_T]dz$, where q_v is the virtual potential temperature of the air parcel, q_T is the environmental potential temperature and g is acceleration due to gravity. The integral is taken from the level of free convection to the level of neutral buoyancy. The mean values of CAPE, which is a measure of stability in the

atmosphere as far as vertical displacements are concerned [62], in different months for
Chennai (13°N, 80.2°E), a station located very close to Gadanki, obtained through India
Meteorology Department (Web site http://www.weather.uwyo.edu/upperair/
sounding.html.) are presented in Figure 11b. The CAPE also shows a prominent peak
during the period April to May along with a small secondary peak in September. Intense
convection is closely related to thunderstorm activity. Climatologically averaged
thunderstorm activity in different latitude belts over the Indian subcontinent was
investigated in detail in earlier study [63]. The annual variation of the mean number of
thunderstorm days for the latitude belt 10–15°N (Figure 11c) shows two prominent peaks in
May and September–October period. Thick convective clouds associated with deep
convection and high thunderstorm activity leads to the formation of thick convective clouds.
The main reason for low lidar observation statistics during the monsoon period is the
presence of these convective clouds which impede the observations. Though these clouds
will have a large spatial extent and persist for several days, some gap region which is
devoid of thick clouds (favouring lidar observation) starts developing after a few days The
outflow from adjacent convective anvils spreading over to this gap region leads to the
formation of STCs.

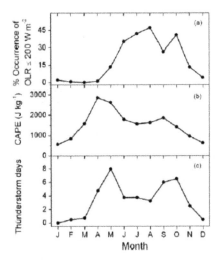

Figure 11. Annual variation of the (a) frequency distribution of outgoing long-wave radiation (<200
Wm-2) around Gadanki between 12.5°–15°N and 77.5°–80°E along with (b) convective available
potential energy (CAPE) obtained from Radiosonde observations at Chennai during the period 1998–
2003. (c) Month-to-month variation of mean number of thunderstorm days (TSD) for the latitude belt
10–15°N during the period 1970–1980 [63].

Being originated from convective outflow [43,44], especially during the summer monsoon
period, particles of STCs in the UT region during this period will be relatively large and
highly non-spherical [8,45]. Presence of these large non-spherical particles leads to an
increase of δ in this region, as is observed. Subsequent uplift of some of these particles along

with tropospheric air across the tropopause leads to an increase in δ along with the integrated backscatter (Iβp) in the region just above the cold point tropopause. As the particles and precursor gases are not directly injected into the stratosphere but diffuse slowly across the tropopause they confine to a small region just above the cold point tropopause. Note that during the period July to August when CAPE as well as thunderstorm activity shows a small decrease, δ (Figure 12) and Iβp (Figure 13) in the 18–20 km also shows a decrease. However, the STCs during this period shows large spatio-temporal variations [37] introducing large day-to-day variability in Iβp and δ in the UT and LS region as indicated by the large error bars during the May-October period.

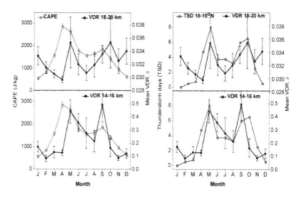

Figure 12. Mean annual variation of Volume Depolarization Ratio in the altitude region 14-16 km (UT) and 18-20 km (LS) superposed on the annual variation of CAPE and number of thunderstorm days. Vertical bars represent the respective standard error.

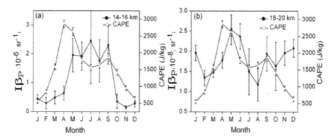

Figure 13. Mean annual variation of Integrated Particulate Back scatter (Iβp) in the UT km (a) and LS (b) superimposed on the annual variation of CAPE for the period 1998-2003.Vertical bars represent the standard error.

Though during the December–February period CAPE is relatively small and thunderstorm activity also is less, a broad peak in Iβp at 18–20 km is observable during this period (Figure 13).This feature though appears to be contradicting to the above hypothesis proposed for the summer monsoon period, the real mechanism responsible for the transport of tropospheric air into the stratosphere during winter is different from that during the monsoon period.

During winter the intrusion of tropospheric air into the lower stratosphere occurs through
vertical ascent of air driven by the strength of the ascending branch of the Hadley cell
circulation in the troposphere and Brewer- Dobson circulation (B–D) in the UTLS region.
The B–D circulation is primarily driven by mechanical forcing (westward directed wave
drag) arising from the dissipation of planetary-scale waves (breaking of Rossby waves) in
the extra tropical stratosphere [64]. Because of filtering by the large-scale stratospheric
winds, vertical propagation of planetary waves into the extra-tropical stratosphere occurs
primarily during winter and this seasonality in wave forcing accounts for the winter
maximum in the B–D circulation [65,66]. In association with these processes, the cold point
tropopause at tropics is pushed to a higher altitude during winter and they play a major role
in exchange of mass from troposphere to stratosphere.

As can be seen from Figure 10a the tropopause is cooler and higher during the January–
March period and warmer and lower during August–October period which is in accordance
with what is reported for tropical locations [67-70]. These Seasonal changes in the cold point
tropopause altitude also contribute to the mass influx to stratosphere [71-73]. Transport of
constituents like water vapor, aerosols, trace gases etc across the tropopause which acts as a
permeable membrane through which continuous mixing between tropospheric and
stratospheric air takes place is well demonstrated by various investigators [74-77]. Over and
above the penetration of deep convection, this annual adjustment of tropopause altitude is
an important mechanism for the stratosphere troposphere exchange.

9. Contribution of STC to particulate scattering in the UTLS region

Fine ice crystals of STCs originating either though the outflow from deep convective anvils
or through freeze drying of moist air lifted up to the tropopause by normal convection
contribute significantly for scattering in the UTLS region. In addition, other aerosol particles
originating from the surface (mainly through bulk to particle conversion and those of
vegetative origin)as well as those formed in the upper troposphere through gas- to-particle
conversion (mainly of various industrial gases) also contribute for the particulate loading in
the UTLS region. It would be worth examining the relative contributions of the two
components in the UTLS region to make a quantitative assessment of the contribution of
STCs to the scattering properties in the UTLS region. Figure 8a shows the contour plots of
month-to-month variation of mean β_p derived from lidar data during the period 1998–2003
(including both STC-contaminated and STC-free). Note that, as this study period was mostly
devoid of major volcanic eruptions [24] the stratospheric aerosol loading can be considered
to be in its background level.

While β_p in the UT is a maximum during the May –September period and minimum during
October–November, in the lower stratosphere it is minimum during summer (July and
August) and maximum during winter. However the winter high in the LS region was
attributed to [7] the transport of tropospheric air (containing aerosols and precursor gases)
in conjunction with the tropical upwelling and B– D circulation while the observed high
during May–June is due to the upward influx of particles (including ice crystals of STCs)

from the UT region. As the uncertainty associated with lidar-derived β_P is small compared to α_P [42], for the lidar based study the altitude profile of β_P and Iβ_P for a desired altitude region (layer integrated) is used for studying the annual pattern of particulate scattering in the UTLS region. Figure 14 shows the contour plots depicting the annual variation of β_P at different altitudes for the STC-contaminated case and the STC-free case.

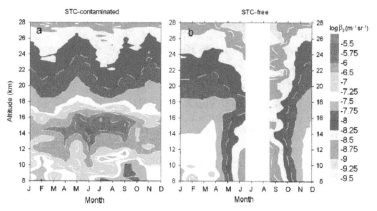

Figure 14. Mean annual variation of particulate backscattering coefficient (β_P) from lidar data at Gadanki for the STC-contaminated (a) and STC-free (b) cases during the period 1998–2003.

The cloud cover being quite large during the summer monsoon period no lidar data were available without STC during the July and August months, which lead to a data-gap for the STC-free case presented in Figure 14b. As can be seen, the variations of β_P in Figure 14a is very similar to those in Figure 8a generated by considering both the STC-contaminated and STC-free profiles even though the absolute magnitude in the UT region is slightly small in the latter case especially during the October–November period. High values of β_P are observed in the UT region for the STC-contaminated case during summer monsoon period. Figure 14b shows that, β_P in the UT region is very small for the STC-free case compared to that of STC-contaminated case for the same period. The values of β_P in the UT region are high during winter and spring and low during summer and autumn for the STC-free case. This annual pattern of β_P is significantly different from that of STC-contaminated case. From this, it is quite reasonable to infer that the prominent peak of β_P in the UT region observed during the summer monsoon period (in Figures 8a and 14a) is due to the influence of STC. However, the enhancement in β_P in the UT region due to STC-contamination is relatively small during the winter months

As seen from Figure 5 the occurrence of STC is mostly confined to the uppermost part of the troposphere, above ~10 km. Because of their presence these STCs directly contribute to the particulate scattering in this region. Over and above, these prevailing STCs in the upper troposphere can also modify the scattering property of particulates (aerosols) above the cloud-top as well as below the cloud-base. These effects in the UT and LS regions can be inferred by examining the Iβ_P at four different altitude regions, 8–10 km (UT$_1$), 12–16 km

Distribution of Particulates in the Tropical UTLS over the Asian Summer Monsoon Region and Its
Association with Atmospheric Dynamics

131

(UT$_2$), 18–21 km (LS$_1$) and 21–25 km (LS$_2$). Among these, the LS$_1$ and LS$_2$ regions are above
the STC-top while UT$_1$ is below the STC-base. The month-to-month variation of mean Iβ_P for
the three cases; viz, (i) including all profiles (ii) considering only SYC-contaminated profiles
and (iii) Considering only STC-free profiles, are presented in Figure15. In the absence of STC
the Iβ_P in the UT$_2$ region is significantly small. The STC contribution to Iβ_P in this region
works out to be around 93±5%. The Iβ_P shows an annual variation with relatively high
values during the winter/dry months and low during the summer monsoon period for the
STC-free case. When STCs prevail, the Iβ_P in the UT$_2$ region increases significantly especially
during the summer monsoon period. Even though the enhancement in Iβ_P due to the
presence of STC is seen during the winter/dry months also, its magnitude is relatively less. It
would be worth in this context to note that, during the monsoon period, the UT$_2$ region is
dominantly influenced by dense STCs originating from the outflow of convective anvils [78].
The particle associated with these STCs will be relatively large and highly non-spherical
[8,45] and hence their contribution to β_p and δ will be significantly large [7].

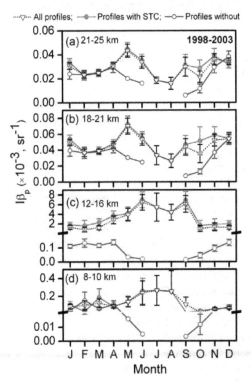

Figure 15. Mean annual variation of layer integrated particulate backscattering coefficient (Iβ_P) in LS$_2$
region (a), LS$_1$ region (b), UT$_2$ region (c) and UT$_1$ region (d), for the three cases (including all profiles,
only STC-contaminated profiles and only STC-free profiles) from lidar data at Gadanki, for the period
1998–2003.

It is quite interesting to note that the presence of STC in the UT$_2$ region significantly enhances the Iβ_P in the lower stratosphere also especially during the May–October period. In the LS$_1$ and LS$_2$ regions Iβ_P shows a double peak structure with prominent maxima during May–June and October–January for the STC-contaminated cases (in the UT region), while it is very small from May to October for the STC-free case. A significant amount from the abundant ice crystals (of STC) present in the UT region during the summer monsoon period will be lifted up along with air-mass across the tropopause. Consequently, Iβ_P in the lower stratosphere also increases. However, unlike the case during winter, this increase in β_P (or Iβ_P) is rather confined to a small region just above the tropopause (the LS$_1$ region). All the three curves in Figure 15 for the LS$_1$ and LS$_2$ regions almost overlap each other during the December to April period. This indicates that the STCs in the UT$_2$ region during this period did not contribute significantly to Iβ_P in the stratosphere. In general, the influence of STC to the mean backscattering coefficient (Iβ_P /Δh, where Δh is the slab thickness) in LS$_1$ region is relatively more than that in LS$_2$, particularly during the summer monsoon period. Thus, presence of dense STC in the UT$_2$ region (during the summer monsoon period) enhances the particulate scattering in the lower stratosphere, even though this region is practically free from STC.

As lidar data yields reliable β_P values only from the region above the altitude of 'full beam overlap' the effect of STCs below the cloud-base can be examined by studying the scattering property of the medium over a narrow altitude region 8–10 km (UT$_1$), which is mostly free from STCs (Figure 6b). The mean annual variation of Iβ_P in the UT$_1$ region for the three cases, (i) when the UT$_2$ region is STC-free, (ii) the UT$_2$ region is STC-contaminated and (iii) with both these cases combined, is presented in Figure 15d. Except for the STC-free case, though the Iβ_P in this region also shows an annual variation similar to that for the UT$_2$ region, the amplitude of this variation is significantly small. Notwithstanding the fact that the values of Iβ_P during the November to April period for the three cases are comparable, for the STC-free condition it shows a decrease during the May–October period. This shows that the presence of dense STCs in the UT$_2$ region enhances the particulate scattering in the region below the STC-base and the magnitude of this enhancement is much smaller than that in the UT$_2$ region (where the Iβ_P increases by a factor >20). But when these STCs are thin (optically as well as geometrically), especially during the November to April period, this contribution is almost negligible. Thus, the prevailing STCs in the UT$_2$ region during the summer monsoon period significantly enhance the scattering from the region above the cloud-top as well as below the cloud-base. The enhancement in particulate scattering in the LS region above the cloud-top could be mainly due to the lofting of STC-particles across the tropopause, which joining with the prevailing LS aerosols modify the volume scattering properties in this region. Though the Brewer–Dobson circulation is weak during summer the troposphere–stratosphere exchange during this period could be aided by an increase in wave activity depending on the prevailing atmospheric condition. Strong convection prevailing over the Indian landmass during the summer–monsoon period, which can transport abundant moisture to the upper troposphere inducing cirrus formation, is also a major source for gravity waves [79].

A study on the influence of STC in the particulate extinction in UT and LS region using SAGE-II data over the Indian longitude sector (70-90°E) from 30°S to 30°N also revealed similar results However, as the occurrence of STC is less frequent in the southern hemispheric off-equatorial region, the increase in τ_P in the UT and LS regions due to the influence of STC is relatively small compared to that in the equatorial region [80].

10. Mean latitude variation of the altitude structure of τ_P over the Indian longitude sector

The latitudinal structure of the annual pattern of τ_P in the tropical UTLS region over the Indian longitude Sector is examined using the altitude profiles of particulate extinction at 532 nm obtained from SAGE-II data archive in the latitude region 30°S to 30°N for the period 1998-2005. The profiles are grouped in different latitude bands each having a width of 5°. Contours presenting the annual variation of τ_P in different altitudes in the UTLS region are presented in Figure16. The mean annual variation of the altitude of the lapse rate tropopause for each of these latitude bands is superposed on the respective contours.

For the latitudinal region between 0-15°N, the tropopause is cooler (higher) during the December-May period and warmer (lower) during July-October period, in accordance with that reported for tropical locations by prior investigators [67,68]. For the latitudinal sector 15-20°N, the tropopause is higher during April-June and lower during the July-August period. At latitudes > 20°N, a pronounced maximum in tropopause altitude can be observed during the boreal summer and minimum during boreal winter. Though the tropopause altitude varies with latitude, time of the day and day of the year, on an average, this altitude mostly lies in the range 16-18 km.

In general, α_P is relatively large (>10^{-7} m^{-1}) in the UT region for the region north of 20°S. In both the hemispheres α_P shows two peaks. The summer peak (in the respective hemispheresis more prominent, compared to the winter peak South of 20°S, α_P in the UT region is relatively small and does not show any pronounced annual variation. The summer-winter contrast in the UT region is almost insignificant beyond 25°S. the winter peak in τ_P becomes relatively weak in the region north of 15°N and becomes almost insignificant beyond 25°N. The summer-winter contrast in α_P is well pronounced beyond 20°N. In contrast to the UT region, the values of α_P in the LS region are relatively small (<10^{-7} m^{-1}), at least by a factor of two. The annual variation of extinction shows a weak summer winter contrast in the LS region for all the latitudinal sectors between 0-30°N. Due to the influence of STCs, which occur at random in different altitudes, the standard error of α_P (expressed as percentages of mean α_P) is generally very large in the UT region. The mean standard error in the attitude region 10-15 km ranges from 20% to 60%. This error decreases progressively with increase in altitude. In the LS region, the error is very small and mostly confined to values in the range 5-15%. In the transition region 15-20 km, the mean error is of the order of 30% which is less than that in the UT region.

Figure 16 shows that the annual variation of α_P in the UT region over the southern hemisphere is distinctly different from that over the northern hemisphere. In this hemisphere high values of α_P (in the UT region) remains fairly confined to latitudes north of 20°S. This difference can mainly attributed to the corresponding difference pattern of deep convection in these two hemispheres. While the southward migration of ITCZ (during the boreal winter) is fairly confined to latitudes north of 20°S, northward migration (during the boreal summer) extends beyond ~28°N. More over, the convection over the northern hemispheric land mass is much stronger than that over the southern hemispheric oceanic regions. Because of these features, the increase in α_P in the UT region is also confined to the geographical region north of 20°S. The value of α_P in the LS region is relatively small and show similar variations in both the hemispheres. This indicates that the features of aerosol transport in the equatorial (±15°latitude) LS region are fairly symmetric with respect to equator.

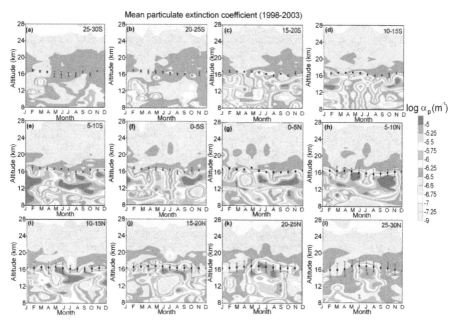

Figure 16. Contour plots showing the mean annual variation of α_P in the UTLS region obtained from SAGE-II in six latitude bands of width 5°from 30°S to 30°N averaged for the longitude region 70°E to 90°E for the period 1998-2005. Month-to-month variation of the mean tropopause altitude for each sector, with vertical bars representing its standard error, is superimposed on the respective contours. For the northern hemisphere the tropopause altitudes are derived from the altitude profiles of temperature obtained from the daily Radiosonde measurements carried out by the India Meteorology Department (IMD) at different locations in each band. For the latitudes south of 5°N, where no Radiosonde measurements are available, the tropopause altitude is obtained from the NCEP data provided along with SAGE-II version 6.2-data archive.

11. Tropospheric convection over the Indian region

The strength of tropospheric convection which can be indexed based on the CAPE could in turn also be related to the thunderstorm activity. The climatology of monthly mean thunderstorm days (TSD) for the Indian region was studied [63] using the data for the period 1970-1980. The monthly mean values of CAPE at different latitude sectors for the period 1998-2003 over this region is estimated from the respective daily values in each month obtained from the website of IMD (www.weather.uwyo.edu /upperair/ sounding.html). The annual variation of these two parameters for different latitude bands are shown in Figure 17. On an average, the annual pattern of TSD matches well with that of CAPE in all these latitude belts except for the fact that the secondary peak in October is less prominent in CAPE for latitudes north of 15°N. Both these parameters show a significant positive correlation with coefficient exceeding 0.7. Two prominent peaks observed during April-May and October in the near equatorial region merges to become a broad peak in the latitude region 20-25°N (off-equatorial) and subsequently becomes a well defined sharp peak around June-July in the latitude region 25- 30°N. Examining the annual variation of α_p in the light of the annual variation of CAPE/TDS it can be seen that the occurrence of the high values of α_p (during May and October in Figure 16g,h,i and those during June, July and August in Figure 16k,l), coincides with the peak in thunderstorm activity (or CAPE) in the respective latitudinal belt. During high convective activity while high values of α_p are observed very close to the tropopause in the latitude belt 0-15°N, significant high values of extinction are observed just above the tropopause in the latitude region between 15-30°N. Intense thunderstorm activity along with deep convection during boreal summer is highly favorable for the formation of dense STCs very close to tropopause leading to an increase in α_p in the UT region. Penetration of particles from these STCs to higher altitudes increases α_p in the LS_1 region, as is observed in Figure 16.

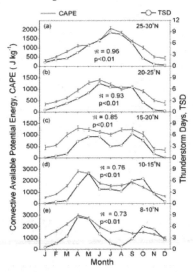

Figure 17. Annual variation of CAPE at different latitude bands over the Indian region along with the mean climatology of thunderstorm days [63]. The vertical bars represent the standard error associated with CAPE. The correlation coefficient ($\mathfrak{R}$) between CAPE and thunderstorm days and its level of significance (p) are also shown in respective panels.

12. Influence of dynamics in the latitude distribution of particulates in the UTLS region

To delineate the latitude variation of α_P in the UTLS region over the Indian longitude sector (70-90°E), the annual variation of monthly mean τ_P for a latitude band of 5°width in the latitude region 30°S to 30°N, for the three altitude regions 12-16 km (UT), 18-21 km (LS1) and 21-30 km (LS2) are examined separately. Figure 18 shows the contour plots of the mean τ_P in the UT, LS1 and LS2 regions with month along x-axis and latitude along y-axis. In the UT region, τ_P shows a general decrease with increase in latitude from equator, with its gradient showing a pronounced variation from month-to-month. The summer-winter contrast (with relatively low values during winter and high values during summer) is well discernable in the UT region (Figure 18a). Relatively high values of τ_P observed in the UT region between 15°S-15°N during May to February period are mostly due to presence of dense STCs resulting from the outflow of convective anvils. These are the periods when convective activity in the troposphere is very strong (associated with the southwest and northeast monsoons) in this region. Above 15°N, relatively high values of τ_P are more-or-less confined to the June-August period when the monsoon trough usually reaches its extreme north over the continent. However, beyond 15°S, the values of τ_P decreases significantly with increase in latitude, compared to that observed in the northern hemisphere.

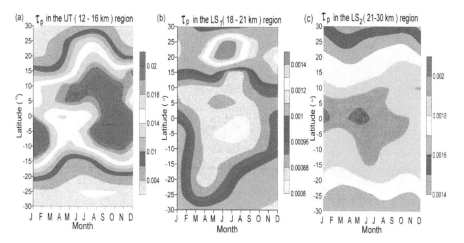

Figure 18. Contour plots showing the latitudinal dependence of the annual variation of τ_P obtained from SAGE-II data in the altitude regions 12–16km (UT), 18–21km (LS1) and 21–30 km (LS2) during the study period

The latitudinal variation of the annual pattern of τ_P in the UT region shows fairly good correspondence with CAPE and mean thunderstorm activity (Figure.17) over the northern hemisphere. The annual variation of TSD in the latitude region 8-15°N shows two peaks; one during the April-May period and then another during October. Further north, these two peaks get closer and merge to form a single peak in the latitude belt 20-30°N during the June-September period. This corresponds well with the latitudinal variation of the annual pattern of τ_P in the UT region (Figure 18a). Increase in convective activity (as well as convective outflow) over the Indian region associated with the northward migration of ITCZ along with the development of deep convection over the Bay of Bengal (particularly over its northern parts) during this period aid the formation of abundant dense STCs in the UT region. The annual variation of high altitude cloud amount over the Indian region also shows the presence of large deep convective cloud systems reaching very high altitudes and even penetrating the tropopause has been reported [81]. As the strength of the tropospheric convection decreases significantly during winter and the ITCZ shifts to the southern hemisphere sector, the probability of occurrence of dense STC over the southern hemisphere Indian region increases significantly. The deep convective clouds over this region get confined to a small geographical region between equator and 10°. Most of the STCs occurring over the Indian region during this period will be of in-situ origin [78]. These clouds will be either ultra-thin or sub-visual type cirrus with very low values of optical depth (<0.03).

The latitude variation of the annual pattern of τ_P in the LS region for the two bands LS$_1$ and LS$_2$ are presented in Figure 18b and Figure 18c respectively. In general, the mean τ_P in the LS$_1$ (18-21 km) region is in the range of 0.0008-0.003, which is one order in magnitude less than that in UT. In this altitude region, τ_P clearly shows relatively high values in the off-equatorial (north of 15°) regions and low values in the equatorial regions. Studies [82] on stratospheric aerosol optical depth during the decay phase of the volcanic aerosol at mid-latitudes using SAGE-II data have shown the presence of sinusoidal variation in the aerosol optical depth superimposed on an exponential decay with maximum and minimum occurring during the local winter and summer respectively. Over tropics the amplitude of these oscillations are significantly small and hence gets submerged in the disturbances caused by subsequent minor volcanic eruptions. As the period selected for the present analysis is volcanically quiescent, these oscillations are well discernable over the tropics. In the region 10-15°N, the annual variation of τ_P shows a relatively high value during winter and low value during summer. Note that, this variation is very similar to the annual variation of particulate backscatter observed in the lidar data from Gadanki [7]. Above 15°N, high values of τ_P are observed in the LS$_1$ region during the May-August period centered around 20-25°N. The particulate extinction just above the tropopause also shows a significant enhancement in the latitude region between 20-25°N (Figure 18b). This could be due to the penetration of the top of the high altitude semitransparent cirrus clouds above the cold point (tropopause). The two peaks observed in CAPE (and TSD) near equator (Figure 17) during April and October becomes more prominent and gets closer with increase in latitude (towards north) and merges to become a strong broad peak centered around July. This shows that the convective activity in the 20-30°N is very strong and the outflow occurs very close to the tropopause. The frequency of occurrence of cloud top altitude (observation from CALIPSO) shows a maximum value of 17±0.5 km in this latitude region during the June to September period which is at least 1 km larger than the maximum value observed at other latitudes (between 30°S to 30°N) at any period

[81]. Thus convection over the Indian land mass during the summer monsoon period in the 20-30°N latitude band is the strongest one in the entire latitude region 30°S to 30°N at any time during the year. This is a characteristic feature for the Indian longitude region. Strong convection plays a major role in transporting particulates from the upper tropospheric cirrus cluster to lower stratosphere causing a pronounced increase in τ_P in the LS1 region. Note that, such a feature in convection and hence in τ_P (in the LS_1 region) is not observed in the southern hemisphere. In LS_2, τ_P shows a distinct latitude variation with relatively high values near the equator up to 15°®in both the hemispheres and low values over the off-equatorial regions (>15°). Relatively high values of τ_P are observed in the LS_2 region during the January-June period in the latitude region 15°S-10°N and low values during the rest of the period. This pattern slowly reverses with increase in latitude. Beyond 10°N, τ_P shows a pronounced winter peak with low values during the March-May period. The annual variation of τ_P in the LS_2 region over the southern hemisphere is quite similar to that in the northern hemisphere. High values of τ_P observed between 15°S and 15°N in the LS_2 region confirms the presence of a Tropical Stratospheric aerosol Reservoir (TSR) during the study period. Earlier studies carried out by several investigators [29,83] revealed the presence of this band structure (with high aerosol loading) in the 21- 30 km altitude region over the equatorial region during volcanically perturbed period. Examining the aerosol climatology in the LS region using the SAGE-II data both during the volcanically perturbed period as well as during the near background conditions, *Bauman et al.* [84] reported maximum aerosol optical depth near the tropics and minimum between 15-45°®latitudes. *Trepte and Hitchman* [28] were the first to propose the existence of a low-latitude maximum in lower stratospheric aerosol optical depth which they referred to as the 'tropical aerosol reservoir'. They also examined the post-volcanic aerosol distribution in tropics and observed that the 18-21 km (LS_1) region experiences a rapid pole ward transport, while in the upper regime (LS_2) aerosol lofting and subsequent accumulation occurs within 20°S-20°N. This can fairly well explain the observed low values of τ_P in the lower regime (LS_1 region) and high values in the upper regime (LS_2 region) in the equatorial region between 15°S to 15°N in the present study. The variation of τ_P in the UT region in both these hemisphere are more-or-less complementary to each other indicating similar seasonal dependence in both the hemispheres, except for the fact that decrease in τ_P with increase in latitude (from equator) towards south is much faster than that in northern hemisphere. This is quite expected because the southern hemispheric sector is mostly occupied by ocean. The seasonal influence is rather insignificant in the LS_1 and LS_2 regions.

13. Influence of moderate volcanic eruptions on α_P in the LS region during 1998-2005

The background stratospheric aerosol layer usually referred to as the Junge layer [85] consists of liquid droplets composed of a mixture of sulfuric acid and water. This layer will be quite prominent subsequent to major volcanic eruptions, such as El Chich'on (Mexico, 1982) and Mount Pinatubo (the Philippines, 1991), which are powerful enough to inject large amount of SO2 into the stratosphere [86]. After oxidation, sulfate aerosols are formed at these altitudes. These particles are subsequently distributed globally depending on the latitude of the eruption. Removal of these aerosols is rather difficult. It takes several months to years (depending on size) to scavenge these volcanic aerosols. The last major eruption (Mount Pinatubo) took place in 1991 and the stratospheric aerosol layer returned to its "background" level around 1997. There is no major increase in stratospheric aerosol loading after 1997 [13,25]. However, increase in the anthropogenic SO2 emission has been proposed as a plausible mechanism responsible for the observed small increasing trend in stratospheric background level in the recent past [26]. An overview of the current understanding on stratospheric aerosol science can be found in *Thomason and Peter* [24].

Though, in general, most of the particles generated through the gas-to-particle conversion process in the stratosphere will be small and nearly spherical in nature, there could be a few larger size particles in the lower stratosphere associated with moderate and intense volcanic eruptions, leading to a pronounced enhancement in δ and α_P in this region. These volcanic perturbations are clearly distinguishable from STCs in the UT region based on the amount of enhancement its temporal structure as well as the duration of enhancements. Moreover, while δ (and α_P) of STCs vary significantly at shorter time scales [57,87] the stratospheric cloud formed through volcanic emissions will be stable for a longer period. In addition, the values of δ associated with STC will be significantly larger (as they are mostly composed of non-spherical ice crystals) those of volcanic clouds. The vertical structure of δ for the volcanic cloud will remain fairly stable at short time scales typically over a night. Thus a long lasting enhancement of δ in the LS region observed by the lidar is an indicator for assessing the volcanic impact on stratospheric aerosols. In order to illustrate this in detail and depict the difference in the nature of perturbations due to STC and volcanic cloud, a contour map of δ for a typical night (25 November 2002) during the eruption period of Reventador (started in November 2002 and lasted up to January 2003) is presented in Figure 19a. In order to accommodate the large variations, the contouring of δ in this figure is performed in two bands; one from 0.04 to 0.2 in steps of 0.02 and the other above 0.2 at 0.2 interval. Two enhanced layers (of δ) one between 12 and 16 km and the other around 19km, are distinctly seen in this figure. The value of δ and its temporal variations are very large for the lower layer while these are very small for the upper layer. The lower layer disappears after mid-night while the upper layer continues to persist up to the end of lidar observations. This shows that the lower layer is an STC while the upper one is the volcanic cloud. This volcanic cloud will persist on subsequent night also while STC may or may not be present. To illustrate this the sequences of lidar profiles observed on a few nights during the eruption period of Reventador are presented in Figure 19b.The features of the layer located around 19 km (in the LS region) is rather steady in all these profiles while those of the layer below ~17 km (due to STC), is highly variable. The stratospheric layer is strongest in December and started decaying in January 2003 and became almost insignificant by March 2003, while the STC layers appear at random.

The perturbation in the lower stratosphere over the equatorial and off-equatorial regions during the period 1998-2005, when the volcanic activity was relatively small, are examined in detail [88] using the altitude profiles of α_P over the tropics from SAGE-II data archive (Figure 20a) and the altitude profiles of α_P and δ obtained from lidar data (Figure 20b and 20c) at Gadanki. Zonal-averaged monthly mean α_P at different altitudes in the lower stratosphere for four different latitude belts (over tropics) are used mainly to have sufficient number of profiles in each belt as well as owing to the fact that in the lower stratosphere the spatial variability along longitude could be minimal (because of efficient mixing in the zonal direction and strong horizontal transport prevailing in the region). As expected the aerosol extinction decreases with increase in altitude with significant loading confined to the altitude region 18–27 km. Short-lived enhancement in α_P are distinctly seen in the lower stratosphere (Figure 20a) in different years. These signatures could be attributed to influence of various minor volcanic eruptions during the study period. Those eruptions which could make discernable impact in the stratosphere are marked at the start of the respective eruption along the X-axis of each image and further details of these eruptions are summarized in Table 1. Only those eruptions which occurred between 30°S and 30°N are included in Table 1. The VEI for different eruptions [89] is obtained from the Web site of Smithsonian Institution–Global Volcanism Program (GVP). The eruptions of Ulawun (eruption d), Ruang (eruption e), Reventador (eruption f), and Manam (eruption j) are relatively strong with VEI ~4 (Table 1), and the signature of these eruptions are well discernable in the lower stratospheric (18–20 km) aerosol extinction (Figure 20a). These perturbations can be distinctly seen both over the equatorial as well as off-equatorial regions.

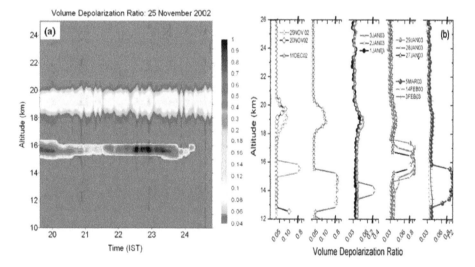

Figure 19. Altitude-time cross-section of δ during the night of 25th November 2002 (a). Altitude profiles of δ on different nights (averaged for one hour around 22:00 IST) for the period November 2002 to March 2003, during the active and post -active phases of Mt. Reventador (b).

Distribution of Particulates in the Tropical UTLS over the Asian Summer Monsoon Region and Its
Association with Atmospheric Dynamics

141

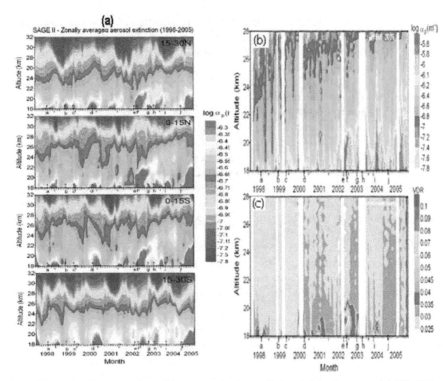

Figure 20. (a) Zonally averaged monthly mean extinction coefficient (α_P) for the equatorial (0°S–15°S
and 0°N–15°N) and off-equatorial (15°N–30°N and 15°S–30°S) regions estimated from SAGE-II data, (b)
Altitude-time cross section of aerosol extinction coefficient (α_P) from lidar and (c) Altitude-time cross
section of the Volume Depolarization Ratio from lidar (at Gadanki) during the period 1998–2005. The
letters along the abscissa represent the eruption of each volcano, the name and other details of which
are listed in Table 1. The major ticks correspond to December of each year.

Identification Letter	Volcano	VEI	Duration		Location		
			Start	End	Latitude (deg)	Longitude (deg)	Altitude (m)
a	Guagua-Pichincha	3	August 1998	May 2001	−0.17	−78.6	4784
b	Mayon	3	June 1999	March 2000	13.3	123.7	2462
c	Tungurahua	3	October 1999	January 2009	−1.5	−78.4	5023
d	Ulawun	4	September 2000	November 2000	−5.0	151.3	2334
e	Ruang	4	September 2002	September 2002	2.3	125.4	725
f	Reventador	4	November 2002	January 2003	−0.08	−77	3562
g	Anatahan	3	May 2003	July 2003	16.4	145.8	790
h	Lokon-Empung	3	September 2003	September 2003	1.4	124.8	1580
i	Soufrière	3	March 2004	May 2004	16.7	−62.2	915
j	Manam	4	October 2004	May 2009	−4.1	145.0	1807

Table 1. Details of volcanic eruptions in the tropics during 1998–2005 having significant stratospheric
impact

Figure 20b shows the temporal variation of the altitude structure of α_P obtained from lidar during the period 1998– 2005 in the form of a contour plot. This figure also depicts the signatures of various minor volcanic eruptions similar to that depicted in the zonal mean values obtained from SAGE-II data, in the latitude sector 0–15°N (Figure 20a). Small-scale features are more pronounced in the lidar data (Figure 20b) mainly because of the fact that it corresponds to a point observation while that in Figure20a is the zonal average. In volcanically quiescent periods the sulfur bearing gases, SO2 and OCS emitted from the earth's surface are transported across the tropopause [90], photolyzed, and oxidized to sulfuric acid before condensation to form sulfuric acid and water droplets in the stratosphere [24,91], which are, by their liquid nature, spherical. The depolarization caused by these particles will be very small and hence the resultant δ in the stratosphere will be very close to that of the molecules. In general, it is in the range 0.03 to 0.04. But, during major volcanic eruption abundant amount of precursor gases will be injected into the lower stratosphere along with a few fine particulates. Because of this influx of particles and gases there will an increase in the number density of particles as well as there will be an increase in size of these particles. The size spectrum of the stratospheric particles also shifts toward the larger size regime following the volcanic eruption [92]. As some of those particles that are directly injected into the lower stratosphere during volcanic eruption could be non-spherical in nature, an increase in δ in the stratospheric aerosol would be expected.

Figure 20c shows a contour plot of δ at different altitude in the lower stratosphere over Gadanki during the period 1998–2005. This plot generated adopting the same procedure as that used for generating Figure 20b from α_P profiles, clearly shows a few short-lived δ enhancements in the lower stratosphere. The sporadic increase in δ is associated with the eruption of a few moderately intense volcanic eruptions. The duration of the increase in α_P also matches well that of δ. The disturbances caused by eruption of volcanoes Ruang (eruption e) and Reventador (eruption f) are relatively stronger (δ ranging from 0.05 to 0.2 in the altitude region 18–21 km during November 2002 to February 2003). In a few cases the enhancements in α_P and δ does not match exactly. This could be due to the fat that the volcanic locations are at different distance from Gadanki as well as the prevailing transport process could be different at various occasions

14. Long term variations of τ_P in the LS region during the period 1998-2005

Zonal mean values of τ_P in the altitude region 18–28 km averaged for the four latitude belts, each of width 15°, in each month are used to examine the temporal variations in both the hemispheres. Time series plots of these for the equatorial (0–15°S and 0–15°N) and off-equatorial regions (15–30°N and 15–30°S) are shown in Figures 21a and 21b, respectively, along with a similar plot of τ_P obtained from lidar at Gadanki in Figure 21c. In the equatorial region the temporal variations in τ_P are very similar in both the hemispheres. The mean level of τ_P shows an abrupt increase after 2002 [88] though the oscillations around this mean level is fairly similar to those before 2002. Over the off-equatorial regions the mean level of τ_P shows a rather steady (gradual) increase from 1998 to 2005. A sharp increase in τ_P is observed towards the trailing edge of the data in the year 2005. During the period 1998–2002, the value of τ_P in the equatorial region is a minimum and is close to ~0.0025. The mean τ_P in the altitude region 18–28 km obtained from lidar at Gadanki also shows similar feature as that observed from SAGE-II over the equatorial region. In general the value of τ_P in the LS region obtained from lidar data is larger than that obtained from SAGE-II. This could be due to the fact that the lidar observation is a single point measurement while the SAGE-II data used in this analysis is zonally and meridionally averaged for the equatorial and off-equatorial regions.

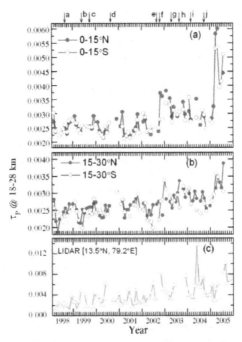

Figure 21. Time series of zonal mean monthly average τ_P in the altitude region 18–28 km obtained from SAGE-II for (a) the equatorial region (0°–15°N, 0°–15°S) and (b) the off-equatorial region (15°–30°N and 15°–30°S), along with (c) the mean τ_P obtained from lidar at Gadanki.

All the plots in Figure 21 show a general increasing trend in stratospheric particulate optical depth during the period 1998–2005 in addition to the periodic variations. On the basis of lidar observations during the period 2000–2009, a similar increase in the integrated stratospheric backscatter coefficient (in the altitude region 20–25 km) at the rate of 4.8% and 6.3% per year (with respect to its value in 2002) was reported [26] at Hawaii (19.5°N, 155.6°W) (Mauna Loa Laboratory) as well as at Boulder (40°N) (Colorado). The increase in τ_P at Hawaii and Boulder was attributed to the increase in global coal consumption since 2002, mainly from China, and subsequent increase in emission of SO2. Anthropogenic aerosols produced through gas-to-particle conversion of precursor gases like sulfates and ammonia transported to the upper troposphere [5] through intense convection [93] in the tropics and subsequently across the tropopause also could possibly be a contributing factor for this increase. An increase in tropical upwelling (Brewer- Dobson circulation) because of global warming also was suggested to be a plausible mechanism for the observed increasing trend [93,94] in stratospheric β_P after 2002.

Even though on an average the stratospheric particulate loading is in its background level during the period 1998–2005, it was influenced particularly by a few moderate volcanic eruptions mainly after September 2002. While the period before September 2002 was absolutely quiet (with low particulate loading), the later period was mildly disturbed. The variation in stratospheric particulate loading need not solely be represented by a corresponding variation in tephra emissions Their could be some other causative mechanisms, such as increase in anthropogenic emissions as well as the increase in tropical upwelling, which could influence the stratospheric particulate loading.

15. Periodic variation of τ_P in the LS region during the period 1998-2005

In addition to the general increasing trend, the values of τ_P, Figure 21 shows a seasonal cycle with winter maximum and summer minimum modulated by a long-period oscillation. These oscillations could primarily be due to the influence of large-scale atmospheric waves. The time series data of τ_P is spectrum analyzed to bring out the characteristics of the prevailing periodic variations. Before subjecting the data to spectrum analysis, the linear trend is removed from the original data. The residual part is Fourier analyzed and the resulting amplitude spectra for different latitude bands (0–15°N, 0–15°S, 15–30°N and 15–30°S) are presented in Figures 22a and 22b. These amplitude spectra reveal the presence of a strong annual component (~12 months) along with a quasi-biennial component (~30 month) both in the equatorial and off-equatorial regions [88]. The spectral amplitude of QBO is as strong (significant) as that of annual oscillation (AO). Figure 22c shows the amplitude spectrum obtained from the lidar derived values of τ_P (at Gadanki). Even though Figure 22c shows more significant peaks in the short-period regime, the spectral amplitude is more pronounced for semi-annual (SAO) and annual (AO) components. This spectrum also shows a secondary peak around 46 months followed by troughs at 23 and 92 months. The spectral amplitude for 30 month periodicity is larger than those at the troughs on either side of this secondary peak. Though the period for the peak amplitude (46 months) is much larger than that expected for the stratospheric QBO, on the basis of the inference derived from Figures 22a and 22b, as well as owing to the fact that the spectral amplitude at 30 months is not a minimum, the characteristics of the 30 month periodicity is examined in the later part to delineate its altitude structure in the lidar data. The SAGE-II data did not show the signature of SAO, which could be due to the inherent smoothing out of this component while taking the zonal mean. The higher spectral amplitude of AO compared to that of QBO is quite expected (as it is true for the wind field also).

16. Quasi-biennial oscillations variation in τ_P and zonal in the LS region

For this study the high resolution Radiosonde data of zonal wind in the lower stratosphere from an equatorial station (where the quasi-biennial oscillation, QBO, signature is expected to be maximum), Singapore (1°22"N, 103°55"E), obtained from Web site http://www.geo.fu-berlin.de/en/met/ag/strat/produkte/qbo/ are used. The time series data of monthly mean zonal wind at Singapore at 20 hPa and 30 hPa levels during the study period are presented in Figure 23. This figure shows that the quasi-biennial oscillation in zonal wind (QBOu) is in the easterly phase during the 1998, 2000–2001, 2003 and 2005. The westerly phase during 1999 and the easterly phase during 2000–2001 are relatively broad. Different phases of QBOu are indentified from this time series to study the influence of QBOu in lower stratospheric aerosols.

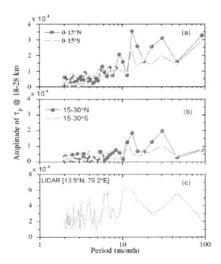

Figure 22. Amplitude spectra of τ_P in the altitude region18–28 km during the period 1998–2005 derived from the SAGE-II data for (a) the equatorial region and (b) the off-equatorial-region, along with (c) that obtained from lidar data at Gadanki

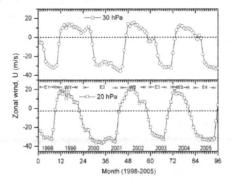

Figure 23. Time series of monthly mean zonal wind at Singapore at 20 and 30 hPa levels during the study period. E_1, E_2, E_3, and E_4 are the periods in which QBOu was in its easterly phase, and W_1, W_2, and W_3 are those in which QBOu was in its westerly phase.

16.1. Latitude variation of τ_P in LS region over the Indian longitude sector in two different phases of QBOu

The latitude variation of τ_P in the altitude range 21–28 km in the band 0–30°N (averaged for every 5°) over the Indian longitude sector (70–90°E) is examined during the consecutive easterly and westerly phases of QBOu in 1998 and 1999 separately. Figure 24 shows the latitudinal variation of τ_P during these two phases of QBOu. This mean τ_P is obtained by averaging the particulate optical depth in individual months when the QBOu phase has reversed completely (the wind speed has reached its highest value). While the value of τ_P is relatively high during the westerly phase of QBOu in the equatorial region up to 15°N, relatively high values are observed during the easterly phase of QBOu in latitudes beyond 15°N. The lidar derived τ_P is relatively low during the easterly phase of QBOu compared to that during its westerly phase. This is in good agreement with that from the latitude variation of τ_P for these 2 years derived from SAGE-II data.

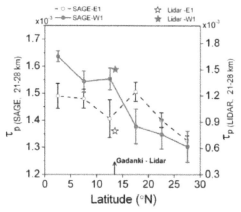

Figure 24. Latitude variation of mean τ_P in the altitude region 21–28 km obtained from SAGE-II averaged over the longitude sector 70°–90°E during the easterly phase (E1) and westerly phase (W1) of QBOu in 1998 and 1999, respectively (vertical bars represent the standard error). The open star represents the value of τ_P during the easterly phase of QBOu (1998), and the red star represents that during the westerly phase (1999) for the altitude region 21–28 km obtained from lidar at Gadanki (marked on the X- axis by a vertical arrow), the scale of which is shown on the right-hand side of the plot.

The value of τ_P shows a pronounced decrease with latitude during westerly phase of QBOu compared to that during its easterly phase. Interestingly, during the easterly phase of QBOu, while the latitude variation of τ_P is relatively small in the equatorial region (0– 15°N), beyond 20°N it decreases sharply with increase in latitude. By examining the altitude profile of backscatter ratio obtained from lidar data at Mauna Loa at Hawaii (19.5°N, 155.6°W), Barnes and Hofmann [11] reported an enhancement of backscatter ratio (enhanced aerosol loading) in the altitude region 21–30 km during the easterly phase of QBOu. This feature agrees well with the above observation derived from the latitude variation of τ_P in the Hawaii.

16.2. Latitude variation of zonal mean τ_p in the LS region over the tropics in different phases of QBOu

On the basis of the measured of particulate loading in the stratosphere, out of the 8 year period considered for the present analysis, the period 1998–2002 was absolutely quiescent while the period 2003–2005 was mildly disturbed. The variation of zonal mean τ_p in the latitude region 30°S to 30°N (averaged for every 5°) during the alternative easterly and westerly phases of QBOu are presented in Figures 25a and 25b. As seen from Figure 23 during the study period (8 years) the QBOu completes approximately 3.5 cycles, with two easterly (E1 and E2) and two westerly (W1 and W2) phases during the period 1998–2002 and two easterly (E3 and E4) and one westerly (W3) phase during the mildly disturbed period of 2003–2005.

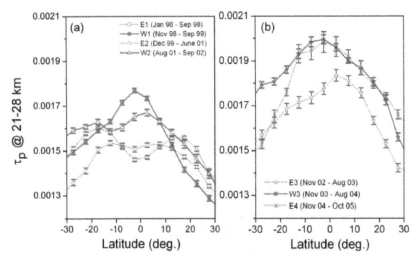

Figure 25. Latitude variation of zonal mean τ_p in the altitude region 21–28 km from 30°S to 30°N obtained from SAGE-II during the alternate easterly and westerly phases of QBOu during the (a) absolute quiet period (1998–2002) and (b) mildly disturbed period (2003–2005). Vertical bars show the corresponding standard error

During the former half (very quiet period), the mean τ_p in the equatorial region shows a general enhancement during westerly phase of QBOu (W1 and W2) while in the off-equatorial region of northern hemisphere it shows an enhancement during the easterly phase of QBOu [88]. When the stratospheric QBOu was in its westerly phase the mean values of τ_p in both the hemispheres decreases steeply with increase in latitude on either side of equator. During the easterly phase of QBOu, though the value of τ_p decreases with increase in latitude beyond 15°N, it is fairly uniform in the equatorial region (15°S to 15°N) with a bite-out over the equator. During the latter half (mildly disturbed period), the mean τ_p decreases rapidly on either side of equator with increase in latitude in both the phases of

QBO$_U$. Figure 25b shows a general enhancement in τ_P during W3 period in the equatorial and off-equatorial regions of both the hemispheres compared to that during the E3 period. However, during E4, τ_P in general is relatively large in the equatorial and northern hemispheric off-equatorial region compared to that during E3 and W3. Note that, in the above analysis two consecutive easterly and westerly cycles of QBO$_U$ are considered and the inferences arrived based on the latitudinal structure of τ_P in these two phases. The inference may apparently be contradicting if one considers the other two pairs westerly of the first pair and easterly of the next (eg W3 and E4) particularly during the mildly disturbed phase of 2003–2005 (Figure 25b). This is mainly due to the fact that the background τ_P has a strong increasing trend during this half in addition to the periodic variations. This can override the periodic variations associated with QBO$_U$. However, this contradiction is totally absent in the former half (1998–2002) when the background τ_P was fairly steady. During this period τ_P in the equatorial region is high in both the westerly phases (W1 and W2) compared to that during the easterly phases E1 and E2, irrespective of sequence of pairs considered. This shows that the influence of QBO$_U$ in τ_P can be clearly delineated only during the very quiescent volcanic periods, when the background stratospheric aerosol loading remains in its steady background level.

16.3. Altitude structure of α_P in the lower stratosphere in different phases of QBO$_U$

Figure 26a shows the mean profile of α_P in the altitude region 18–28 km obtained from lidar data at Gadanki for the two periods January to September 1998 and November 1998 to September 1999 when the QBO$_U$ was in its easterly and westerly phases, respectively. Figure 26a clearly shows that the value of α_P is consistently larger during the westerly phase of QBO than the corresponding values during its easterly phase. This shift in the altitude profile toward a higher value side during the westerly phase is not due to the influence of any trend because the mean level of the stratospheric particulate optical depth in the 0°–15°N region remains fairly the same during the period 1998–2002 and an increase in this level occurred only after 2002 (Figure 21).

For a more detailed study on the effect of QBO$_U$ on stratospheric aerosols, the zonal mean altitude profile of α_P in different phases of QBO during the study period is examined for the equatorial and off -equatorial regions separately. The mean profiles thus obtained for the latitude region 0–10°N and 20–30°N (averaged zonally, as well as along the respective latitude bands) are shown in Figures 26b and 26c, respectively. Over the equatorial regions the altitude profile of aerosol extinction coefficient in the altitude region 22–27 km during the easterly phase of QBO$_U$ are consistently lower than that during the subsequent westerly phase. In the off-equatorial region, the aerosol extinction coefficient in the altitude region 25–32 km. is found to be relatively low during the westerly phase It may be noted that the earlier studies also have shown that aerosol extinction in the lower stratosphere can be influenced by the phase of the quasi biennial oscillation [11,28,29,95] in this region even though the difference in the nature of the latitudinal dependence over the equatorial and off-equatorial regions were not addressed in detail.

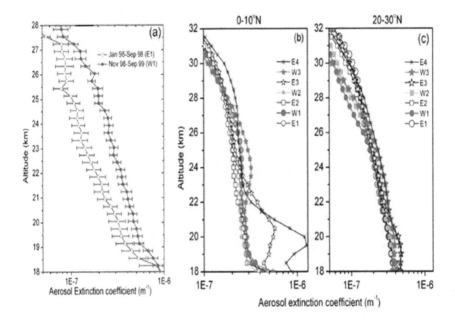

Figure 26. Altitude profiles of mean extinction from lidar at Gadanki for the E_1 and W_1 phases of QBO_U
for the period 1998-1999, along with the altitude profiles of mean extinction a from SAGE-II during the
four alternate phases of QBO_U from 1998-2005, averaged over the latitude regions (a) 0°N–10°N and (b)
20°N–30°N, representative of equatorial and off-equatorial region

Association between the altitude structure of QBO in particulate extinction (QBO_a) and QBO_U
for the entire study period is examined by subjecting the time series of zonal mean monthly
average α_P (obtained from SAGE-II) and zonal mean zonal wind (obtained from NCEP
reanalysis) at each altitude bin from 20 to 30 km to Fourier analysis after removing the linear
trend. To illustrate the QBO features, the spectral components corresponding to the periods
other than 22-45 month are removed from the respective trend removed time series data using
the corresponding amplitudes and phases. The altitude structure of the residual amplitudes
thus obtained for α_P and U over the equatorial and off-equatorial regions are presented in
Figure 27. The left side panels show the residual amplitude for α_P and the right side panels the
same for zonal wind. Both these panels show clearly the presence of a QBO. While the phase of
the biennial oscillation in wind shows a clear downward propagation with time as expected,
the phase of QBO_a changes many times with increase in altitude. In the lower altitudes (<22.5
km) its phase propagation matches with that of wind while at higher altitudes it is opposite.
Above 28km again both are in the same phase. A similar feature can be seen in the off-
equatorial region also where the signature of QBO in U is rather weak. This change in the
phase of QBO_a with altitude is analogous to that reported for the stratospheric ozone [96] and
was attributed to the secondary meridional circulation (SMC).

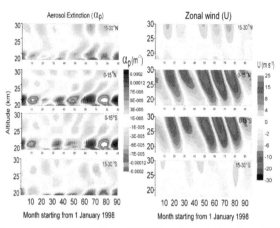

Figure 27. Amplitude spectra of QBO in α_P (from SAGE-II) and zonal wind (from NCEP) for the altitude region 18 to 30 km showing the time history of the biennial component for the period 1998 - 2005 for the equatorial and off-equatorial regions.

17. QBO in α_P and secondary meridional circulation

To study the altitude structure of the QBO_a, the time series of zonal-averaged monthly mean aerosol extinction coefficient at different altitudes (at 0.5 km interval) are subjected to Fourier analysis after removing the linear trend. This analysis of α_P at different levels showed a prominent peak in its amplitude corresponding to the periodicity of QBO (30 month) and AO (12 month). Figure 28 shows the altitude profile of the amplitude and phase of QBO_a from 18 to 32 km in the equatorial (0–15°N and 0–15°S) and off-equatorial (15–30°N and 15–30°S) regions. In general, the amplitude of QBO_a decreases with increase in altitude (as that observed in the raw aerosol extinction data). Over the equatorial region, the altitude variation of QBO_a amplitude (Figures 28b and 28c) shows three prominent peaks in the region 18–22 km, 23– 27 km and 28–32 km with peak amplitude centered around 20 km, 25 km and 30 km, respectively. This feature is remarkably symmetric about the. equator, in both the hemispheres. Another interesting feature to be noted is that the phase remains fairly constant around these peaks. In the equatorial region, the phase of QBO_a around 25 km is ~15 months and that at around 30 km and 20 km are ~28 and 0 months, respectively. Thus the phase difference of QBO_a in the upper and lower regime with respect to that at 25 km is around 13 and 15 months, which corresponds to 156° and 180°, respectively (for the 30 month periodicity one month in phase corresponds to 12°). This shows that the QBO_a around 25 km is almost out of phase with that in the upper (28–32 km) and lower (18–22 km) regime [88].

Over the off-equatorial region the altitude structure of the amplitude and phase of QBO_a are found to be different from that over the equatorial region. Figures 28a and 28d show that the amplitude of QBO_a is significant in the lower and upper altitude regions with a minimum value around 24.5 km in the northern hemisphere and around 21 km in the southern hemisphere. While the amplitude of QBO_a shows a broad maximum in the altitude region 24–28 km with a peak around 26 km in the southern part, correspondingly it shows a maximum around 29 km in the northern part.

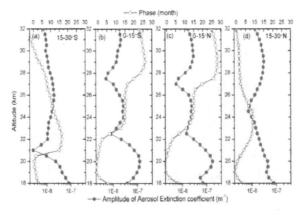

Figure 28. Altitude structure of the QBO amplitude and phase in aerosol extinction for the altitude region 18–32 km in the equatorial region [(c) 0°N–15°N and (b) 0°S–15°S] and off-equatorial region [(d) 15°N–30°N and (a) 15°S–30°S].

Over the equatorial region, the observed features of QBO_a in the three altitude regions could be attributed to the influence of secondary meridional circulation (SMC) induced by the vertical shear of QBO in stratospheric zonal wind. Through a detailed analysis of the QBO modulation of the meridional wind in the stratosphere, Ribera et al. [87] demonstrated the existence of discrete zones of meridional wind convergence and divergence over the equator. These convergence (divergence) zones during the westerly (easterly) phase of QBO_U are located at the lower and upper limits of the maximum zonal wind shear and maximum temperature anomaly layers. The existence of convergence and divergence zones, which are directly related to the rising and sinking motions, forms two circulation cells (SMC) in the consecutive vertical levels, quasi-symmetric about the equator.

To examine the cycle-to-cycle variation of the QBO in aerosol extinction coefficient and its phase structure at the three altitude regions the monthly mean values of α_p at the central altitudes (of the three regions) 20 km, 25 km and 30 km for the equatorial region 0–15°N and 0–15°S are estimated during the period 1998 to 2005 and this data is subjected to wavelet analysis (after de-trending). Contour plots in Figure 29 shows the time history of the amplitudes for different periodicities (ranging from 2 to 45 months) in α_p for the study period. The spectral characteristics at the respective altitudes are similar in both the hemispheres. The annual component repeats coherently in all the three altitudes indicating that they follow the same pattern in the entire altitude range 18–32 km. But the temporal pattern of QBO differs. When the QBO in α_p is in its positive phase (maximum amplitude) in the lower (20 km) and upper (30 km) regimes, it is in the negative phase (minimum amplitude) in the middle regime (25 km). The QBO_a at 25 km is out of phase with that in the upper and lower regimes during the entire study period. However, the amplitudes of AO and QBO in α_p at 30 km are relatively small in the first 40 months compared to rest of the period. This analysis confirms the temporal repeatability of QBO_a phase structure in these three altitude regions inferred from the mean α_p at different altitudes averaged for the entire period of analysis shown in Figure 28. The phase of QBO_U (Figure 23) matches with that of QBO_a in the middle regime (25 km) while it is in opposite phase with those in the lower (20 km) and upper (30 km) regime.

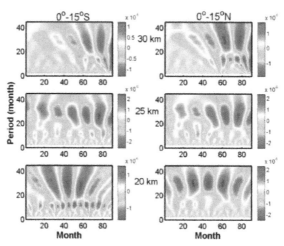

Figure 29. Wavelet analysis of zonal averaged monthly mean aerosol extinction at 20, 25, and 30 km for the equatorial region 0°N–15°N and 0°S–15°S during the period 1998–2005.

Studies by Fadnavis and Beig [96] on the influence of SMC induced by QBO in the spatiotemporal variations of ozone over the tropical and subtropical regions however showed maximum amplitude of the QBO signal in ozone concentration at the two pressure levels, 30 hPa and 9 hPa, where the QBO manifests in opposite phase This feature is in good agreement with the amplitude and phase of the QBO signal observed in aerosol extinction coefficient at 25 km and 30 km. From Figures 28b and 28c it can be seen that while the maximum amplitude of QBO signal in aerosol extinction coefficient centered around 20 and 30 km coincides with SMC cells centered at ~80 hPa and ~9 hPa as illustrated by Plumb and Bell [98] and Punge et al. [99], the maximum QBO signal in aerosol extinction coefficient in the altitude region 23–27 km centered around 25 km coincides with the level of maximum zonal wind shear (centered ~24 hPa) where the divergence and convergence zones (circulation cells) intersect. Similarly, Dunkerton [100] studied the QBO anomalies in ozone, methane and water vapor and observed a double peak structure in their amplitudes. Numerical simulations of QBO anomalies also showed influence of SMC in modulating the distribution of tracers leading to the formation of two peaks in their number density near 24 and 32 km [101-103]. However, the observed maximum amplitude of QBO in α_p centered around 20 km is not reproduced in these model simulations.

Distribution of Particulates in the Tropical UTLS over the Asian Summer Monsoon Region and Its
Association with Atmospheric Dynamics

153

18. Summary

The altitude structure of aerosol extinction in the tropical UTLS region and its variability in different time scales is examined with particular stress for the Indian longitude sector (70-90°E) , using the data from a dual polarization lidar (at 532 nm wavelength) located at Gadanki along with data obtained from SAGE-II onboard ERBS. Organized convection associated with the Asian summer monsoon (ASM) and highly dynamic ITCZ makes this region unique in the global scenario. Altitude profiles of particulate extinction for a period of eight years from 1998-2005 (volcanically quiescent period) obtained from SAGE-II in the latitude region 30°S to 30°N and the altitude profiles of particulate backscatter , particulate extinction and volume depolarization for the same period derived from lidar data at Gadanki are is used for this purpose. The lidar data are used mainly to study the influence of local features such as STCs, local convection etc in the properties of particulates in the UTLS region while the satellite data are used to study the spatial features in the tropical UTLS from 30°S to 30°N. A comparison of aerosol extinction from SAGE-II with that from lidar on a profile basis as well as on monthly mean basis over a small geographical grid (size 6° in latitude and 13° in longitude) centered at Gadanki showed a very good agreement between the two in the major features such as mean altitude structure as well as the mean annual pattern.

Lidar studies show a significant increase is particulate scattering in the upper troposphere with a high value of δ over the Indian subcontinent during the ASM period when the tropospheric convective activity is the highest. This feature is clearly associated with particle formation due to homogenous/heterogeneous nucleation and condensation and subsequent freezing to form non spherical ice crystals (associated with STC). The particulate backscatter coefficient in the LS region is maximum during winter and minimum during the ASM period. The winter peak is closely associated with the increase in vertical mass flux in conjunction with the tropical upwelling and Brewer Dobson circulation.

Occurrence of STCs in the UT region is very common in this geographical region. The frequency of occurrence of STC is the largest during the ASM period, when most of the observed STCs are optically dense and geometrically thick. The most favoured altitude for STC is 14-16 km. Most of the observed STCs are optically and geometrically thin. Thin STCs are generally observed at higher altitudes, very close to the cold point. Most of the dense are found to be associated with organized convection while the thin STCs could be of *in situ* origin. The values of δ in these clouds vary from 0.03 to 0.6 with low values occurring more frequently than high values. Though cloud depth (thickness) generally varies from 0.4 to ~4.0 km, in majority of the cases it is less than 1.7km. While the cloud optical depth in general increases with increase in cloud temperature, the depolarization shows a decrease. For values of temperature < 198 K, the cloud thickness and depolarization show a sharp decrease. These clouds contribute significantly to the particulate scattering in the UT region. High value of β_P observed in the UT region during the ASM period is mainly contributed by the STC particulates. They also contribute to scattering in the lower stratosphere, very close to the cold point, mainly through the penetration on particles from these STCs across the tropopause aided by the upward propagating inertia-gravity waves. This feature which is characteristic for the Indian monsoon region is almost absent in the southern hemisphere where the occurrence of organized very deep convection is minimal. However, the Brewer-Dobson circulation plays a significant role in the transport of UT particles in to the lower stratosphere during the winter season.

The convective activity prevailing in the troposphere significantly influences the microphysical properties of particulates in the UTLS region. The particulate scattering and optical depth in the UT region shows a general decrease with increase in latitude on either side of the equator with a well pronounced summer-winter contrast. However, organized convection during the ASM period enhances the particulate loading in the UT in the northern latitudes even beyond 25°N. While the particulate optical depth in the 18–21 km region (lowest part of the stratosphere) is relatively low in the equatorial region, it shows an increase in the off-equatorial region mainly due to this enhancement in particulate concentration above the cold point, particularly over the Indo-Gangetic Plain, during this period. At a higher altitude (21–30 km) it shows a different pattern, with high values near the equator and low values in the off-equatorial region. This confirms the existence of a stratospheric aerosol reservoir. This spatial distribution could be attributed to horizontal advection in the lower regime (rapid transport from near equatorial region to higher latitudes) as well as lofting to higher altitudes over the equatorial region (B-D circulation).

Spectral analysis of zonal mean particulate optical depth in the stratosphere (18-32km) revealed the existence of a strong QBO both in the equatorial and off-equatorial regions. The phase of the QBO signal in particulate extinction (QBO$_a$) around 25 km is found to be in opposite phase with that in the upper (28-32 km) and lower regime (18-22 km), illustrating the existence of a secondary meridional circulation (SMC) produced due to vertical shear of QBO phase in zonal wind (QBO$_U$). While the particulate optical depth in the lower stratosphere is relatively large during the westerly phase of QBO$_U$ in the equatorial region, relatively high values are observed during the easterly phase of QBO$_U$ in the off-equatorial region. During the westerly phase of stratospheric QBOU, the mean particulate optical depth rapidly decreases with increase in latitude on either side of equator in both the hemispheres. During the easterly phase, this remains fairly steady between ±15° latitude, with a small bite-out around the equator, and decreases steadily for latitudes beyond 15°.

Author details

S.V. Sunilkumar and K. Parameswaran
Space Physics Laboratory, Vikram Sarabhai Space Centre, Thiruvananthapuram, India

Bijoy V. Thampi
Laboratoire de Météorologie Dynamique, IPSL, Place Jussieu, Paris, France

Acknowledgement

We thank the NASA Langley Research Center (NASA-LaRC) and the NASA Langley Aerosol Research Branch for providing the SAGE-II data through the web site ftp://ftp-rab.larc.nasa.gov/pub/sage2/v6.20. The authors are thankful to the technical and scientific staff of National Atmospheric Research Laboratory (NARL), Gadanki for their dedicated efforts in conducting the Lidar observations. One of the authors, Dr. K. Parameswaran would like to acknowledge CSIR for providing grant through the Emeritus Scientist scheme.

19. References

[1]   Jensen EJ, Toon OB, Pfister L, Selkirk, HB. Dehydration of the upper troposphere and lower stratosphere by subvisible cirrus clouds near the tropical Tropopause. Geophysical Research Letter 1996;23 825-828.

[2]   Luo BP, et al. Dehydration potential of ultrathin clouds at the tropical Tropopause. Geophysical Research Letter 2003;30(11) 1557 doi:10.1029/2002GL016737.

[3]   Hartmann DL. Holton JR, Fu Q. The heat balance of the tropical tropopause, cirrus, and stratospheric dehydrationGeophysical Research Letters 2001;28(10) 1969–1972.

[4]   Riese M, Friedl-Vallon F, Spang R, Preusse P, Schiller C, Hoffmann L, Konopka P, Oelhaf H, von Clarmann Th., Höpfner M. GLObal limb Radiance imager for the atmosphere (GLORIA): scientific objectives. Advances in Space Research 2005;36, 989–995.

[5]   Froyd KD, Murphy DM, Sanford TJ, Thomson DS, Wilson JC, Pfister L, Lait L. Aerosol composition of the tropical upper troposphere, Atmospheric Chemistry and Physics 2009;9 4363-4385.

[6]   de Reus M, Borrmann S, Bansemer A, Heymsfield AJ, Weigel R, Schiller C, Mitev V, Frey W, Kunkel D, Kürten A, Curtius J, Sitnikoev NM, Ulanovsky A, Ravegnani F. Evidence for ice particles in the tropical stratosphere from in-situ measurements. Atmospheric Chemistry and Physics 2009;9, 6775–6792.

[7]   Thampi BV, Sunilkumar SV, Parameswaran K. Lidar studies of particulates in the UTLS region at a tropical station over the Indian subcontinent. Journal of Geophysical Resarch 2009 ;114, D08204, doi:10.1029/2008JD010556.

[8]   McFarquhar GM, Heymsfield AJ, Pinhirne J, Hart B. Thin and subvisual tropopause tropical cirrus: Observations and Radiative Impacts. Journal of Atmospheric Sciences 2000;57, 1841-1853.

[9]   Haladay T, Stephens G. Characteristics of tropical thin cirrus clouds deduced from joint CloudSat and CALIPSO observations. Journal of Geophysical Research 2009;114, D00A25, doi:10.1029/2008JD010675.

[10] Sato M, Hansen J, McCormick M, Pollack J. Stratospheric Aerosol Optical Depths 1850–1990. Journal of Geophysical Research 1993;98(D12), 22,987-22,994.

[11] Barnes JE, Hofmann DJ. Variability in the stratospheric background aerosol over Mauna Loa Observatory. Geophysical Research Letter 2001;28, 2895–2898.

[12] Jäger H. Long-term record of lidar observations of the stratospheric aerosol layer at Garmisch-Partenkirchen. Journal of Geophysical Research 2005;110, D08106, doi:10.1029/2004JD005506.

[13] Deshler T, Anderson-Sprecher R, Jäger H, Barnes JE, Hofmann DJ, et al. Trends in the nonvolcanic component of stratospheric aerosol over the period 1971-2004. Journal of Geophysical Research 2006;111, D01201, doi:10.1029/2005JD006089.

[14] Grant WB et al. Aerosol-associated changes in tropical stratospheric ozone following the eruption of Mount Pinatubo. Journal of Geophysical Research 1992;99, 8197-8211.

[15] Angell JK, Korshover J, Planet WG. Ground-based and satellite evidence for a pronounced total-ozone minimum in early 1983 and responsible atmospheric layers. MonthlyWeather Review 1985;113, 641-646.

[16] Hofmann DJ, Solomon S. Ozone destruction through heterogeneous chemistry following the eruption of El Chichón. Journal of Geophysical Research 1989;94, 5029-5041.

[17] Deshler T, Johnson BJ, Hofmann DJ, Nardi B. Correlations between ozone loss and volcanic aerosol at latitudes below 14 km over McMurdo Station, Antarctica. Geophysical Research Letter 1996;23, 2931–2934, doi:10.1029/96GL02819.

[18] Koike M, Kondo Y, Matthews WA, Johnston PV, Yamazaki K. Decrease of stratospheric NO2 at 44°N caused by Pinatubo volcanic aerosols. Geophysical Research Letter 1993;20(18), 1975-1978.

[19] Coffey MT, Mankin WG. Observations of the loss of stratospheric NO2 following volcanic eruptions. Geophysical Research Letter 1993;20(24), 2873-2876.

[20] Johnston PV, McKenzie RL, Keys JG, Matthews WA. Observations of depleted stratospheric NO2 following the Pinatubo volcanic eruption. *Geophysical Research Letter* 1992;*19*, 211- 213.

[21] McCormick MP, Thomason LW, Trepte CR. Atmospheric effects of the MtPinatubo eruption. *Nature* 1995;*373*, 399-404.

[22] Thomason LW, Burton SP, Luo BP, Peter T. SAGE II measurements of stratospheric aerosol properties at non-volcanic levels. Atmospheric Chemistry and Physics 2008;8, 983–995.

[23] Thomason LW, Kent GS, Trepte CR, Poole LR. A comparison of the stratospheric aerosol background periods of 1979 and 1989–1991. Journal of Geophysical Research 1997;102, 3611-3616.

[24] Thomason LW, Peter T., editors. Assessment of stratospheric aerosol properties (ASAP). SPARC Report 4; 2006. WMO/TD 1295, Technical Report 2006 WCRP-124, World Climate Research Program, Geneva, Switzerland.

[25] Deshler T, Hervig ME, Hofmann DJ, Rosen JM, Liley JB. Thirty years of in situ stratospheric aerosol size distribution measurements from Laramie, Wyoming (41°N), using balloon-borne instruments. Journal of Geophysical Research 2003;108(D5), 4167, doi:10.1029/2002JD002514.

[26] Hofmann DJ, Barnes JE, O'Neill M, Rudeau MT, Neely R. Increase in background stratospheric aerosol observed with lidar at Mauna Loa Observatory and Boulder, Colorado. Geophysical Research Letter 2009;36, L15808, doi:10.1029/2009GL039008.

[27] Graf H.-F, Langmann B, Feichter J. The contribution of Earth degassing to the atmospheric sulfur budget. Chem Geol. 1998;147, 131-145.

[28] Trepte CR, Hitchman MH. Tropical stratospheric circulation deduced from satellite aerosol data. Nature 1992;355, 626–628, doi:10.1038/355626a0.

[29] Hitchman MH, Mckay M, Trepte CR. A climatology of stratospheric aerosol. Journal of Geophysical Research 1994;99(D10), 20,689-20,700.

[30] Choi W, Grant W, Park J, Lee KM, Lee H, Russell III J. Role of the quasi-biennial oscillation in the transport of aerosols from the tropical stratospheric reservoir to midlatitudes. Journal of Geophysical Resresearch 1998;103(D6), 6033-6042.

[31] Barnes JE, Hofmann DJ. Lidar measurements of stratospheric aerosol over Mauna Loa. Geophysical Research Letter 1997;24, 1923–1926.

[32] McCormick MP. SAGE II: An overview. Advances in Space Research 1987;7(3), 219–226.

[33] Mauldin III LE, Zaun NH, McCormick MP, Guy JH, Vaughan WR. Stratospheric Aerosol and Gas Experiment II instrument: A functional description. Optical Engineering 1985;24, 307–312.

[34] Chu WP, McCormick MP, Lenoble J, Brogniez C, Pruvost P. SAGE II inversion algorithm. Journal of Geophysical Research 1989;94(D6), 8339–8351.

[35] Wang P-H, McCormick MP, Minnis P, Kent GS, Yue GK, Skeens KM. A method for estimating vertical distribution of the SAGE II opaque cloud frequency. Geophysical Research Letter 1995;22(3), 243–246.

[36] Wang P-H, Minnis P, McCormick MP, Kent GS, Skeens KM. A 6-year climatology of cloud occurrence frequency from Stratospheric Aerosol and Gas Experiment II observations (1985–1990). Journal of Geophysical Research 1996;101(D23), 29,407–29,429.

[37] Sunilkumar SV, Parameswaran K, Krishna Murthy BV. Lidar observations of cirrus cloud near the tropical tropopause: General features. Atmospheric Research 2003;66, 203-227.

[38] Sunilkumar SV, Parameswaran K. Temperature dependence of tropical cirrus properties and radiative effects. Journal of Geophysical Research 2005;110, D13205, doi:10.1029/2004JD005426.

[39] Parameswaran K, Thampi BV, Sunilkumar SV. Latitudinal dependence of the seasonal variation of particulate extinction in the UTLS over the Indian longitude sector during volcanically quiescent period based on lidar and SAGE-II observations. Journal of Atmospheric and Solar Terrestrial Physics 2010;72, 1024-1035.

[40] Fernald, FG. Analysis of atmospheric lidar observations: Some comments. Applied Optics 1984;23, 652– 653, doi:10.1364/AO.23.000652.

[41] Bodhaine BA, Wood BN, Dutton EG, Slusser JR. On Rayleigh optical depth calculations. Journal of Atmosphere and OceanTechnology 1999;16, 1854-1861.

[42] Sunilkumar SV, Parameswaran K, Thampi BV, Interdependence of tropical cirrus properties and their variability. Annales Geophysicae 2008;26, 413-429.

[43] Pfister et al.Aircraft observations of thin cirrus clouds near the tropical Tropopause. *Journal of Geophysical Research* 2008;106(D9), 9765-9786.

[44] Garrett TJ, Zulauf MA, Krueger SK. Effects of cirrus near the tropopause on anvil cirrus dynamics. *Geophysical Research Letter* 2006;33, L17804, doi:10.1029/2006GL027071.

[45] Chen JP, McFarquhar GM, Heymsfield AJ, Ramanathan V. A modeling and observational study of the detailed microphysical structure of tropical cirrus anvils. Journal Geophysics Research 1997;102(D6), 6637-6653.

[46] Sassen K, Cho BS. Subvisual thin cirrus lidar data set for satellite verification and climatological research. Journal of Applied Meteorology 1992 ;31, 1275-1285.

[47] Yue GK, Fromm MD, Shettle EP. Intercomparison of aerosol extinction measured by Stratospheric Aerosol and Gas Experiment (SAGE) II and III. Journal of Geophysical Research 2009;114, D07205, doi:10.1029/2008JD010452.

[48] Antuna JC, Robock A, Stenchikov GL, Thomason LW, Barnes JE. Lidar validation of SAGE II aerosol measurements after the 1991 Mount Pinatubo eruption. Journal of Geophysical Research 2002;107 (D14), 4194. doi:10.1029/2001JD001441.

[49] Lu J, Mohnen VA, Yue GK, Jäger H. Intercomparison of multiplatform stratospheric aerosol and ozone observations. Journal of Geophysical Research 1997;102 (D13), 16,127–16,136.

[50] Lu C-H, Yue GK, Manney GL, Jäger H, Mohnen VA. Lagrangian approach for stratospheric aerosol and gas experiment (SAGE) II profile intercomparisons. Journal of Geophysical Research 2000;105, 4563–4572.

[51] Winker DM, Trepte CR. Laminar cirrus observed near the tropical tropopause by LITE. Geophysical Research Letter 1998;25, 3351-3354.

[52] Mergenthaler JL, Roche AE, Kumer JB, Ely GA. Cryogenic limb array etalon spectrometer observations of tropical cirrus. Journal Geophysical Res 1999;104, 22183-22194.

[53] Massie ST, Lowe P, Tie X, Hervig M, Thomas G, Russell J. Effect of the 1997 El Nino on the distribution of upper tropospheric cirrus. Journal of Geophysical Research 2000;105, 22725-22741.

[54] Roca R, Viollier M, Picon L, Desbois M. A multisatellite analysis of deep convection and its moist environment over the Indian Ocean during the winter monsoon. Journal of Geophysical Research 2002;107(D19), 8012, doi:10.1029/2000JD000040.

[55] Rajeev K, Parameswaran K, Meenu S, Sunilkumar SV, Thampi BV, Suresh Raju C, Krishna Murthy BV, Jagannath KS, Mehta SK, Narayana Rao D, Rao KG. Observational assessment of the potential of satellite-based water vapor and thermal IR brightness temperatures in detecting semitransparent cirrus. Geophysical Research Letter 2008;35, L08808, doi:10.1029/2008GL033393.

[56] Heymsfield AJ, McFarquhar GM. Mid-latitude and tropical cirrus-Microphysical properties. In: Lynch D, Sassen K, Starr DO, Stephens G. (eds.) Cirrus. Oxford University Press;2002. p78-101.

[57] Sassen K. The lidar backscatter depolarization technique for cloud and aerosol research. In Mishenko ML, Hovenier JW, Travis LD. (eds.) *Light Scattering by Nonspherical Particles: Theory, Measurements and Geophysical Applications.* Academic Press, New York; 2000.pp393-416.

[58] Noel V, Winker DM, McGill M, Lawson P. Classification of particle shapes from lidar depolarization ratio in convective ice clouds compared to in situ observations during CRYSTAL-FACE. *Journal of Geophysical Research* 2004;*109*, D24213, doi:10.1029/2004JD004883.

[59] Roca R, Ramanathan V. Scale dependence of monsoonal convective systems over the Indian Ocean. *Journal of Climate* 2000;*13*, 1286-1297.

[60] Wong S, Dessler AE. Regulation of H_2O and CO in tropical tropopause layer by the Madden-Julian oscillation. *Journal of Geophysical Research* 2007;*112*, D14305, doi:10.1029/2006JD007940.

[61] Gettelman A, Seidel DJ, Wheeler MC, Ross RJ. Multi-decadal trends in tropical convective available potential energy. *Journal of Geophysical Research* 2002;*107*(D21), 4606, doi:10.1029/2001JD001082.

[62] Stull RB. *Meteorology Today for Scientists and Engineers.* West publishing company, StPaul, Minneapolis; 1995.

[63] Manohar G, Kandalgaonkar S, Tinmaker M. Thunderstorm activity over India and the Indian southwest monsoon. *Journal of Geophysical Research* 1999;*104*(D4), 4169-4188.

[64] Haynes PH, McIntyre ME, Shepherd TG, Marks CJ, Shine KP. On the "downward control" of extratropical diabatic circulations by eddy-induced mean zonal forces. *Journal of Atmospheric Sciences* 1991;48, 651-678, doi:10.1175/1520-0469.

[65] Holton JR, Haynes PH, McIntyre ME, Douglass AR, Rood RB, Pfister L. Stratosphere-troposphere exchange. Review of Geophysics 1995;33, 403-439.

[66] Yulaeva E, HoltonJR, Wallace JM. On the cause of the annual cycle in tropical lower stratospheric temperatures. *Journal of Atmospheric Sciences*1994;*51*, 169-174.

[67] Reid GC, Gage KS. On the annual variation of height of the tropical Tropopause. *Journal of Atmospheric Sciences* 1981;38, 1928– 1937.

[68] Krishna Murthy BV, Parameswaran K, Rose KO. Temporal variations of the tropical tropopause characteristics. Journal of Atmospheric Sciences 1986;43:914–923.

[69] Highwood EJ, Hoskins BJ. The tropical Tropopause. *Quarterly Journal of Royal Meteorological Society* 1998;*124*, 1579-1604.

[70] Seidel DJ, Ross RJ, Angell JK, Reid GC. Climatological characteristics of the tropical tropopause as revealed by radiosondes. *Journal of Geophysical Research* 2001;106(D8), 7857-7878.

[71] Staley D. On the mechanism of mass and radioactivity transport from stratosphere to troposphere. *Journal of Atmospheric Sciences* 1962;*19*(6), 450-467.

[72] Reiter ER, Glasser ME, Mahlman JD. The role of the tropopause in the stratospheric-tropospheric exchange process. *Pure Applied Geophysics* 1969;75, 185-218.

[73] Reiter ER. Stratospheric-tropospheric exchange processes. *Review of Geophysics and Space Physics* 1975;*13*(4), 459-474.

[74] Rosenlof KH, Tuck AF, Kelly KK, Russell III JM, McCormick MP. Hemispheric asymmetries in water vapor and inferences about transport in the lower stratosphere. *Journal of Geophysics Research* 1997;*102*, 13,213-13,234.

[75] Dessler AE. The effect of deep, tropical convection on the tropical tropopause layer. *Journal of Geophysical Research* 2002;*107*(D3), 4033, doi:10.1029/2001JD000511.

[76] Gettelman A, Kinnison DE, Dunkerton TJ, Brasseur GP. Impact of monsoon circulations on the upper troposphere and lower stratosphere. *Journal of Geophysical Research* 2004;*109*, D22101, doi:10.1029/ 2004JD004878.

[77] Park M, Randel WJ, Kinnison DE, Garcia RR, Choi W. Seasonal variation of methane, water vapor, and nitrogen oxides near the tropopause: Satellite observations and model simulations. *Journal of Geophysical Research* 2004;*109*, D03302, doi:10.1029/2003JD003706.

[78] Sunilkumar, SV, Parameswaran K, Rajeev K, Krishna Murthy BV, Meenu S, Mehta SK, Asha Babu. Semitransparent Cirrus clouds in the Tropical Tropopause Layer during two contrasting seasons. Journal of Atmospheric and Solar Terrestrial Physics 2010; 72,745-762,doi:10.1016/j.jastp.2010.03.020.

[79] Das SS, Kumar KK, Uma KN. MST radar investigation on inertia-gravity waves associated with tropical depression in the upper troposphere and lower stratosphere over Gadanki (13.5°N, 79.2°E). Journal of Atmospheric and Solar Terrestrial Physics 2010;72, 1184-1194, doi:10.1016/j.jastp2010.07.016.

[80] Thampi BV, Parameswaran K, Sunilkumar SV. Semitransparent cirrus clouds in the upper troposhere and their contribution to the particulate scattering in the tropical UTLS region. Journal of Atmospheric and Solar Terrestrial Physics 2012;74,1-10, doi:10.1016/j.jastp2011.09.005.

[81] Meenu S, Rajeev K, Parameswaran K, Nair AKM. Regional distribution of deep clouds and cloud top altitudes over the Indian subcontinent and the surrounding oceans. Journal of Geophysical Research 2010 ;115, D05205, doi:10.1029/2009JD011802.

[82] Yue GK, McCormick MP, Chiou EW. Stratospheric aerosol optical depth observed by the Stratospheric Aerosol and Gas Experiment II: Decay of the El Chichon and Ruiz volcanic perturbations. Journal of Geophysical Research 1991;96(D3), 5209–5219.

[83] Trepte et al., 1994

[84] Bauman JJ, Russell PB, Geller MA, Hamill P. A stratospheric aerosol climatology from SAGE II and CLAES measurements: 2Results and comparisons, 1984-1999. *Journal of Geophysical Research* 2003;*108 (D13)*, 4383, doi:10.1029/2002JD002993.

[85] Junge CE, Changnon CW, Manson JE. Stratospheric aerosols. *Journal of Meteorology* 1961;*18*, 81-108.

[86] Robock A. Volcanic eruptions and climate. Review of Geophysics 2000;38, 191-219.

[87] Parameswaran K, Sunilkumar SV, Krishna Murthy BV, Satheesan K, Bhavanikumar Y, Krishnaiah M, Nair PR. Lidar Observations of cirrus cloud near the tropical tropopause:

temporal variations and association with tropospheric turbulence. *Atmospheric Research* 2003;*69*, 29-49.

[88] Sunilkumar SV, Parameswaran K, Thampi BV, Ramkumar G. Variability in background stratospheric aerosols over the tropics and its association with atmospheric dynamics. *Journal of Geophysical Research* 2011;116, D13204, doi:10.1029/2010JD015213.

[89] Newhall CG, Self S. The volcanic explosivity index (VEI): An estimate of explosive magnitude for historical volcanism. Journal of Geophysical Research 1982 ;87,1231–1238.

[90] Hamill P, Jensen EJ, Russell PB, Bauman JJ. The life cycle of stratospheric aerosol particles. Bulletin of American Meteorological Society 1997;7, 1395–1410.

[91] Rosen JM. The boiling point of stratospheric aerosol. Journal of Applied Meteorology 1971;18, 1044-1046.

[92] Deshler T. A review of global stratospheric aerosol: Measurements, importance, life cycle, and local stratospheric aerosol. Atmospheric Research 2008; 90,223-232,doi:10.1016/j.atmosres.2008.03.016.

[93] Randel WJ, Wu F, Voemel H, Nedoluha GE, Forster P. Decreases in stratospheric water vapor after 2001: links to changes in the tropical tropopause and the Brewer-Dobson circulation. Journal of Geophysical Research 2006;111(12), doi:10.1029/2005JD006744.

[94] Randel WJ, Wu F, Oltmans SJ, Rosenlof K, Nedoluha GE. Interannual changes of stratospheric water vapor and correlations with tropical tropopause temperatures. Journal Atmospheric Sciences 2004;61, 2133–2148.

[95] Bingen C, Fussen D, Vanhellemont F. A global climatology of stratospheric aerosol size distribution parameters derived from SAGE II data over the period 1984–2000. Journal of Geophysical Research 2004;109, D06202, doi:10.1029/2003JD003511.

[96] Fadnavis S, Beig G. Spatiotemporal variation of the ozone QBO in MLS data by wavelet analysis. Annales Geophysicae, 2008;26, 3719–3730.

[97] Ribera P, Penã-Ortiz C, Garcia-Herrera R, Gallego D, Gimeno L, Hernàndez E. Detection of the secondary meridional circulation associated with the quasi-qiennial oscillation. Journal of Geophysical Research 2004;109, D18112, doi:10.1029/2003JD004363.

[98] Plumb RA, Bell RC. A model of the quasi-biennial oscillation on equatorial beta plane. Quarterly Journal of Roya; Meteorological Society 1982;108, 335–352.

[99] Punge HJ, Konopka P, Giorgetta MA, Müller R. Effects of the quasi-biennial oscillation on low-latitude transport in the stratosphere derived from trajectory calculations. Journal of Geophysical Research 2009;114, D03102, doi:10.1029/2008JD010518.

[100] Dunkerton TJ. Quasi-biennial and subbiennial variation of stratospheric trace constituents derived from HALOE observations. Journal of Atmospheric Sciences 2001;58, 7–25.

[101] Chipperfield MP, Gray LJ, Kinnersley JS, Zawodny J. A two-dimensional model study of the QBO signal in SAGE-II NO2 and O3. Geophysical Research Letter 1994;21, 589–592.

[102] Jones DBA, Schneider HR, McElroy MB. Effects of the quasi-biennial oscillation on the zonally averaged transport of tracers. Journal of Geophysical Research 1998;103, 11,235–11,249.

[103] McCormack JP, Siskind DE, Hood LL. Solar–QBO interaction and its impact on stratospheric ozone in a zonally averaged photochemical transport model of the middle atmosphere. Journal Geophysical Research 2007;112, D16109.

Chapter 6

Changes of Permanent Lake Surfaces, and Their Consequences for Dust Aerosols and Air Quality: The Hamoun Lakes of the Sistan Area, Iran

Alireza Rashki, Dimitris Kaskaoutis, C.J.deW. Rautenbach and Patrick Eriksson

Additional information is available at the end of the chapter

1. Introduction

Changes in the frequency and extent of natural inundation occurring on large permanent and ephemeral lake systems may lead to significant fluctuations in regional dust loading on both a seasonal and an inter-annual basis [1]. As surface water diversion increases, arid-land surfaces that were previously wet or stabilized by vegetation are increasingly susceptible to deflation by wind, resulting in desertification and increase in dust outbreaks [2-4]. Desiccation of lake beds, whether due to drought or to water diversion schemes, as in the Aral Sea in Turkmenistan [5,6], Owens lake in California [7,8], lake Eyre in Australia, Hamoun lakes in Iran [9-12], can lead to increased dust storm activity. Thus, some dust may be derived from dried lake beds and can be highly saline, while the finest aerosols can be injurious to health. Anthropogenic sources were previously considered as important dust contributors [13], but more recent estimates of only 5-7% of total mineral dust from such sources gives major importance to natural sources [14]. Each year, several billion tons of soil-dust are entrained into the atmosphere playing a vital role in solar irradiance attenuation, and affects marine environments, atmospheric dynamics and weather [15-20].

Atmospheric aerosols affect the global climatic system in many ways, i.e. by attenuating the solar radiation reaching the ground, modifying the solar spectrum, re-distributing the earth-atmosphere energy budget and influencing cloud microphysics and the hydrological cycle. Mineral dust plays an important role in the optical, physical and chemical processes in the atmosphere, while dust deposition adds exogenous mineral and organic material to terrestrial surfaces, having a significant impact on the Earth's ecosystems and biogeochemical cycles. The impact of dust aerosols in the Earth's system depends mainly on particle characteristics such as size, shape and mineralogy [21], which are initially determined by the terrestrial sources from which the soil sediments are entrained and from their chemical composition [22]. Size distribution is a key parameter to characterize the aerosol chemical, physical, optical properties and their effects on health. The lower and upper size limits of dust aerosols are from a few nanometers to ~100μm and aerosol properties change substantially over this size range. Several studies demonstrated that airborne Particulate Matter (PM) has an impact on climate [23], biogeochemical cycling in ecosystems [24], visibility [25] and human health [26-28]. Over recent years in the public health domain the PM concentration has become a topic of considerable importance, since epidemiological studies have shown that exposure to particulates with aerodynamic diameters of < 10 μm (PM10) and especially < 2.5 μm (PM2.5) induces an increase of lung cancer, morbidity and cardiopulmonary mortality [e.g. 29-35]. Thus air pollution appears to have an adverse effect on respiratory and cardiovascular systems [36], which might result in an acute reduction of lung function, aggravation of asthma, increased risk of pneumonia in the elderly, low birth weight and high death rates in newborns [37].

Some dust contaminants (soluble and chelatable metallic salts, pesticides, etc) affect human health when they are transported over densely populated areas [38], retained in residences and other occupied structures [39], and they also impact the nutrient loading of waters flowing from adjacent watersheds [40] and terminal bodies of water by direct and indirect deposition [41-42]. PM is a complex mixture of substances suspended in to the atmosphere in solid or liquid state with different properties (e.g. variable size distribution or chemical composition amongst others) and origins (anthropogenic and natural). Owing to this mixture of substances, the chemical composition of PM may vary widely as a function of emission sources and the subsequent chemical reactions which take place in the atmosphere [43-45].The chemical mass balance is the most commonly used method for assessing PM source contributions [46], while statistical methods, such as factor analysis and multi-linear regression [47], have also produced interesting results regarding dust source identification. Elemental and mineralogical analyses have also been used to identify the source regions of dust deposited in Arctic ice caps [48] and on other depositional surfaces [49, 42].

One spectacular example of such dust effects is the Hamoun (dry) lakes in the eastern part of Iran that has attracted scientific interest during recent years, since it constitutes a major dust source region in southwest Asia, often producing intense dust storms that cover the Sistan region in eastern of Iran and the southwest of Afghanistan and Pakistan and influencing the air quality, human health and ecosystems as well as aerosol loading and climate from local to regional scales [50-52, 12]. Particles from the Hamoun dust storms might also cover farm and grasslands to result in damage to crops and fill the rivers and water channels with aeolian material. Over recent years, tens of thousands of people have suffered from respiratory diseases and asthma during months of devastating dust storms in the Sistan basin, especially in the cities of Zabol and Zahak and the surrounding villages [53]. According to the Asthma Mortality Map of Iran, the rate of asthma in Sistan is, in general, higher than in other regions [54].

The aim of this Chapter is to: (1) assess the seasonal and annual variability of dust storms that originate from the Hamoun Basin, (2) identify the amount of dust loading from the dry lakes, (3) assess the contribution of the lake beds to regional dust emissions for specific locations, (4) assess the dust chemical and mineralogical composition to provide useful information regarding the status of dust storms and also human health and, (5) assess dust concentrations and air quality of the regions affected by the dust storms.

2. Sistan region

The Sistan basin (Helmand basin) lies between the Hindu Kush Mountains in Afghanistan and the mountain ranges flanking the eastern border of Iran and is the depository that receives the discharge of the Helmand (Hirmand) river in the lower Sistan Basin (Fig. 1). It is a large and remote desert basin, extremely arid and known for its windstorms, extreme floods and droughts. The closed basin receives the waters of the Helmand river, the only major perennial river in western Asia between the Tigris-Euphrates and Indus rivers [9]. The Helmand and its tributary streams drain the southern Hindu Kush Mountains of Afghanistan and flow into an otherwise waterless basin of gravel plains and sandy tracts before terminating in Sistan (also Seistan, British spelling), a depression containing the large inland delta of the Helmand river and a series of shallow, semi-connected playas at the western edge of the basin (Fig. 1).

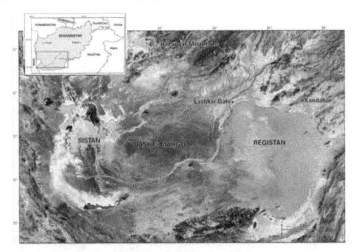

Figure 1. Landsat-5 image showing the lower Hirmand Basin and Sistan region. [9].

In 1949 the United States initiated a new program for the improvement of underdeveloped areas of the world. The damming of the Helmand river in southern Afghanistan became one of the showcase projects of U.S. foreign aid in the "Third World" after World War II. Dams were built on the Helmand river and its main tributary (Arghandab river) during the 1950s. The main project goals were to provide hydroelectric power and increase agricultural productivity through irrigation and land reclamation. The Arghandab dam, located northwest of the city of Kandahar, was completed in 1952 with a height of 145 feet (44.2 meters) and storage capacity of 388,000 acre-feet (478.6 million cubic meters). The larger Kajakai dam on the Helmand was completed a year later with a height of 300 feet (91.4 meters) and length of 919 feet (280 meters) and storage capacity of 1,495,000 acre-feet (1,844 million cubic meters). About 300 miles (482.8 kilometers) of concrete-lined canals were built to distribute the reservoir waters [9]. The Hirmand river is the longest river in Afghanistan (ca. 1150 km; catchment > 160,000 km²) and the main watershed for the Sistan basin, finally draining into the natural swamp of the Hamoun lakes complex [10].

Severe dust plumes usually extend from Sistan into southern Afghanistan and southwestern Pakistan obscuring the surface over much of the region (Fig 2). Severe droughts during the past decades, especially after 1999, have caused desiccation of the Hamoun lakes leaving a fine layer of sediment that is easily lifted by the wind [55], thus modifying the basin to one of the most active sources of dust in southwest Asia [56, 50, 57]. Therefore, the Hamoun dry-lake beds exhibit large similarities with the other two major dust source regions of the world that comprise of dried lakes and topographic lows, i.e. Bodele depression in Chad [3] and lake Eyre in Australia [4]. The strong winds blow fine sand off the exposed Hamoun lake beds and deposit it to form huge dunes that may cover a hundred or more villages along the former lakeshore. As a consequence, the wildlife around the lake has been negatively impacted and fisheries have been brought to a halt.

Fig. 2 (left panel) shows a severe dust storm over the Sistan region as observed from the Terra-MODIS satellite's true color image on 15 June 2004. The intense dust plumes form a giant U-shape extending from Sistan into southern Afghanistan and southwestern Pakistan that obscures the surface over much of the region. The pale color of the dust plume is consistent with that of dried wetland soils. The dust is blowing off the dry lake beds that become the Hamoun wetlands during wet years.

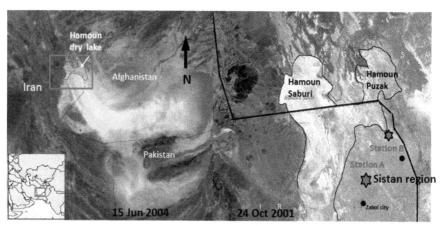

Figure 2. MODIS image of dust deflation over southern Afghanistan originating from the dry Hamoun lakes in Sistan on June 15, 2004.

3. Hamoun lakes

The Hamoun lakes are situated roughly at the termination of the Hirmand river's inland delta. The Hamoun lakes complex (Hamoun-e-Puzak, Hamoun-e-Sabori and Hamoun-e-Hirmand and Baringak) are located in the north of the Sistan region, which is also the largest fresh water ecosystem of the Iranian Plateau and one of the first wetlands in the Ramsar Convention [58]. Water in the Hamoun lakes is rarely more than 3 meters deep, while the size of the lakes varies both seasonally and from year to year. Maximum expansion takes place in late spring, following snowmelt and spring precipitation in the mountains. In years of exceptionally high runoff, the Hamoun lakes overflow their low divides and create one large lake that is approximately 160 kilometers long and 8–25 kilometers wide with nearly 4,500 square kilometres surface area. Overflow from this lake is carried southward into the normally dry Gaud-i Zirreh (Fig. 1), the lowest playa (463-meter altitude) in the Sistan depression. Furthermore, mountain runoff varies considerably from year to year; in fact, the Hamoun lakes have completely dried up at least three times in the 20th century [9]. The maximum extent of the Hamoun lakes following large floods is shown in Fig. 4, where a continuous large lake has been created covering an extended area of ~4,500 km^2 with a volume of 13000 million m^3 in Sistan and southwestern Afghanistan. This figure corresponds to spring of 1998 after snowmelt in the Afghanistan mountains that transferred large quantities of water into the Hamoun Basin. As a consequence, livelihoods in the Sistan region are strongly interlinked with and dependent on the wetland products and services, as well as on agricultural activities in the Sistan plain. Fishing and hunting represent an important source of income for many households and, therefore, the local and regional economy is strongly dependent on weather conditions, precipitation and land use – land cover changes. The political boundary between Iran and Afghanistan splits the Hamoun system, further complicating management possibilities in the area. Most (90%) of the watershed is located in Afghanistan and practically all of the wetlands' water sources originate there. The Iranian part is desert, and produces runoff only in rare cases of significant local rainfall [10].

In view of the Hirmand and the surrounding rivers that supply most of the sediments to the Hamoun lakes, a brief encapsulation of the relevant geology of the catchment area in Afghanistan is given. Afghanistan has a very complex geology, encompassing two major relatively young orogenies, Triassic and subsequent Himalayan, resulting in amalgamation of crustal blocks and formation of concomitant ophiolites and younger clastic and carbonate sedimentary rocks as well as basaltic lavas and, more recently, extensive alluvial and eolian detritus [59] . The Sistan region and Hamoun dry lake beds are mainly composed of Quaternary lacustrine silt and clay material as well as Holocene fluvial sand, silt and clay (Fig. 3). These materials have been carried to the basin by the rivers, while along their courses Neogene fluvial sand, eolian sand, silt and clay are the main constituents. Note also the difference in the soil-dust composition between two major desert areas, Registan and Dasht-i-Margo in Afghanistan. The former is composed of Neogene coarse gravels and the latter of Quaternary eolian sand. More details about the geology in the Sistan region can be found in [60].

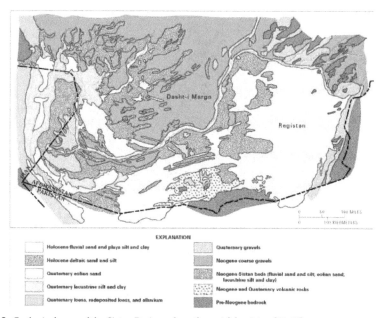

Figure 3. Geological map of the Sistan Basin and southern Afghanistan [61-63].

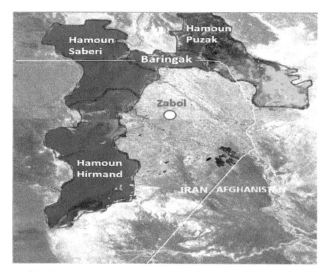

Figure 4. Position of the Hamoun Lakes in Iran and Afghanistan, showing a maximum inundation period.

4. Droughts in the Sistan

The Sistan Basin has recently experienced an unusually long 10-year drought starting in 2000 [10]. Combined with war and severe political disruption over the past 2 decades, the 10-year drought has created conditions of widespread famine that affected many people in eastern Iran and southwestern Afghanistan. A suggested, climatic forcing mechanism has been proposed for the recent drought by Barlow and others (2002). A prolonged ENSO (El Niño-Southern oscillation) cold phase (known as La Niña) from 1998 to 2001 and unusually warm ocean waters in the western Pacific appear to have contributed to the prolonged drought. The unusually warm waters (warm pool) resulted in positive precipitation anomalies in the Indian Ocean and negative anomalies over central Afghanistan [64], thus contributing to the drying of the Hamoun Basin. The contrast between a relatively wet year in 1976 and the nearly dry Hamoun lakes in 2001 is shown in Figs. 2 and 4. Millions of fish and untold numbers of wildlife and cattle died. Agricultural fields and approximately 100 villages were abandoned, and many succumbed to blowing sand and moving dunes [65].

Most of the Sistan population lives near the Hamoun lakes and is employed in agricultural, fishery, handicrafts and other jobs. To counter the effects of droughts, the Iranian government prepares facilities such as food and flour supplies, medicine and health services and employment in the region to prevent the forced emigration of people, but the continuous and extreme droughts have forced some people to leave the Sistan region. Long droughts at the end of the 1960s, middle of the 1980s, and from 1999 to 2010 affected the Sistan region significantly and resulted in desiccation of the Hamoun lakes, making the surrounding lands saline and disturbing their soil fertility, while some places became barren (see Fig. 5). The most important findings in Fig. 5 are: (1) in 1976, the Hamoun lakes were still thriving. Dense reed beds appear as dark green, while tamarisk thickets fringing the margins of the upper lakes show up as pink shades in the satellite images (Fig. 5). Bright green patches represent irrigated agricultural lands, mainly wheat and barley. The lakes flood to an average depth of half a meter, denoted by lighter shades of blue, while dark blue

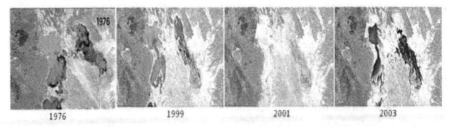

1976   1999   2001   2003

Figure 5. Satellite (Landsat) images of the Hamoun Basin in spring of different years. Hamoun lakes are fed primarily by water catchments in neighbouring Afghanistan. In 1976, when rivers in Afghanistan were flowing regularly, the lake's water level was relatively high. Between 1999 and 2011, however, drought conditions caused frequent dryness of the Hamoun lakes that almost disappeared in 2001 after a 3-year intense drought period [65].

to black indicates deeper waters, which, however, do not exceed four meters. (2) By 2001, the Hamoun lakes had vanished since central and southwest Asia were hit by the largest persistent drought anywhere in the world. The only sign of water in this scorched landscape of extensive salt flats (white) is the Chah Nimeh reservoir in the southern part of Sistan (not shown on the satellite image), which is now only used for drinking water. Degraded reed stands in muddy soil are visible as dark green hues at the southern end of Hamoun-i Puzak. In 2003 the Hamoun Basin was covered with water again, but with significantly lower coverage than in the mid-1970s [65].

5. Data set and experimental methods

In order to address the scientific topics mentioned above multiple ground-based instrumentation and several sets of satellite images (mostly Landsat and MODIS) were used for illustration purposes and for detection of water level in Hamoun and for monitoring of land use – land cover changes, seasonality of dust storms and associated sediments, air quality perspectives, chemical and mineralogical composition of dust over the Sistan region and Hamoun Basin. Within this framework, satellite images from different years were used to identify changes in the lake's surface. Information about dust storm occurrence was obtained from Zabol meteorological station, 5 km from the Hamoun lakes (see Fig. 4). The ground-based measurements were primarily used to compare effects of the Hamoun surface on dust aerosols.

More specifically, the amount of dust loading during dust storms was measured using passive dust traps fixed at two monitoring towers (respectively, at four and eight meters above ground level in altitude), with one meter distance between the adjacent individual traps; the 4 m tower had four traps and the 8 m tower had 8 traps (Fig. 6). The two towers were established in open sites (station A and station B, denoted by red stars in Fig. 2 right panel) close to the dry-bed lake dust source region during the period August 2009 to July 2010. The dust sampler used in the campaign was developed by the Agricultural and Natural Research Center of Sistan (Fig. 6), and is a modified version of the SSDS sampler [66-67]. At the observation sites, the samplers collect airborne dust sediment. The traps were mounted on a stable bracket parallel to the wind direction. The samplers consist of a tube with a diameter of 12 cm. The sediment-laden air passes through a vertical 2.5 cm x 6 cm sampler opening in the middle. Inside the sampler, air speed is reduced and the particles settle in a collection pan at the bottom, while the air discharges through an outlet with a U shape. After each measurement, the samplers were evacuated to make them ready for measuring the following dust events. The collected samples were oven dried at 105 °C for 24 hours, and then, dried samples were weighed using an electronic scale in order to obtain total mass quantities at each sampling height and for each dust storm. The samples were also transported to a laboratory for chemical and mineralogical analysis.

Furthermore, soil samples were collected from topsoil (0–5 cm depth) at several locations in the dry-bed Hamoun lakes and downwind areas. These samples were analyzed to investigate the chemical and mineralogical characteristics of dust, relevance of inferred sources and contributions to air pollution.

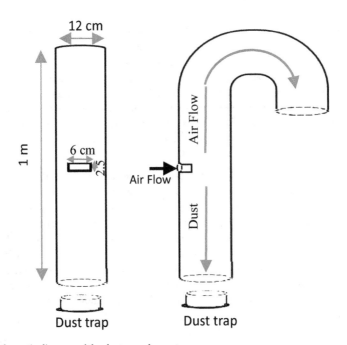

Figure 6. Schematic diagram of the dust sampler system.

These samples were analyzed for major and trace elements and for minerals by applying X-Ray Fluorescence (XRF) and X-Ray Diffraction (XRD) techniques, respectively. The samples were prepared for XRD analysis using a back loading preparation method. They were analyzed using a PANalytical X'Pert Pro powder diffractometer with X'Celerator detector and variable divergence and receiving slits with Fe filtered Co-Kα radiation. The phases were identified using X'Pert High score plus software. The relative phase amounts (weights %) were estimated using the Rietveld method (Autoquan Program). Mineral analysis by XRD is the single most important non-destructive technique for the characterization of minerals such as quartz, feldspars, calcite, dolomite, clay, silt and iron oxides in fine dust. Mineral phase analysis by XRD is one of the few techniques that are phase sensitive, rather than chemically sensitive, as is the case with XRF spectrometry. Quantitative mineralogical analyses using the XRD technique have been performed by a number of scientists over the globe [e.g., 68-70, 44, 71-72].

The sample preparation for XRF is made up of two methods, pressed powders and fusions. The former samples were prepared for trace element analyses and the latter for major element analyses. Each milled sample (<75μm) was combined with a polyvinyl alcohol, transferred into an aluminum cup and manually pressed to ten tons. The pressed powders were dried at 100°C for at least 30 minutes and stored in a desiccator before analyses were conducted. For the fusion method, each milled sample (<75μm) was weighed out in a 1/6 sample to flux (Lithium tetraborate) ratio. These samples were then transferred into mouldable Pt/Au crucibles and fused at 1050°C in a muffle furnace. Aluminum cooling caps were treated with an iodine-ethanol mixture (releasing agent) and placed on top of the crucibles as they cooled. Some samples needed to be treated with an extra 3g of flux if they continued to crack. Finally, all geochemical samples were analyzed using the Thermo Fisher ARL 9400 XP+ Sequential XRF. The Quantas software package was used for the major element analyses and the WinXRF software package was used for the trace element analyses. The concentrations of the major elements are reported as oxides in weight percentages, while the trace element concentrations are reported as elements in parts per million (ppm).

Furthermore, in order to provide analysis of the air quality, PM_{10} concentration measurements were obtained by using an automatic Met One BAM 1020 beta gauge monitor (Met One, Inc.,) over Zabol. The instrument measures PM_{10} concentrations (in $\mu g.m^{-3}$) with a temporal resolution of one hour. The measurements were carried out at the Environmental Institute of Sistan located at the outskirts of Zabol during the period September 2010 to September 2011 (total of 373 days). The recording station is close to the Hamoun basin and is placed in the main pathway of the dust storms of the Sistan region. The hourly measured PM_{10} data were daily-averaged, from which the monthly values and seasonal variations were obtained. For further assessing the air quality over Zabol, the PM_{10} concentrations were used to calculate an Air Quality Index (AQI) [52].

6. Meteorology and climatology over Sistan

The climate over Sistan is arid, with low annual average precipitation of ~55 mm occurring mainly in the winter (December to February) and evaporation exceeding ~4000 mmyear[-1] [58]. During summer, the area is under the influence of a low pressure system attributed to the Indian thermal low that extends further to the west as a consequence of the south Asian monsoon system. These low pressure conditions are the trigger for the development of the Levar northerly wind, commonly known as the "120-day wind" [73], causing frequent dust and sand storms, especially during summer (June to August) [74, 56] and contributing to the deterioration of air quality [52]. Therefore, one of the main factors affecting the weather conditions over the region is the strong winds rendering Sistan as one of the windiest deserts in the world. These winds blow continuously in spring and summer (from May to September), and on some days during winter, and have significant impacts on the landscape and the lives of the local inhabitants.

The annual variation of mean Temperature (T), Relative Humidity (RH) and atmospheric Pressure (P) over Zabol (a large city in the Sistan region) during the period 1963 to 2010 is shown in Fig 7. The monthly mean T exhibits a clear annual pattern with low values in the winter (9 to 12 °C) and high (~35 °C) in summer, following the common pattern found in the northern Mid-latitudes. During the summer period the maximum T often goes up to 46 or 48 °C causing an extremely large diurnal variation, which is a characteristic of many arid environments. RH illustrates an inverse annual variation with larger values in winter (50 to 57%) and very low values in summer (~25%), which are about 10 to 15% during daytime. During the period October to April P values are generally high (1020 to 1024 hPa in winter), which is above the standard mean sea level value of 1013.25 hPa. P values decrease during summer (~ 996 hPa in July) as a result of the Indian thermal low that develops over the entire south Asia during summer monsoon months. This has a direct impact on the intensity of the winds over the region, which has a monthly mean of as high as 12 m.s⁻¹ during June and July with frequent gusts of above 20 to 25 m.s⁻¹. In contrast, during late autumn and winter months the wind speed is confined to ~3 to 4 m.s⁻¹ [12].

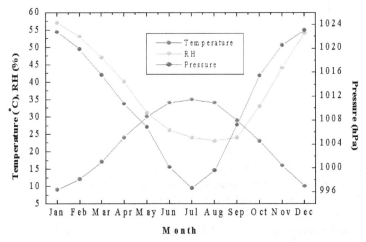

Figure 7. Monthly mean variation of air Temperature, Relative Humidity and atmospheric pressure in Zabol during the period 1963 –2010.

7. Temporal changes of Hamoun dry lake beds and dust storms

The dust storms over a region cause several climatic implications, environmental and human concerns [16, 75,18, 76, 36] and can be examined via multiple instrumentation and techniques. Among others, the analysis of the visibility records can constitute a powerful tool for monitoring of the seasonal and inter-annual variation of the dust storms, since the main result of such phenomena is the limitation of visibility and deterioration of air quality. The annual variation of visibility (as the main indicator for the dust storms) over Sistan follows a clear annual pattern, with large values in winter, usually above 10 km, and very low in summer (< 4 km on average) as analyzed from meteorological observations taken at Zabol (Fig. 8). A power-decreasing curve relation associated with 93% of the variance was observed between wind speed and visibility [12]. This inverse relation indicates that the wind speed does not act as a ventilation phenomenon over Zabol, as usually occurs in coastal urban environments with local sea-breeze cells [77], but rather as a factor responsible for the deterioration of visibility, since the intense Levar winds are the cause of the dust outbreaks over Sistan. Therefore, the major dust storms over the region are associated with intense winds of northwesterly direction that are responsible for the deterioration of visibility to lower than 100 m in many cases.

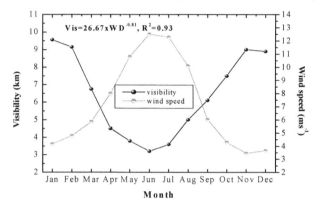

Figure 8. Monthly mean variation of the visibility (km) and wind speed (ms-1) in Zabol during the period 1963–2009.

Although the visibility exhibits a clear annual pattern (Fig. 8) suggesting that the summer season is the favourable period for the occurrence of frequent and intense dust storms, long-term data series over Zabol (1963-2009) show that it contains considerable year-to-year variations (Fig. 9b). Focusing on recent years, the days with visibility <= 2 km have been dramatically increased from about 20 during 1995 to 1999 to >100 during 2000 to 2001. This is attributed to a severe drought period that dried the largest part of the Hamoun wetlands (see Fig. 5) and favored the alluvial uplift, as well as the frequency and mass intensity of dust storms that affected the visibility over Sistan (Fig. 9b). However, in the 2000s the days with very low visibility seem to have a decreasing trend, but remaining above the standards of the climatological mean. It is, therefore, concluded that the regional and synoptic meteorology (mainly precipitation) is strongly linked to land use – land cover changes over Hamoun and then, to dust outbreaks over Sistan region.

In contrast, the annual variation of the wind speed (Fig. 9a) exhibits an opposite pattern with higher intensities during summer (June and July) and lower in winter. As far as the wind direction is concerned, it is found from the Zabol data series that the northwestern direction clearly dominates, being more apparent in summer, while high percentages for intense winds are also associated with a northwesterly flow (Fig. 9a). The probability for intense winds to blow from other directions is low; summer winds are much more intense with ~27% of wind speeds above 11 ms-1, while calm conditions are limited to 3% against 19% and 20% for autumn and winter, respectively. The higher frequency and intensity of northwestern winds is the reason for the frequent dust storms that affect Zabol. It is to be noted that Zabol is located at the downwind direction of dust storms that normally originate from Hamoun (Fig. 2).

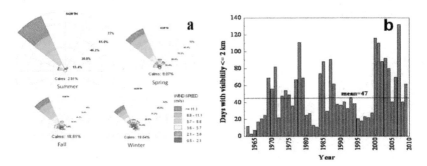

Figure 9. Flow chart of the seasonal wind speed and direction in Zabol during the period 1963-2010. The percentage of calm periods is shown at the bottom of each wind rose. The thickest bar represents wind speeds in excess of 12 m/s; (b) year-to-year variation of the visibility recordings at Zabol meteoroological station.

The water levels in the Hamoun lakes change considerably from year to year as has been discussed above. Table 1 summarizes the percentage of water surface in July in the Hamoun lakes, as well as the annual precipitation and number of dusty days during the period 1985-2005. Yearly variations of Hamoun lakes water surface identified four periods from 1985 to 2005: [10]:

1. A low-water period from 1985-1988: the Hamoun dried out or shrunk to a very small size almost every year, but there was some inflow every year.
2. A high-water period from 1989-1993: there was considerable inflow for five years, during which time the Hamoun only shrunk below the previous period's maximum levels for a very short time.
3. A medium-water period from 1994-1999: a dynamic balance of inflow and outflow maintained a reasonably high minimum water volume every year.
4. A dry period from 2000-present: the inflow ceased and a catastrophic drought ensued except for a flood in 2005 that immediately dried up before 2006.

On the other hand, Table 2 summarizes the correlation coefficients between the percentage of dried beds in July, precipitation and number of dusty days, i.e. the parameters that are included in Table 1. The analysis shows that precipitation has a direct effect on water levels (r=0.63 for Hamoun Saberi). On the other hand, in years with high precipitation the lakes had high water surface. Hamoun Saberi is also affected by the Farah river that has a closer watershed, but this correlation is low for both Hamoun Hirmand and Hamoun Puzak (r= 0.35 and r=0.54 respectively). The correlation between dusty days and percentage of dried Hamoun beds (100-percent of water surface) shows high correlation coefficient values regarding Hamoun Saberi and Baringak (r=0.88 and r=0.82 respectively) and lower correlation for Hamoun Hirmand (r= 0.63). The high correlation for the Hamoun Saberi and Baringak indicates that Sistan dust storms are directly affected by the north and northwestern winds flowing through the Saberi. The year-to-year variation of the dusty days and the percentage (%) of dried bed lakes in Baringak and Hamoun Saberi (Fig. 10) indicates a co-variation of the examined parameters, thus suggesting that the land use – land cover changes play a major role in the occurrence of dust storms over Sistan region.

Year	Baringak	Saberi	Hirmand	Puzak	precipitation	Dusty days
1985	0	30	7	35	25.6	88
1986	35	65	12	53	72.8	30
1987	25	40	2	53	8.7	91
1988	15	50	6	50	69.5	62
1989	90	92	60	54	26.1	38
1990	90	98	70	72	96.1	37
1991	80	95	90	80	85.8	41
1992	80	98	80	93	80.9	33
1993	75	95	70	60	52.4	46
1994	43	60	20	60	116.6	39
1995	48	66	7	62	76.2	21
1996	62	90	47	70	84.3	18
1997	60	85	25	60	76.4	23
1998	80	100	73	60	61.4	22
1999	72	90	25	60	87.7	28
2000	0	12	0	0	26.8	116
2001	0	0	0	0	7.2	110
2002	0	5	0	0	37.5	88
2003	0	0	0	0	32.3	92
2004	0	0	0	0	51.1	80
2005	80	90	18	32	129.5	41

Table 1. Yearly variability of percentage of water surface in Hamoun lakes in July, annual precipitation and the dusty days (visibility <= 2km) over Sistan region

	Baringak	Hamoun Saberi	Hamoun Hirmand	Hamoun Puzak	Precipitation	Dusty days
Baringak	1					
Hamoun Saberi	0.96**	1				
Hamoun Hirmand	0.84**	0.80**	1			
Hamoun Puzak	0.80**	0.89**	0.74**	1		
Precipitation	-0.59**	-0.63**	-0.35	-0.54	1	
Dusty days	0.82**	0.88**	0.60**	0.81**	-0.730**	1

** Correlation is significant at the 0.01 level

Table 2. Correlations between percent of Hamoun dried beds in July and dusty days (1985-2005).

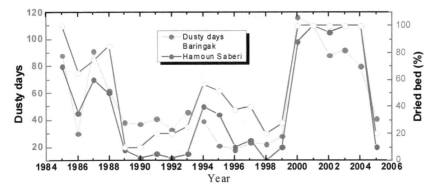

Figure 10. Yearly variability of the dusty days (visibility <= 2km) over Sistan region with association to percentage of Hamoun dried beds (1985-2005). The lower coverage of the Hamoun Basin by water (high percentage of dried beds) corresponds to higher number of dusty days over Sistan region.

8. Dust loading measurements

Dust activity is a function of several parameters, such as topography, rainfall, soil moisture, surface winds, regional meteorology, boundary layer height and convective activity [78-79.

Data on dust loading are available at only a few places around the world [e.g.80-84] and those presented here are the first for the Sistan region. Hence, obtaining measurements of horizontal dust flux will significantly increase our understanding of wind erosion and dust influences. Apart from the natural emissions of dust, [85] identified two ways in which human activities can influence dust emissions: (a) by changes in land use, which alter the potential for dust emission, and (b) by perturbing local climate that, in turn, alter dust emissions. As has been discussed above, both ways are considerably active over the Sistan region and Hamoun Basin.

The dust loading measured at the two stations close to the Hamoun basin for several dust events during the period August 2009 to July 2010 is plotted in Fig. 11. In the same graph, meteorological data from the Zabol station that give information about the duration of dust events (for the examined days as well as on the preceding or succeeding days, i.e. about 2-3 days before the peak-day of the dust storm) and daily mean and maximum wind speeds, are also plotted. The results of the average dust loading measured at eight heights at station B and at four heights at station A reveal considerable variation, ranging from ~0.10 to ~2.5 kgm^{-2}. In general, the highest dust loading is observed for dust events occurring in summer, but intense dust storms can also take place in winter, since the Hamoun basin is an active dust source region throughout the year. The dust loading is highly correlated with the duration of the dust storms, as shown from their correlation, with the linear regressions being statistically significant at the 0.99% confidence level (Fig. 12). Apart from the strong linkage to the duration of dust storms, the dust loading at both stations also seems to have a dependence on the daily mean and maximum wind speeds (not presented). However, this dependence was found to be more intense and statistically significant (at the 95% confidence level) at station B, which is located closer to the dust source, whereas for station A the correlation was not found to be statistically significant. This finding emphasizes the strong effect of the wind speed on dust erosion and transportation, as well as on dust loading, at least for areas close to dust sources. However, the results show that the main factor that controls the dust loading at both stations is the duration of the dust storms, and secondly the wind speed. The role of the wind might have been found to be more critical if measurements were taken at the sampler stations instead of using the meteorological data from Zabol. The analysis showed that the total dust loading for the 19 events of measurements at station A is 16.9 kg m^{-2} corresponding to 0.88 kg m^{-2} per event, whereas at station B the measurements yielded 15.8 kg m^{-2} (17 events), corresponding to 0.93 kg m^{-2} per event. The larger dust loading at station B is attributed to the smaller distance from the source region.

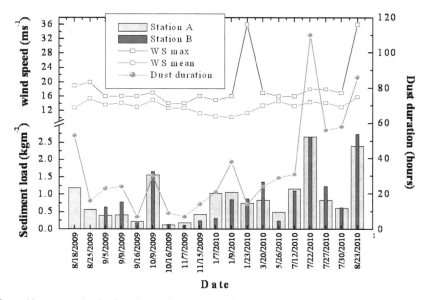

Figure 11. Average dust loading (kgm^{-2}) during various dust events in the Sistan region as measured at the 4m (station A) and 8m (station B) monitoring towers. The duration of dust events (hours), as well as the mean and maximum wind speeds on the dusty days were obtained from the Zabol meteorological station.

Figure 13 illustrates the height variation in dust loading during the dust storms measured at station A (19 days, up to 4m in height) and station B (17 days, up to 8m in height). Contrasting height variations measured during intense dust storms occurred between the two stations, while similar variations correspond to moderate and low dust storm events. More specifically, the dust loading shows an increase (decrease) with height in station A (station B), revealing a difference in the dust transport mechanisms. This finding can be explained by considering the fact that station B is located closer to the Hamoun dust source region, meaning that uplift and newly transported dust concentration is higher near the surface. On the other hand, at station A that is located about 20 km away, the dust loading presents larger values up to 3 m since the near-ground dust particles have already been deposited near the source, and as the distance increases so does the dust-plume height. The diurnal variability of the dust loading at the two stations (not presented) showed increased mass concentrations during daytime that can be explained by enhanced convection and turbulent mixing in a deepened boundary layer. Furthermore, the local winds are stronger during daytime due to thermal convections.

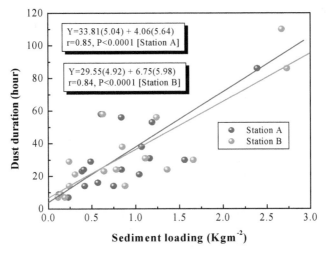

Figure 12. Correlation between dust loading measurements and duration of dust storm events for 19 days at station A and 17 days at station B.

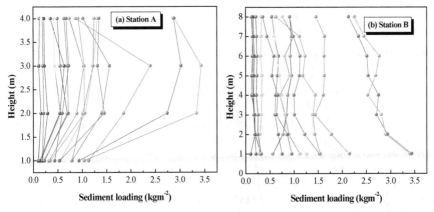

Figure 13. Height variation of the dust loading at stations A (a) and at station B (b) for several dust storm days. Green colors are loadings for winter, yellow for spring, red for summer and blue for autumn.

9. PM₁₀ measurements

In order to provide a first ever in-situ analysis of the air quality over Sistan, PM₁₀ concentration measurements were obtained by using an automatic Met One BAM 1020 beta gauge monitor (Met One, Inc.,) at Zabol [12]. The instrument measures PM₁₀ concentrations (in $\mu g.m^{-3}$) with a temporal resolution of one hour. The measurements were carried out at the Environmental Institute in Sistan located at the outskirts of Zabol during the period September 2010 to August 2011. The recording station is close to the Hamoun basin and is placed in the main pathway of the dust storms in the Sistan region. The hourly measured PM₁₀ data were daily-averaged, from which the monthly values and seasonal variations were obtained (Table 3). For further assessing the air quality over Zabol, the PM₁₀ concentrations were used to calculate the Air Quality Index (AQI).

	Monthly Mean PM10 (µg.m-3)	Daily minimum PM10 (µg.m-3)	Daily maximum PM10 (µg.m-3)
January	196	29	597
February	147	13	787
March	262	21	2698
April	224	97	515
May	322	71	1276
June	627	100	1875
July	847	110	2007
August	807	155	2448
September	564	88	1046
October	531	100	2339
November	200	66	737
December	476	84	3094
Winter	273	13	3094
Spring	270	21	2698
Summer	716	100	2448
Autumn	484	66	2339

Table 3. Monthly mean, daily maximum and daily minimum PM₁₀ concentrations in Zabol during the period September 2010 to August 2011.

The results show extremely large PM₁₀ concentrations at Zabol (see Fig. 14). Even the mean values are much higher than the most risky and dangerous maximum levels provided by the U.S. Environmental Protection Agency (397 $\mu g.m^{-3}$). Throughout the year, and especially during the period June to October, the area suffers from severe pollution since even the lower PM₁₀ values are above 100 $\mu g.m^{-3}$, while the maximum ones are usually above 1000 $\mu g.m^{-3}$. On the other hand, extreme PM₁₀ measurements associated with severe dust events may also occur in other months, for example like December. Daily PM₁₀ concentrations during major dust storms are about 10 to 20 times above the standard levels. Regarding the monthly mean PM₁₀ concentrations, the results show extremely large values (>500 $\mu g.m^{-3}$) during the period June to October, reaching up to 847 $\mu g.m^{-3}$ in July.

Changes of Permanent Lake Surfaces, and Their Consequences for Dust Aerosols and Air Quality: The Hamoun Lakes of the Sistan Area, Iran

181

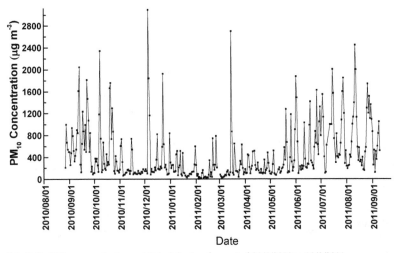

Figure 14. Daily PM_{10} concentrations at Zabol during the period 28/8/2010 to 10/9/2011.

The frequency of occurrence of PM_{10} concentrations for each season over Zabol is depicted in Fig. 15. In summer ~60% of the PM_{10} values were higher than 425 $\mu g.m^{-3}$, while the lower PM_{10} values occur in winter and spring with larger frequency in the 55-154 $\mu g.m^{-3}$ interval. A very significant finding is the very low frequency for PM_{10} concentrations below ~400 $\mu g.m^{-3}$ in summer, suggesting an extremely turbid atmosphere with frequent dust storms and near absence of clear or relatively clear conditions over Sistan during summer. Autumn also presents high frequency in the >425 $\mu g.m^{-3}$ interval that might be due to continuation of the Levar winds in September favouring the dust storms over Sistan.

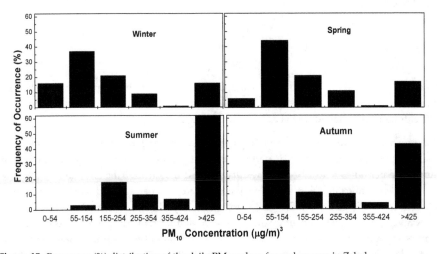

Figure 15. Frequency (%) distribution of the daily PM_{10} values for each season in Zabol.

10. Air quality index (AQI)

In order to identify the impact of air pollution on human health, air pollution indices are commonly used, of which the AQI is the most well known [86-88]. As a consequence, the AQI is a powerful prenatainarry tool to ensure public health protection [86].

The AQI is divided into six categories, varying from 0 to 500, with different health impacts [86] as listed in Table 4. The two first AQI categories (good and moderate, <155 PM_{10} $\mu g.m^{-3}$) have no impact on health, while the last AQI category (hazardous, >424 PM_{10} $\mu g.m^{-3}$) is associated with a serious risk of respiratory symptoms and aggravation of lung disease, such as asthma, for sensitive groups and with respiratory effects likely in the general population [89, 87]. The AQI for Zabol was calculated for the period September 2010 to July 2011.

Health Quality	AQI	PM10 (µg.m-3)	Days	(%)
Good	0-50	0-54	21	5.7
Moderate	51-100	55-154	106	28.6
Unhealthy for sensitive people	101-150	155-254	66	17.8
Unhealthy	151-200	254-354	36	9.7
Very unhealthy	201-300	355-424	12	3.2
Hazardous	301-500	425<	129	34.9

Table 4. Health quality as determined by the Air Quality Index (AQI), PM_{10} and number of days with severe pollution in Zabol during the period September 2010 to July 2011.

Based on the technological rules related to AQI, the following formula was used to derive the PM_{10} concentration from AQI [90, 88]:

$$I = \frac{I_{high} - I_{low}}{C_{high} - C_{low}}(C - C_{low}) + I_{low}$$

Where I is the concentration of PM_{10}, I_{low} and I_{high} are AQI grading limited values that are lower and larger than I (AQI index), respectively, and C_{high} and C_{low} denote the PM_{10} concentrations corresponding to I_{high} and I_{low}, respectively.

Provisional studies focusing on air quality and dust over Iran have already been carried out. For example, amongst others, [91] performed a comparative study of air quality in Tehran during the period 1997 to 1998. The results revealed that in 1997 the air quality on 32% of the days was unhealthy, and on 5% of the days it could be regarded as very unhealthy, whereas in 1998 the unhealthy and very unhealthy days increased to 34% and 6%, respectively. [92] studied the air quality in Tehran and Isfahan and offered solutions for its improvement using the AQI. It was found that on 329 days of the year in Tehran, and on 34 of the days in Isfahan, the AQI departed beyond 100. [93] also studied AQI in Tehran reporting that on 273 days in 2001 the values were higher than those set for the air quality standards; 13% of the days were considered as very unhealthy and 0.27% were classified as dangerous. [52] found that 15 % of the days were unhealthy for sensitive people in the city of Zahedan thatwas affected by Sistan dust storms, while 2 % were associated with a high health risk or were even hazardous.

Comparing the present results with those of the above-mentioned studies, it is concluded that the Sistan region experiences much higher PM concentration levels. Assessment of air quality in Zabol shows that 243 days out of 370 (65%) exhibit air pollution levels of above the air quality standards (>155 $\mu g.m^{-3}$), a fraction that is much higher than that (26.5%) reported for Zahedan city located also in Sistan about 200 km south of Zabol [52]. The most significant finding is the 129 days (34.9%) that are characterized as hazardous (Table 4), which in combination with the adverse effects on human health, make it clear that environmental conditions in the Sistan region are rather poor for human well-being. On the other hand, only 5.7% of the days are associated with low pollution levels when the air quality is considered satisfactory and air pollution poses little or no risk. Several studies have shown that ambient air pollution is highly correlated with respiratory morbidity, mainly amongst children [94, 36, 95]. The results gathered from hospitals in the Sistan region showed that during dust storms respiratory patients increased significantly, especially those affected by chronic obstructive pulmonary disease and asthma. The percentage of these diseases increases in summer (June and July) [53]. Apart from the dust storms, re-suspended dust within the urban environment is a strong source of PM_{10} concentrations, while urban-anthropogenic and industrial activities are considered to have a much lower effect on the air pollution over Zabol.

The mean diurnal variation of PM_{10} concentrations for each season in Zabol indicates a clear pattern for all seasons except winter, with the maximum of the diurnal variation being observed in the middle of the day (~08:00-11 LST) while in winter PM_{10} values reach a maximum in the afternoon hours to early morning (~16:00 – 02:00 LST). In general, solar heating and vertical mixing of pollutants are the main factors for the reduction of PM_{10} levels at local noon to early afternoon hours. However, the maximum PM_{10} concentrations normally occur between 08:00 and 11:00 (LST) over Sistan. The diurnal PM_{10} variability in all seasons, except winter, is closely associated with the intensity of the wind speed measured at the Zabol meteorological station (see Fig. 16). This wind, being northerly in direction, carries large quantities of dust from the Hamoun dry lake bed. The mean diurnal wind speed variation is similar for all seasons; however, the wind favors the increase of aerosol load in summer and autumn (maximum PM_{10} for higher wind speeds). Note that the Hirmand River and some other ephemeral channels provide little water in winter and spring to the Hamoun lake beds. Therefore, in early summer, the Hamoun lakes are wet and at the end of summer and early autumn are always completely dried out. On the other hand, the Levar winds continue also in September and so high wind speeds cause huge dust storms.

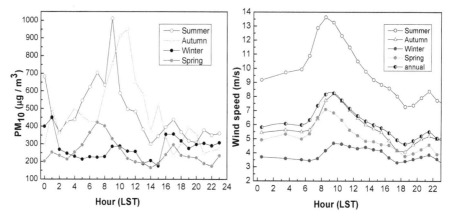

Figure 16. Mean hourly variation of the PM₁₀ (left panel) and wind speed (right panel) for each season in Zabol.

11. Mineralogical characteristics of dust

In order to understand the influence of dust on the atmospheric environment, climatic system and health and to establish effective remedial policies and strategies, it is regarded as necessary to investigate the compositional (chemical and mineralogy) characteristics of airborne and soil dust over Sistan. To the best of our knowledge there are currently no published studies about the geochemical characteristics and dust mineralogy in this region. Moreover, nearby locations, Bagram and Khowst in Afghanistan, were selected for analyzing the mineralogical dust composition, major and trace elements within the framework of the Enhanced Particulate Matter Surveillance Program (EPMSP) campaign [44]. Furthermore, mineralogical and geochemical characteristics of dust were recently examined at Khuzestan province in southwestern Iran [72].

In this Chapter, an overview of the geological-geochemical characteristics of airborne and soil dust in the Sistan region is given for airborne and soil samples collected during the period August 2009 to August 2010. The chemical constituents during major dust storms over the region are analyzed at two locations (Fig. 2), also investigating the relationship between the chemical constituents of the dust storms and those of the inferred (Hamoun) source soils. The mineralogy percentage composition averaged at all heights for each day is shown in Fig. 17a, b for stations A and B, respectively. The chemical formulas of the main mineralogical components are given in [96], as well as the chemical reactions of dust with atmospheric constituents and trace gases during the dust life cycle. The mineralogical composition corresponds to screened samples with diameter <75 μm and can constitute an indication of both regional geology and wind transported dust that is deposited in local soils [44].

Emphasizing the dust mineralogy at station A, it is seen that the airborne dust is mainly composed of quartz, which is the dominant component (26-40%) for all the days of observations. Calcareous particles, mainly consisting of calcite, are the second dominant mineralogical component over the site with average mass percentage of 22%, while micas (muscovite) contribute 13% and plagioclase (albite), 11%. The remaining components contribute much less to the dust mass, while chlorite (6.3%) is apparent in all dust samples for all days. The others, i.e. dolomite, enstatite, gypsum, halite, etc are present only in some samples with various percentages. It is quite interesting to note that quartz is much more common over Sistan than the feldspars (plagioclase, microcline and orthoclase).

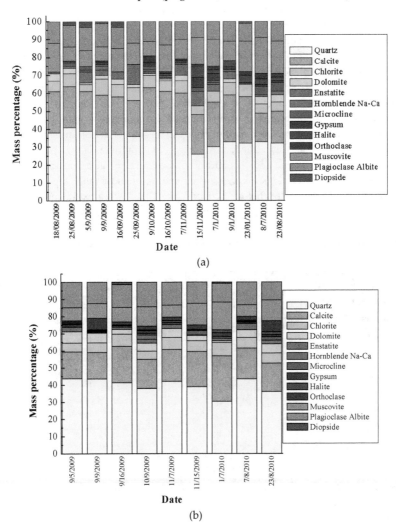

Figure 17. a. Mineralogical composition as obtained from XRD analysis for airborne dust samples collected on different days in station A. **b.** Same as in Figure 17a, but for the station B.

The mineralogical analysis for the 9-days recorded data at station B (Fig. 16b) shows more or less similar results to those obtained for station A and, therefore, any discussion will be given on their comparison (Fig. 19). The mineralogical composition has the same descending order as in station A, i.e. quartz (39.8±4.4%), calcite (18.8±3.5%), plagioclase (albite) (12.7±1.4%) and muscovite (10.1±3.2%). On the other hand, dust deposition may influence biogeochemical cycling in terrestrial ecosystems, while dust accumulation in soils can influence texture, element composition and acid neutralizing capacity [97-98]. Furthermore, the chemical and mineralogical composition of soil dust provides useful information about its provenance [99], radiative forcing implications [100] and human health effects [101]. For these reasons, in addition to the airborne dust samples, soil samples were collected at 16 locations around Sistan and Hamoun, at depths ranging from 0 to 5 cm from the soil crust. The results of soil sample mineralogy are summarized in Fig. 18. From an initial consideration of these results, it is established that the soil samples exhibit similar mineralogy to the airborne dust at both stations, thus suggesting similarity in sources for both airborne and soil dust. On the other hand, some soils in the Sistan region have been primarily formed from dust transported from the Hamoun lakes, presenting large similarities in mineralogy and chemical composition to airborne dust. However, atmospheric chemical reactions involving dust and aerosols of other types can alter the chemical characteristics of dust before its deposition [102]. Therefore, the mineralogy of the soil samples may differ significantly in comparison to the results obtained for airborne dust at stations A and B, since some of the soil samples (11 samples) were collected in the Hamoun dried lakes and others (five samples) around stations A and B.

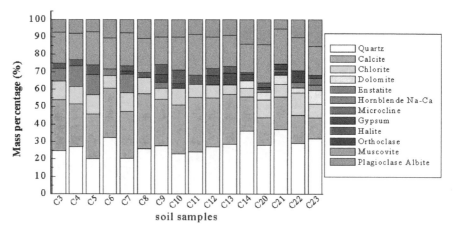

Figure 18. Mineralogical composition as obtained from XRD analysis for soil samples collected at various locations in the Hamoun Basin.

Figure 19 summarizes the results from the mineralogical analysis of samples taken at the two stations and from the soil samples, allowing a quantitative comparison between them. The vertical bars correspond to one standard deviation from the mean for both airborne and soil samples. The distance from the source region from whence dust is deposited also influences the particle size distribution, mineralogy and chemical composition of dust. Therefore, generally speaking, at local scales quartz clearly dominates with fractions up to ~50%, while as the distance from the dust source increases, feldspars (plagioclase, microcline) and phyllosilicate minerals (illite and kaolinite) present increased fractions [103, 42]. However, in our study the dust samples were all obtained within the same area and, therefore, are mineralogically similar. Nevertheless, station B, which is located closer to the Hamoun basin, the source of dust exposures, exhibits higher percentages of quartz, while station A (near to Zabol city) exhibits higher concentrations of calcite and muscovite compared to station B. On the other hand, the soil samples exhibit a lower mean percentage for quartz (27.7±4.7) and higher percentages for calcite, chlorite, halite and muscovite compared to the airborne samples.

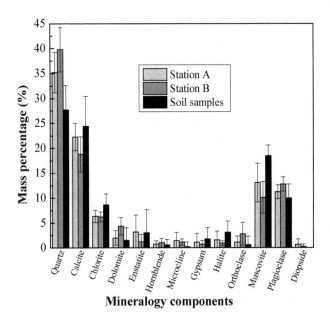

Figure 19. Average mineralogy components for airborne dust samples at stations A and B and for soil samples obtained at various locations in Hamoun Basin. The vertical bars express one standard deviation from the mean.

These mineralogical airborne dust and soil compositions, derived essentially from the Hamoun source region, reflect the composition of the material available from this provenance as well as the relevant grain size characteristics, enabling the wind storms to entrain this material into the lower atmosphere. While most of the minerals (quartz, feldspars of various types, muscovite) can easily be tied to basement-type lithology of generally gneissic-granitic character, others (chlorite, pyroxenes and hornblende) rather suggest mafic parent rocks, as can be inferred from basic mineralogical analysis [e.g.,104] . However, the calcite, dolomite, halite and gypsum suggest evaporate minerals, although both calcite and dolomite can also reflect alteration products of primary acid or mafic rock constituents. The inferred evaporate minerals reflect local derivation of salt from desiccating water bodies in the Hamoun lakes, originally formed from altered transported components via the Hirmand river system. Thus, the semi-quantitative mineral determinations for the airborne dust over the Sistan region support derivation of the particles from well weathered and well eroded (transported) argillaceous alluvium from the extensive Hirmand river system draining Afghanistan and terminating in the Hamoun Basin. The general geology of Afghanistan encompasses extensive terrains of both acidic and mafic rocks, while similar mineralogical composition of dust (i.e. dominance of quartz, but lower percentage of calcite) was found at the Bagram and Khowst sites located in eastern Afghanistan [44]. More specifically, they found that these sites are underlain by loess (wind deposited silt), sand, clay and alluvium containing gravel. As shown in Fig. 2, as well as in other studies [44, 51, 12], nearly the whole of Afghanistan is affected by the dust storms originating from Hamoun, since the dust plume usually follows a counter-clockwise direction, carrying wind-blown dust towards eastern Afghanistan. Similarly to our findings, the airborne dust at selected locations in southwestern Iran was found to be composed mainly from quartz and calcite, suggesting detritus sedimentary origin, followed by kaolinite and a minor percentage of gypsum [72]. Furthermore, [44] found that airborne dust samples derived from poorly drained rivers and lakes in central and southern Iraq contain substantial calcite (33–48%), quartz and feldspar with minor chlorite and clay minerals. Previous studies [105-106], have shown that silicate minerals (quartz, feldspars) and phyllosilicates (illite, kaolinite, smectite/montmorillonite clays, chlorite) dominate aeolian dust. Dust samples may also contain substantial amounts of carbonates, oxides, gypsum, halite and soluble salts, but the quantity and percentage of these minerals are quite variable from site to site.

12. Elemental composition of dust

Knowledge of the chemical composition of airborne dust is necessary for clarifying the likely source regions and is important for quantitative climate modelling, in understanding possible effects on human health, precipitation, ocean biogeochemistry and weathering phenomena [50]. Chemical analysis of dust provides valuable information about potentially harmful trace elements such as lead, arsenic and heavy metals (Co, Cr, Cu, Ni, Pb). On the other hand, the major-element and ion-chemistry analyses provide estimates of mineral components, which themselves may be hazardous to human health and ecosystems and which can act as carriers of other toxic substances. The chemical analysis of dust samples at both stations was performed via XRF analysis for the major oxides (Figs. 20a, b).

In general, the analysis reveals that all samples at both stations contain major amounts of SiO_2, mainly in the mineral quartz, variable amounts of CaO in the mineral calcite, plagioclase feldspar and to a limited extent in dolomite, as well as substantial Al_2O_3 concentrations. More specifically, average major elements of airborne dust at both stations indicate a predominant SiO_2 mass component (46.8 – 47.8%) with significant CaO (12-12.2%)

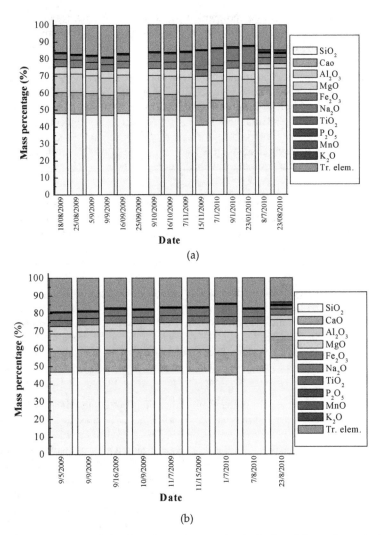

Figure 20. a. Major elements (oxides) for airborne dust samples obtained on different days at Station A analysed by means of XRF. **b.** Same as in Fig. 20a, but for the station B.

and Al_2O_3 (10.4-10.8%) contributions; a few percent of Na_2O (4.2-5.4%), MgO (4.3%) and total iron as Fe_2O_3 (3.8-4.1%), as well as trace amounts (<1%) of TiO_2, K_2O, P_2O_5 and MnO, while the remaining major elements (Cr_2O_3, NiO, V_2O_5, ZrO_2) were not detected by XRF analysis (Figs. 20a, b). When compared to various average shale analyses in the literature (Geosynclinal Average Shale and Platform Average Shale from [107] ; Average Shale from [108]; North American Shale Composite from [109], the Sistan dust is significantly depleted in SiO_2, Al_2O_3, K_2O and total Fe and significantly enriched in CaO, Na_2O and MgO. The MgO is largely contained in dolomite and, to a lesser extent, in clay minerals such as palygorskite and montmorillonite [78, 44]. These components can be ascribed to the importance of evaporate minerals such as calcite, dolomite, halite and gypsum (as also suggested by the mineralogical analysis) inferred to have come from the desiccation taking place in the Hamoun dust source region. Furthermore, the elevated values for the trace elements Cl, F and S (Table 5) support the latter postulate as it would be expected from an evaporate-rich source for deflation of dust [e.g.,110]. Similar to the elemental composition of dust over Sistan, [44] determined a high fraction of SiO_2 in silt, less CaO in calcite and slightly more Al_2O_3 in clay minerals at the Khowst site. At both Afghanistan sites (Bagram and Khowst), the SiO_2 was dominant with fractions of about 50-55%, followed by Al_2O_3, CaO and MgO.

By comparing the major elements of different dust storms, some interesting relationships have been found. More specifically, on days (e.g. 15/11/2009, 7/1/2010, 23/1/2010) (Fig. 20a) when airborne dust was relatively depleted in SiO_2, enhanced MgO and, particularly, Na_2O values were recorded. Conversely, when SiO_2 values were higher (e.g. 8/7/2010, 23/8/2010), both MgO and Na_2O contributions dropped. This suggests that certain intense dust storms were richer in evaporate source material (i.e., elevated MgO and Na_2O) coming from Hamoun dried lake beds, while others had more silica, reflecting weathered rock detritus from the Hirmand river and Afghanistan mountains. An explanation of these variable chemical compositions of dust samples is a real challenge, but it is postulated here that they may reflect local desiccation cycles and, possibly, even micro-climatic changes in the Hamoun-lakes dust source region. Excessive desiccation of the lakes would enhance potential evaporate minerals for deflation in drier periods, while in wetter periods, airborne dust would logically have been derived more from weathered fluvial detritus rich in SiO_2.

Figure 21 summarizes the results of the elemental compositions determined by XRF analysis at both stations. For comparison reasons, the mean elemental composition found for several sites in southwestern Iran (Khuzestan province) [71-72] is also shown. The vertical bars express one standard deviation from the mean. Concerning the major elemental oxides over Sistan, both stations exhibit similar results, well within the standard deviations, suggesting that the transported dust over Sistan is locally or regionally produced with similarity in source region. In contrast, the mean elemental composition of airborne dust over Khuzestan province exhibits remarkable differences from that over Sistan, revealing various source regions and dust mineralogy. More specifically, the SiO_2 percentage is significantly lower and highly variable over Khuzestan, which is also characterized by higher contributions of Na_2O, MgO and K_2O compared to Sistan. The dust storms over southwestern Iran may originate from local sources as well as being transported over medium- and long-ranges from different sources located in Iraq as well as in the Arabian Peninsula. A comparative study of the mineralogy and elemental composition of airborne dust at several locations in Iraq, Kuwait and the Arabian Peninsula [44] has shown significantly variable contributions, suggesting differences in overall geology, lithology and mineralogy of these regions. In further contrast, airborne dust over Sistan seems to have its individual characteristics originating from local and well-defined sources.

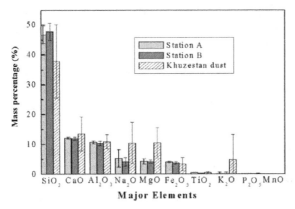

Figure 21. Average X-ray fluorescence (XRF) results for major dust elements in stations A and B. Similar results obtained in Khuzestan Province, southwestern Iran [71-72] are also shown for comparison reasons.

The Earth's crust is dominated by silicon and aluminum oxides. Numerous studies [78, 50 and references therein] reviewing the elemental composition of airborne dust over the globe report that mineral dust is composed of ~60% SiO_2 and 10-15% Al_2O_3. The contribution of other oxides, i.e. Fe_2O_3 (~7%), MgO (~2.5%) and CaO (~4%), are, in general, more variable depending on source location. Furthermore, the review study of [96] showed that airborne dust samples collected over the globe have fairly small variations in elemental composition. The CaO concentrations over Sistan are found to be much higher than those (5.5%) summarized in [96].

13. Trace elements

The average concentrations of trace elements (in ppm) in dust samples collected during major dust storms at stations A and B are summarized in Table 5, as obtained from XRF analysis. The results show that the dominant trace elements over Sistan are F and Cl, with the former being dominant in the vast majority of the dust events at station A. However, on two days (8/7/2010 and 23/8/2010) the Cl concentrations were extremely large, thus controlling the average value; there is a lack of observations at station B on 23/8/2010, thus the lower average Cl concentration. Note that on both these days, the SiO_2 component is large, while MgO and Na_2O are low (Fig. 20a). The dominance of chlorine indicates soil salinization in the Hamoun basin and along the Hirmand river and its tributaries. Furthermore, S exhibits higher concentration at station A, while for the other elements the concentrations between the two stations are more or less similar. The concentrations of potentially harmful and toxic elements, like Cs, Pb and As are, in general, low at both stations; however, Ba, Cr and Zn present moderate concentrations.

On the other hand, the analysis of the major element ratios provides essential knowledge of the dust chemical composition and source region. The ratios of Si/Al at stations A and B are similar (7.8±0.8 and 8.3±0.9, respectively), due to the presence of silicate and aluminosilicate minerals in most dust samples. The ratios of Mg/Al (0.90±0.16, 0.92±0.12), Ca/Al (3.09±0.19, 3.12±0.19) and Fe/Al (0.51±0.02, 0.49±0.01) at the two stations suggest contributions of clays and Ca-rich (calcite) minerals to the chemical compositions of the airborne dust. In contrast, the Fe/Al ratio is low over Sistan and is nearly half of that found for airborne dust over southwestern Iran and several locations over the globe [72], but is comparable to that found over central Asia [111]. It should be noted that this ratio remains nearly invariant, ranging from 0.47 to 0.54, for all the collected dust samples at both stations and can be a good surrogate for the dust source region, since any variation in Fe/Al mainly corresponds to variations in clay minerals and not to coating during dust transportation [50]. In contrast, the Ca/Al ratio exhibits the highest variations from sample to sample (2.80-3.46), since it is influenced by particle size, with higher values as particle-size increases [72]. Synoptically, all the ratio values and the low standard deviations suggest similarity in geochemical characteristics over Sistan and a uniform source of airborne dust.

Trace Elements		
Parts per million (ppm)	Station A	Station B
Cl	28670	15047
F	13938	13456
S	4445	2506
Ba	210	253
Sr	154	125
Zr	83	76
Cr	70	84
V	69	69
Zn	57	51
La	30	32
Rb	24	19
Ni	18	16
Ce	17	16
Cs	14	13
Sc	11	11
Cu	11	11
Pb	10	10

Table 5. Average X-ray fluorescence (XRF) values for trace elements of airborne dust for stations A and B.

14. Conclusions

The present Chapter focused on shedding light on the dust loading, PM concentrations, physical and chemical composition of dust in the Sistan region, southeastern Iran, which constitutes a major dust source region in south west Asia. Sistan region is a closed topographic low basin surrounded by arid and rocky mountains, while its northern part drains the Hilmand river, thus constituting a wetland area known as Hamoun. Hamoun lakes complex have an area about 4500 Km^2 with water volumes of 13025 million m^3 and play the role of a "water cooler" for the region when they are full of water as the severe winds blow across the lakes.

Severe droughts over the past decades, especially after 1999, have caused desiccation of the Hamoun lakes, leaving a fine layer of sediment that is easily lifted by the wind and therefore making the basin one of the most active sources of dust in south-west Asia [56, 50]. The strong "Levar", especially during the summer season, blows fine sands off the exposed lake bed and deposits this detritus within huge dune bed forms that may cover a hundred or more villages along the former lakeshore. As a consequence, the wildlife around the lake has been negatively impacted and fisheries have been brought to a halt, which also implies an impact on society. The drainage of the Hamoun wetlands, in association with the intense Levar winds in summer, is the main factor responsible for the frequent and massive dust storms over the Sistan region. Analysis of water surface in combination with dust storms showed that the Hamoun dried beds, particularly Hamoun Saberi and Baringak, have a dramatic effect on dust storms as sources of aerosols.

Systematic PM concentrations were measured in Zabol city, affected by the Sistan dust storms, covering the period September 2010 to August 2011. The results show that the PM_{10} concentrations were considerably higher than the corresponding European Union air quality annual standard. The analysis of the daily PM concentrations showed that the air quality is affected by dust storms from the Sistan desert, which may be very intense during summer. Hamoun, as an intense dust source region, caused a dramatic increase in PM_{10} concentrations and a deterioration of air quality (65% of the days were considered unhealthy for sensitive people and 34.9% as hazardous).

Dust loading from the Hamoun basin appears to have a significant contributing influence on the development of extreme dust storms, especially during the summer days. This influence firstly seems to depend on the intensity and duration of dust storms, and secondly, on the distance from the source region, the wind speed and altitude. The grain-size distribution of the dust loading was strongly influenced by the distance from the dust source, since grain sizes shifted to larger values towards station B that is closer to the Hamoun basin. Furthermore, the particle size distribution exhibited a shift towards lower values as the altitude increases, with this feature seen to be more obvious amongst larger size particles, while the frequency of particles below 2.5 μm seemed not to be affected by altitude. In general, the regional dust loading and characteristics are subject to significant spatio-temporal variability. This finding necessitates more systematic observations at as many locations as possible around the Hamoun basin in order to improve the understanding of forcing dynamics, transport mechanisms as well as to quantify the dust amounts emitted from the Hamoun basin.

To fully understand mineral dust characteristics and the potential impact on human health, dust mineralogy and geochemical properties were examined in the Sistan region by collecting airborne samples at two stations and soil samples from several locations over

Sistan and the Hamoun basin. The Sistan region is an ideal site to study the nature of dust storms as it receives large amounts of fine alluvial material from the extended Hirmand river system draining much of the Afghanistan highlands, which comprise crystalline basement rocks, Phanerozoic sediments and extensive flood basalts. As a result, large quantities of quartz-rich, feldspar- and mica-bearing silt, as well as mafic material from flood basalt sources and carbonate minerals from dolomites, are transported to the Hamoun wetlands in northern Sistan. Due to droughts at Hamoun and large irrigation projects upstream on the river catchment, extensive desiccation has occurred in the wetlands resulting in large dry lake environments. These have produced large quantities of evaporate minerals to add to the alluvial silts, and the combination of these materials provides the provenance for the airborne dust.

Dust aerosol characterization included chemical analysis of major and trace elements by XRF and mineral analysis by XRD. The results showed that quartz, calcite, muscovite, plagioclase and chlorite are the main mineralogical components of the dust, in descending order, over Sistan, and were present in all the selected airborne dust samples. In contrast, significantly lower percentages for enstatite, halite, dolomite, microcline, gypsum, diopside, orthoclase and hornblende were found, since these minerals occurred only in some of the samples at both stations. On the other hand, SiO_2, CaO, Al_2O_3, Na_2O, MgO and Fe_2O_3 were the major elements characterising the dust, while large amounts of F, Cl and S were also found as trace elements. The mineralogy and chemical composition of airborne dust at both stations were nearly the same and quite similar to the soil samples collected at several locations downwind. This suggests that the dust over Sistan is locally emitted, i.e. from the Hamoun basin, and in a few cases can also be long-range transported to distant regions. On the other hand, individual dust storms showed significant differences between either evaporite-dominated aerosols or those characterized by deflation from alluvial silts. These possibly reflect either localized climatic cyclicity or desiccation cycles. However, in some cases the soil samples showed poor comparisons with aerosol compositions, suggesting that dynamic sorting, soil-forming processes and climatic influences, such as rainfall, altered the mineralogy and chemistry in these partially eolian deposits. Sistan is also an ideal site for studying dust storms and enrichment factors relative to crustal norms; the latter factors suggest that the dust is essentially of crustal rather than anthropogenic origin. SEM analyses of the samples indicated that airborne dust has rounded irregular, prismatic and rhombic shapes, with only the finer particles and a few examples of the coarser dust being spherical.

Author details

Alireza Rashki and C.J.deW. Rautenbach
Department of Geography, Geoinformatics and Meteorology,
Faculty of Natural and Agricultural Sciences, University of Pretoria, Pretoria,
South Africa
Department of Drylands and Desert Management, Faculty of Natural Resources and Environment,
Ferdowsi University of Mashhad, Mashhad,
Iran

Dimitris Kaskaoutis
Research and Technology Development Centre, Sharda University, Greater Noida, India

Patrick Eriksson
*Department of Geology, Faculty of Natural and Agricultural Sciences, University of Pretoria,
Pretoria, South Africa*

15. References

[1] Mahowald, N.M., Bryant, R.G., Corral, J., and Steinberger, L. 2003. Ephemeral lakes and desert dust sources, Geophysical Research Letters,. 30, NO. 2, 1074

[2] Engelstaedter, S., Tegen, I., Washington, R., 2006. North African dust emissions and transport. Earth Science Reviews 79, 73-100.

[3] Koren, I., Y. J. Kaufman, R. Washington, M. C. Todd, Y. Rudich, J. V. Martins, and D. Rosenfeld 2006, The Bodele depression: a single spot in the Sahara that provides most of the mineral dust to the Amazon forest, Environmental Research Letters, 1(1).

[4] Baddock, M.C., Bullard, J.E., Bryant, R.G., 2009. Dust source identification using MODIS: A comparison of techniques applied to the Lake Eyre Basin, Australia. Rem. Sens. Environ. 113, 1511-1528.

[5] Orlovsky, L., Orlovsky, N., Durdyev, A. (2005) Dust storms in Turkmenistan. J Arid Environ 60:83–97

[6] Breckle, S.W., Wucherer,W., Liliya A. Dimeyeva, L.A., Nathalia P. Ogar, N.P. 2012. Aralkum - A Man-Made Desert: The Desiccated Floor of the Aral Sea (Central Asia), Springer, pp: 486

[7] Reheis M (1997) Dust deposition of Owens (dry) Lake, 1991–1994: preliminary findings. J Geophys Res 102:25999–26008

[8] Reheis, M., Budahn, J.R., Lamothe,P.J., and Reynolds, R.L.,, 2009. Compositions of modern dust and surface sediments in the Desert Southwest, United States, Journal Of Geophysical Research, 114, F01028, doi:10.1029/2008JF001009

[9] Whitney, J. W., 2006. Geology, Water, and Wind in the Lower Helmand Basin, Southern Afghanistan U.S. Geological Survey, Reston, Virginia, Retrieved 2010-08-31

[10] United Nations Environment Programme (UNEP). 2006. History of Environmental Change in the Sistan Basin Based on Satellite Image Analysis:1976 – 2005. P: 60

[11] Miri A, Moghaddamnia A, Pahlavanravi A, Panjehkeh N (2010) Dust storm frequency after the 1999 drought in the Sistan region, Iran. Clim Res 41:83-90

[12] Rashki, A., Kaskaoutis, D.G., Rautenbach, C.J.deW., Eriksson, P.G.,Giang, M, Gupta, P., 2012b. Dust storms and their horizontal dust loading in the Sistan region, Iran. Aeolian Research doi:10.1016/j.aeolia.2011.12.001

[13] IPCC, (Intergovernmental Panel on Climate Change), 2001. Climate Change 2001: The Scientific Basis. In Contribution of Working Group I to the Third Assessment Report of the Intergovernmental Panel on Climate. J.T. Houghton et al.; Eds, Cambridge Univ. Press, New York, USA.

[14] IPCC, 2007. Climate Change 2007: Synthesis Report. Contribution of Working Groups I, II and III to the Fourth Assessment Report of the Intergovernmental Panel on Climate Change. In: Core Writing Team, Pachauri, R.K., Reisinger, A. (Eds.), Geneva, Switzerland, p. 104.

[15] Tegen, I., Fung, I., 1994. Modeling of mineral dust in the atmosphere: sources, transport, and optical thickness. J. Geophys. Res. 99, 22897–22914.

[16] Tegen, I., Lacis, A.A., Fung, I., 1996. The influence on climate forcing of mineral aerosols from disturbed soils. Nature, 380, 419-422.

[17] Dunion, J.; Velden, C. 2004. The impact of the Saharan air layer on Atlantic tropical cyclone activity. Bull. Amer. Meteor. Soc. 85, 353-365.

[18] Prasad, A.K., Singh, S., Chauhan, S.S., Srivastava, M.K., Singh, R.P., Singh, R., 2007. Aerosol radiative forcing over the Indo-Gangetic Plains during major dust storms. Atmos. Environ. 41, 6289-6301.

[19] Singh, R.P., Prasad, A.K., Kayetha, V.K., Kafatos, M., 2008. Enhancement of oceanic parameters associated with dust storms using satellite data. J. Geophys. Res., 113, C11008, doi:10.1029/2008JC004815.

[20] Patadia, F., Yang, E.-S., Christopher, S.A., 2009. Does dust change the clear sky top of atmosphere shortwave flux over high surface reflectance regions? Geophys. Res. Lett., 36, L15825, doi:10.1029/2009GL039092.

[21] Mahowald, N., Baker, A., Bergametti, G., Brooks, N., Duce, R., Jickells, T., Kubilay, N., Prospero, J., Tegen, I., 2005. Atmospheric global dust cycle and iron inputs to the ocean. Global Biogeochem. Cycles Vol.19 (No.4), GB4025. doi:10.1029/2004GB002402.

[22] Claquin T, Schulz M, Balkanski Y, Boucher O (1998) Uncertainties in assessing radiative forcing by mineral dust. Tellus Ser B – Chem Phys Meteorol 50: 491–505

[23] Broecker, W.S., 2000. Abrupt climate change: causal constraints provided by the paleoclimate record. Earth Science Reviews 51, 137–154.

[23] Nriagu, J.O., and Pacyna, J.M. 1988. Quantitative assessment of worldwide contamination of air, water and soils with trace metals. Nature, 333: 134–139

[25] Husar, R.B., Prospero, J.M., Stowe, L.L., 1997. Characterization of tropospheric aerosols over the oceans with the NOAA advanced very high resolution radiometer optical thickness operational product. Journal of Geophysical Research 102, 16,889–16,909.

[26] Nriagu, J.O., 1988. A salient epidemic of environmental metal poisoning? Environmental Pollution 50, 139–161.

[27] Dockery, D., Pope, C.A., Xiping, X., Spengler, J., Ware, J., Fay, M., Ferris, B., Spiezer, F., 1993. An association between air pollution and mortality in six US cities. New England Journal of Medicine 329 (24), 1753–1759.

[28] Dockery, D., Pope, A., 1996. Epidemiology of acute health effects: summary of time-series studies. In: Particles in Our Air: Concentrations and Health Effects (Wilson R, SpenglerJD, eds). Harvard University Press, Cambridge, MA, 123-147.

[29] Pope, C. A., 2000. Epidemiology of fine particulate air pollution and human health: Biologic mechanisms and who's at risk?, Environmental Health Perspectives, 108, 713-723.

[30] Schwartz J. 2004. Air pollution and children's health. Pediatrics 113:1037-1043.

[31] Pozzi,R., B.D. Berardis, B.D., Paoletti, L., Guastadisegni, C., 2005. Winter urban air particles from Rome (Italy): effects on the monocytic–macrophagic RAW264.7 cell line, Environ. Res. 99: 344–354.

[32] Chakra, O.R.A., Joyeux, M., Nerriere, E., Strub, M.P., Zmirou-Navier, D., 2007. Genotoxicity of organic extracts of urban airborne particulatematter: an assessment within a personal exposure study, Chemosphere 66:1375–1381.

[33] Brook, R.D., Urch, B., Dvonch, J.T., Bard, R.L., Speck, M., Keeler, G., Morishita, M., Marsik, F.J., Kamal, A.S., Kaciroti, N., Harkema, J., Corey, P., Silverman, F., Gold, D.R., Wellenius, G., Mittleman, M.A., Rajagopalan, S., and Brooky, J.R., 2009. Insights into the mechanisms and mediators of the effects of air pollution exposure on blood pressure and vascular function in healthy humans. Hypertension, 54(3), 659-667.

[33] Sivagangabalan, G., Spears, D., Masse, S., Urch, B., Brook, R.D., Silverman, F., Gold, D.R., Lukic, K.Z., ; Speck, M., Kusha, M., Farid, T., Poku, K., Shi, E., Floras, J., Nanthakumar, K., 2010. Mechanisms of Increased Arrhythmic Risk Associated With Exposure to Urban Air Pollution. Circulation, 122, A17901

[35] Bhaskaran ,k. Wilkinson, p and smeeth, L., 2011. Cardiovascular consequences of air pollution: what are the mechanisms? Heart, 97, 519-520.

[36] Nastos, T., Athanasios. G., Michael, B., Eleftheria, S.R., Kostas, N.P., 2010. Outdoor particulate matter and childhood asthma admissions in Athens, Greece: a time-series study. Environmental Health , 9:45, 1-9.

[37] Wilson, A. M.; Salloway, J. C.; Wake, C. P.; Kelly, T., 2004. Air pollution and demand for hospital services: A review. Environ. Int., 30, 1109-1018

[38] Larney, F. J., Leys, J. F., Muller, J. F., & McTainsh, G. H. (1999). Dust and endosulfan deposition in cotton-growing area of Northern New South Wales, Australia. Journal of Environmental Quality, 28, 692–701.

[39] Lioy, P. J., Freeman, N. C. G., & Millette, J. R. 2002. Dust: A metric for use in residential and building exposure assessment and source characterization. Environmental Health Perspectives, 110, 969–983.

[40] Wood, W. W., & Sanford, W. E. (1995). Eolian transport, saline lake basins and groundwater solutes. Water Resources Research, 31, 3121–3129.

[41] Ganor, E., Foner, H.A., & Gravenshorst, G. (2003). The amount and nature of the dust on Lake Kinneret (the Sea of Galilee), Israel: flux and fractionation. Atmospheric Environment, 37, 4301–4315.

[42] Lawrence, C.R. and Neff, J.C. 2009. The contemporary physical and chemical flux of aeolian dust: A synthesis of direct measurements of dust deposition. Chemical Geology 267: 46-63.

[43] Chow, J. C., Watson, J. G., Ashbaugh, L. L., & Magliano, K. L. (2003). Similarities and differences in PM10 chemical source profiles for geological dust from the San Joaquin Valley, California. Atmospheric Environment, 37, 1317– 1340.

[44] Engelbrecht J.P, McDonald E.V, Gillies, J.A, Jayanty RKM, Casuccio, G., Gertler, A.W. 2009, Characterizing mineral dusts and other aerosols from the Middle East—Part 1: Ambient sampling.Inhalation Toxicology 21:297 -326

[45] Mishra, S.K., Tripathi, S.N., 2008. Modeling optical properties of mineral dust over the Indian Desert. J. Geophys. Res., 113, D23201, doi:10.1029/2008JD010048.

[46] Wilson, W.E., Chow, J.C. Claiborn, C. Fusheng, W. Engelbrecht, J. and Watson, J.G., 2002, Monitoring of particulate matter outdoors. Chemosphere, 49, 1009–1043

[47] Thurston, G.D., Spengler, J.D., 1985, A quantitative assessment of source contributions to inhalable particulate matter pollution in metropolitan Boston. Atmospheric Environment, 19, 9–25.

[48] Biscaye, P. E., & Grousset, F. E. (1998). Ice-core and deep-sea records of atmospheric dust. In A. Busacca (Ed.), Dust aerosols, loess soils, and global change (pp. 101–103). College Agric. Home Econ. Misc. Publ. MISC0190 (1998). Pullman, WA: Washington State Univ.

[49] Shaw, G. E. (1980). Transport of Asian desert aerosol to the Hawaiian Islands. Journal of Applied Meteorology, 19, 1254–1259.

[50] Goudie, A.S., Midelton, N.J., 2006. Desert dust in the global system, Springer. 2006, pp287.

[51] Alam, K., Qureshi, S., Blaschke, T., 2011. Monitoring Spatio-temporal aerosol patterns over Pakistan based on MODIS, TOMS and MISR satellite data and a HYSPLIT model. Atmos. Environ., 45, 4641-4651.

[52] Rashki, A., Rautenbach, C.J.deW., Eriksson, P.G., Kaskaoutis, D.G., Gupta, P., 2012a. Temporal changes of particulate concentration in the ambient air over the city of Zahedan, Iran. Air Quality, Atmosphere & Health. DOI: 10.1007/s11869-011-0152-5

[53] Miri, A., Ahmadi, H., Ghanbari, A., Moghaddamnia, A., 2007. Dust Storms Impacts on Air Pollution and Public Health under Hot and Dry Climate. Int. J. Energy and Environ. 2, 1.

[54] Selinus, 2010, Physics and modelling of wind erosion, springer

[55] Ranjbar, M., and Iranmanesh, F. 2008. Effects of "Drought" on "Wind Eroding and Erosion" in Sistan Region with use of Satellite Multiple Images. WSEAS, ISSN: 1792-4294.

[56] Middleton, N. J., 1986. Dust storms in the Middle East. J. Arid Environ. 10, 83-96.

[57] Esmaili, O and Tajrishy, M, 2006, Results of the 50 year ground-based measurements in comparison with satellite remote sensing of two prominent dust emission sources located in Iran, Proc. SPIE 6362, 636209; http://dx.doi.org/10.1117/12.692989

[58] Moghadamnia, A., Ghafari, M.B., Piri, J., Amin.S., Han. D., 2009. Evaporation estimation using artificial neural networks and adaptive neuro-fuzzy inference system techniques. Adv. Water Resources 32, 88–97.

[59] British Geological Survey; http://bgs.ac.uk/

[60] Tirrul, R., Bell, I. R., Griftis, R. J., Camp, V. E., 1983. The Sistan suture zone of eastern Iran. Bull. of Geological Soc. of America, 94, 134-150.

[61] Wittekindt, Hans, and Weippert, D., compilers, 1973, Geologische Karte von Zentral-und Sudafghanistan: Hannover, Bundesanstalt fur Bodenforschung, 4 sheets, scale 1:500,000

[62] O'Leary, D.W., and Whitney, J.W., 2005a, Geological map of quadrangles 3062 and 2962, Charbuiak (609), Khannesin (610), Gawdezereh (615) and Galach (616), Afghanistan: U.S. Geological Survey Open-File Report 2005–1122A, scale 1:250,000.

[63] O'Leary, D.W., and Whitney, J.W., 2005b, Geological map of quadrangles 3164, Lashkargah (605) and Kandahar (606), Afghanistan: U.S. Geological Survey Open-File Report 2005–1119A, scale 1:250,000.

[64] Barlow, Matthew, Cullen, H., and Lyon, B., 2002, Drought in central and southwest Asia: La Niña, the warm pool, and Indian Ocean precipitation: Journal of Climate, v. 15, p. 697–701.

[65] Partow, Hassan, 2003, Sistan oasis parched by droughts, in Atlas of global change: United Nations Environmental Programme, Oxford University Press, p. 144–145.

[66] Ekhtesasi, M.R., Daneshvar, M.R., Abolghasemi, M., Feiznia, S., and Saremi Naeini, M.A. Measurement and Mapping of Aeolian Sand Flowthrough Sediment Trap Method (Case Study: Yazd-Ardakan Plain), Journal of the Iranian Natural Res., Vol. 59, No. 4, 2007, pp. 773-781

[67] Ekhtesasi, M.R., (2009). National project of monitoring of wind erosion and sand storm in Iran, forests and range and watershed organization of Iran (Persian language)

[68] Rietveld H.M. 1969. A profile refinement method for nuclear and magnetic structures. J. ppl. Crystallogr.2:65–71.

[69] Sturges, W.T., Harrison, R.M. , Barrie L.A., 1989. Semi-quantitative XRD analysis of size fractionated atmospheric particles, Atmospheric Environment, 23 , 1083–1098

[70] Caquineau S, Magonthier MC, Gaudichet A, Gomes L. 1997. An improved procedure for the X-ray diffraction analysis of low-mass atmospheric dust samples. Eur. J. Mineral. 9: 157–166.

[71] Zarasvandi, A., 2009, Environmental impacts of dust storms in the Khuzestan province, Environmental Protection Agency (EPA) of Khuzestan province, Internal Report, 375p

[72] Zarasvandi,A., Carranza, E.J.M., Moore3, F., Rastmanesh, F. Spatio-temporal occurrences and mineralogical-geochemical characteristics of airborne dusts in Khuzestan Province (southwestern Iran), Journal of Geochemical Exploration, 111, 3, 138–151

[73] Hossenzadeh, S.R., 1997 One hundred and twenty days winds of Sistan. Iran Iranian Journal of Research Geography 46,103–127

[74] Goudie, A.S., Middleton, N.J., 2000. Dust storms in south west Asia. Acta Universitatis Carolinae, Supplement 73-83.

[75] Prasad, A. K.; Singh, R. P. & Kafatos, M. (2006), Influence of coal based thermal power plants on aerosol optical properties in the Indo-Gangetic basin, Geophysical Research Letters, 33(5).

[76] Kaskaoutis, D. G., Kambezidis, H. D.,Nastos, P. T., and Kosmopoulos, P. G., 2008. Study on an intense dust storm over Greece. Atmospheric Environment, 42, 6884-6896.

[77] Adamopoulos, A.D., Kambezidis, H.D., Kaskaoutis, D.G., Giavis, G., 2007. A study of particle size in the atmosphere of Athens, Greece retrieved from solar spectral measurements. Atmos. Res. 86, 194-206.

[78] Goudie, A. S., and Middleton.N.J., 2001. Saharan dust storms: nature and consequences. Earth-Science Reviews 56: 179–204.

[79] Knippertz, P., Ansmann, A., Althausen, D., et al., 2009. Dust mobilization and transport in the northern Sahara during SAMUM 2006 - A meteorological overview. Tellus B 61, 12-31.

[80] Zhang, D.E., 1985. Meteorological characteristics of dust fall in China since the historic times. In: Liu, T.S. (Ed.), Quaternary Geology and Environment of China. China Ocean Press, Beijing, pp. 45–56

[81] Offer, Z.Y., Goossens, D. 2001. Airborne particle accumulation and composition at different locations in the Negev desert, Zeitschrift für Geomorphologie, 45 (2001), pp. 101–120

[82] Dong, Z., Man, D., Luo, W., Qian, Q., Wang, J., Zhao, M., Liu, Sh., Zhu, G., Zhu, Sh. 2010. Horizontal aeolian sediment flux in the Minqin area, a major source of Chinese dust storms, Geomorphology 116 58–66

[83] Zhao, M., Zhan, K.J., Qiu, G.Y. Fang, E.T., Yang, Z.H., Zhang, Y.C., Ai De Li, A.D. 2011. Experimental investigation of the height profile of sand-dust fluxes in the 0–50-m layer and the effects of vegetation on dust reduction, Environ Earth Sci, 62:403–410

[84] Zhang, Z., Dong. Z., and Zhao. A. 2011. The characteristics of aeolian sediment flux profiles in the south-eastern Tengger Desert, Sedimentology (2011) 58, 1884–1894

[85] Zender, C.S., Miller, R.L., Tegen, I., 2004. Quantifying mineral dust mass budgets: terminology, constraints, and current estimates. EOS, Transactions, American Geophysical Union 85 (48), 509–512.

[86] Environmental Protection Agency (EPA), 1999. Guideline for reporting the daily air quality-air quality index (AQI). EPA-1999-454/R-99-010.

[87] Mohan, M., Kandya, A., 2007. An analysis of the annual and seasonal trends of Air Quality Index of Delhi. Environ. Monit. Assess. 131, 267–277.

[88] Larissi, I.K., Antoniou, A., Nastos, P.T., Paliatsos, A.G., 2010a. The role of wind in the configuration of the ambient air quality in Athens, Greece. Fres. Environ. Bull. 19, 1989-1996.

[89] Ozer, P., Bechir, M., Laghdaf, O.M., Gassani, J., 2006. Estimation of air quality degradation due to Saharan dust at Nouakchott, Mauritania, from horizontal visibility data. Water Air Soil Pollut, 178:79–87

[90] Triantafyllou, A.G., Evagelopoulos, V., Zoras, S., 2006. Design of a web-based information system for ambient environmental data. Journal of Environmental Management, 80, 230-236.

[91] Mousavi, G., Nadafy, R.K., 2000. Comparative study of air quality in Tehran in 1997 and 1998, The third National Conference on Environmental Health. Kerman. 47-50(In Persian)

[92] Cheraghi, M., 2001. Evaluation and comparison of air quality in Tehran and Isfahan in 1999 and offering solutions to improve It, MSc thesis of Environment, Natural Resources Faculty of Tehran University, 150 Pages (In Persian)

[93] Ardakani, S.Q., 2006. Determine the air quality in Iran in (2004).Journals of Environmental Science and Technology, Volume 8, Number 4, winter, p. 38-33(In Persian)

[94] Bartzokas, A., Kassomenos, P., Petrakis, M., Celessides, C., 2004. The effect of meteorological and pollution parameters on the frequency of hospital admissions for cardiovascular and respiratory problems in Athens. Indoor and Built Environ. 13, 271-275.

[95] Samoli, E., Kougea, E., Kassomenos, P., Analitis, A., Katsouyanni, K., 2011. Does the presence of desert dust modify the effect of PM10 on mortality in Athens, Greece? Sci. Total Environ. 409, 2049-2054.

[96] Usher,C.R., Michel, A.E., VH Grassian, V.H. 2003 Reactions on mineral dust, Chemical Review, 103 (12) , 4883–4940

[97] Larssen, T., Carmichael, G.R., 2000. Acid rain and acidification in china: the importance of base cation deposition. Environmental Pollution 110 (1), 89–102.

[98] Muhs, D.R., Benedict, J.B., 2006. Eolian additions to late quaternary alpine soils, Indian Peaks Wilderness Area, Colorado Front Range. Arctic Antarctic and Alpine Research 38 (1), 120–130.

[99] Yang, X.P., Zhu, B.Q., White, P.D. 2007, Provenance of aeolian sediment in the Taklamakan Desert of western China, inferred from REE and major-elemental data. Quaternary International 175, 71–85.

[100] Sokolik, I.N., Toon, O.B., Bergstrom, R.W., 1998. Modeling the radiative characteristics of airborne mineral aerosols at infraredwavelengths. Journal of Geophysical Research-Atmospheres 103 (D8), 8813–8826

[101] Erel,Y., Dayan,U., Rabi, R., Rudich,Y., Stein,M., 2006. Trans boundary transport of pollutants by atmospheric mineral dust. Environmental Science & Technology 40 (9), 2996–3005.

[102] Dentener, F.J.; Carmichael, G.R.; Zhang, Y.; Lelieveld, J.; Crutzen, P.J. 1996. Role of mineral aerosol as a reactive surface in the global troposphere. J Geophys. Res. 101(17): 22869-22889.

[103] Arnold, E., Merrill, J., Leinen, M., and King, J.: 1998. The effect of source area and atmospheric transport on mineral aerosol collected over the north pacific ocean, Global Planet Change, 18, 137–159

[104] Deer, W. A., Howie, R. A. and Zussman, J.: 1966, An Introduction to the Rock Forming Minerals, Longmans, pp. 528.

[105] Schütz,L., Sebert,M., 1987, Mineral aerosols and source identification, Journal of Aerosol Science, 18 (1) , pp. 1–10

[106] Reheis, M. C., and Kihl, R., 1995, Dust deposition in southern Nevada and California, 1984–1989: Relations to climate, source area, and source lithology: Journal of Geophysical Research, v. 100, p. 8893–8918.

[107] Wedepohl, K.H., 1971. Environmental influences on the chemical composition of shales and clays. In: Ahrens, L.H., Press, F., Runcorn, S.K., Urey, H.C. (Eds.), Physics and Chemistry of the Earth. Pergamon, Oxford, UK, pp. 307–331

[108] Clarke, F. W., 1924. Bull. U.S. geol. Surv., 700, p. 29.

[109] Gromet, L P, Dymek, R.F., Haskin, L.A., and Korotev, R.L. 1984) The "North American shale composite": Its compilation, major and trace element characteristicsG. eochimica et CosmochimicaA cta, 48,2469- 2482

[110] Talbot M.R. and Allen P.A. 1996. Lakes. in Sedimentary Environments: Reading H.G. (ed), Processes, Facies and Stratigraphy, Blackwell: Oxford, 83–124.

[111] Kreutz, K. and Sholkovitz, E. 2000. Major element, rare earth element, and sulfur isotopic composition of a high-elevation firn core: Sources and transport of mineral dust in central Asia. Geochemistry, Geophysics, Geosystems 1(11).1525-2027.

Chapter 7

Interaction Between Aerosol Particles and Maritime Convective Clouds: Measurements in ITCZ During the EPIC 2001 Project

J.C. Jiménez-Escalona and O. Peralta

Additional information is available at the end of the chapter

1. Introduction

Atmospheric particles interact directly with solar radiation extinguishing part of it and decreasing the amount of radiation that reaches the Earth's surface. This effect produces a change in the local radiative balance. On the other hand, it also presents an indirect effect on the interaction with radiation because these particles are an important element in the formation and development of clouds influencing their optical properties and the length of residence.

There are studies that have focused primarily on understanding and explaining the role of atmospheric particles in the formation and evolution of clouds. They have shown enough information able to explain those processes in theory (e.g. Pruppacher and Klett, 1997). And they have been validated with experimental works (e.g. Twomey, 1991; Raga and Jonas, 1993 a, b).

However, other issues of importance that do not yet have much information are the processes that modify the properties of atmospheric particles interacting with the cloud and the effects of changes in the environment. Particles increase their average size in regions of high relative humidity (RH) near the clouds (Baumgardner et al, 1996; Baumgardner and Clarke, 1998). Other studies show that the clouds condensation nuclei (CCN) are relatively higher in regions where a cloud is evaporated compared with places without clouds (DeFelice and Saxena, 1994; DeFelice and Cheng, 1998; Naoki et al, 2001). Also, the composition of atmospheric particles may change resulting from chemical reaction in aqueous state (Hegg et al, 1980; O'Dowd et al, 2000, Alfonso and Raga, 2002). Aerosol particles used as CCN show an increase in size after the cloud drops are evaporated (Hobbs, 1993). Towmey (1974) and Albrecht (1989) showed that changes in particles concentrations in an area influenced by polluted sources modify the local cloud albedo and its life time, inhibiting the rain process (Rosenfeld, 1999).

So, there are many processes that change the properties of atmospheric particles and their interactions with clouds. This chapter is focused on identifying and assessing the main processes involved with particles in the vicinity of maritime convective clouds at the Inter-Tropical Convergence Zone (ITCZ).

1.1. Clouds in the Inter-Tropical Convergence Zone (ITCZ)

The ITCZ is one of the most important weather systems in the Tropics. The area shows a decisive influence on the characterization of different clime and weather conditions in tropical region. The ITCZ is characterized by several interactions between the ocean and the atmosphere, identified as:

- Area of confluence of trade winds from the Northeast and Southeast.
- Area where the Equatorial depression is located due to an increased incidence of solar radiation and the presence of convective phenomena.
- Area of maximum temperature on the sea surface
- Maximum mass convergence zone
- Area that has the band of maximum coverage of convective clouds.

Therefore, the ITCZ is represented as a line of clouds of deep convection extending across the Atlantic and Pacific oceans, located between the 5° and 10° N (Holton, 1992). The band moves depending on the season, always matching in areas with high solar intensity or where the sea surface has higher temperatures. The clouds movement is towards the Southern hemisphere between September and February, and in the opposite direction in the next few months until the end of the summer in the Northern Hemisphere. However, just at the north of the Ecuador, the ITCZ movements are lower (Wallace and Hobbs, 1977). In those areas, the rain intensifies with solar heating. An exception occurs with El Niño-Southern Oscillation (ENSO), and the ITCZ is deflected towards where the ocean surface increases its temperature. A special feature of the region is the presence of a warm water pool, nearby the coast of Mexico, located in the ITCZ between 12° and 8° N from September to November. The presence of high temperatures at the ocean surface promotes a greater amount of moisture and increases convection effects, which result in larger vertical clouds.

1.2. Development of convective clouds

A convective cloud is formed when a mass of moist air acquires buoyancy due to the increase in surrounding temperature; with this process, the presence of atmospheric instability helps to lift the air masses.

Once the air mass reaches the saturation point, water vapor condenses on the CCN. The change of physical state releases latent heat that is absorbed by the air increasing the buoyancy forces. From now on, the cloud experiences a violent vertical development reaching the maximum height generated by the strong temperature gradient between the cloud core and the environment surrounds it. However, the growing process will be affected by dry air entrainment on the cloud's side walls, inhibiting the vertical development of convective cloud (Squires, 1958; Emanuel, 1982).

The air entrainment dilutes and evaporates droplets releasing aerosol particles, which served as CCN, back into the atmosphere and cooling the air surrounding. The process generates downward movements in the cloud and promotes a high turbulence that mixes the air masses.

1.3. Interaction of atmospheric particles with solar radiation

Atmospheric particles play a very important role in the climate system. Their effects on the direct radiative forcing scattering or absorbing sunlight, and facilitating indirectly the formation of clouds are a relevant object of study since there is no adequate knowledge of their significance on clouds creation. Particles radiative forcing is globally comparable to greenhouse gases, but in the opposite direction because it causes a cooling climate (Charlson et al, 1992). Coakley and Grams (1976) consider that particles between the range $0.05 < r < 1$ µm may cause a cooling surface. Research has shown that the radiative forcing of atmospheric particles depends on their composition, size and altitude (Hansen et al, 1980; Pollack et al, 1981). So, it may be considered that a change in one or several properties on the atmospheric particles might affect the local forcing.

The indirect radiative forcing occurs when the aerosol particles are used as CCN creating cloud droplets. The effects are classified in two types:

a. Radiative forcing induced by an increase of anthropogenic particles promoting a higher concentration of droplets that change the cloud's albedo (Twomey, 1974). This effect is also known as the cloud's albedo or Twomey effects.
b. Radiative forcing caused by a higher concentration of anthropogenic particles, causing a decrease in the droplets diameter and more competition for the water vapor available in the atmosphere. This will reduce the precipitation efficiency and modify the cloud's residence time in the atmosphere (Albrecht, 1989). The event is known as cloud lifetime or Albrecht effects.

1.4. EPIC 2001

East Pacific Investigation of Climate Processes in the Coupled Ocean-Atmosphere System 2001 (EPIC 2001) was sponsored by The U.S. Climate Variability and Predictability Research Program (CLIVAR), which has the goal of providing the observational basis needed to improve the representation of certain key physical processes in coupled ocean atmosphere models. In addition to physical processes, EPIC 2001 research was directed toward a better understanding and simulation of the effects of short-term variability in the east Pacific on climate. This variability is particularly important in the region because conditions in the

ITCZ are highly variable on daily to intra-seasonal time scales. The effects of such variability rectify strongly onto climate time scales in this region.

EPIC 2001 was conceived as an intensive process study along and near 95°W during September and October 2001. This longitude was chosen to coincide with the Tropical Atmosphere Ocean project (TAO) mooring array in order to provide an overlap between the process study and long-term monitoring.

In addition to the TAO moorings, two aircrafts, the National Center for Atmospheric Research's (NCAR) C-130 and NOAA's P-3, plus two ships, NOAA's R/V Ron H. Brown and the National Science Foundation's (NSF's) R/V New Horizon, and Galapagos-based soundings, were used to make measurements of the atmosphere and ocean in this region. The aircraft were based from 1 September to 10 October 2001 in Huatulco, Mexico. The ships spent approximately 3 weeks in the vicinity of 10° N, 95° W, and then traversed the 95° W line to the equator. After a short stop in the Galapagos Islands, the Ron H. Brown then proceeded south along 95° W and then to the Woods Hole Oceanographic Institute Improved Meteorological Recorder (IMET) mooring at 20° S, 85° W. Meanwhile the New Horizon reversed its track along 95° W and then returned to port.

2. Methodology

On this study we use a P-3 aircraft belonging to the National Oceanic and Atmospheric Administration (NOAA) and a C-130 Hercules property of the National Science Foundation (NSF) operated by the National Center of Atmospheric Research (NCAR). El Centro de Ciencias de la Atmósfera, at the Universidad Nacional Autónoma de México, (UNAM) installed and operated instrumental at the C-130 to measure some properties of atmospheric particles.

2.1. Instrumentation

We used instruments to measure the physical properties of atmospheric particles. The particles chemical composition is inferred from their optical properties as it is explained later in this chapter. Since the atmospheric particles are micro and sub-micron range sizes, the number of particles per volume may provide information on their origin or formation and might infer possible causes about their changes in concentrations.

We use optical counters to estimate the concentration of particles. Their operating principle is based on the extinction of a beam of known wavelength, having gone through an air sample with a certain amount of particles. Table 1 contains the technical details of the instruments used in the EPIC 2001 flights for the study.

2.1.1. Optical counters

Condensation nuclei particle counter (TSI-3760) is an instrument that increases the size of the particles with a forced growth in an environment artificially saturated with butanol. This increases the efficiency of detection and counting of very small droplets (Twomey, 1991).

Instrument type	Parameter	Range	Accuracy
CN counter (TSI Model 3760)	Light-Absorption Coefficient	10^{-7} to 10^{-2} m^{-1}	±5%
PCASP-100X	Light-scattering Coefficient	1.0×10^{-7} to 10×10^{-3} m^{-1}	±5%
PMS Model FSSP 300	Number concentration of aerosol	0.01 to > 3 μm 0 to 2×10^4 cm^{-3}	Varies with concentration, about 6% at 3300 cm^{-3}.
	Size spectra of aerosols	0.12 to 3.0 mm (15 channels)	± 20% (Diameter) ±16% (Concentration)
PMS Model FSSP 100	Size spectra of aerosols	0.3 to 20 mm (30 channels)	± 20% (Diameter) ±16% (Concentration)
CCN Counter	Size spectra of aerosols and cloud droplets	2 to 47 mm (30 channels)	± 20% (Diameter) ±16% (Concentration)
	Number concentration of CCN	0.2 to 1.0 % supersaturation	10% at 1% supersaturation

Table 1. Aerosol and cloud particle instrumentation on the C130 aircraft

The CCN is a subset of the total concentration of particles, which can form droplets in an environments of over-saturation, as in a cloud (101 - 110% SS). The CCN counter model 100, determines the concentrations of these atmospheric particles. The operating principle is based on measuring the variation within a thermal gradient diffusion chamber, to create an environment of over-saturation. An electrical system controls the temperature of two plates that create the conditions of over saturation. A beam of laser light passes through the chamber and the instrument measures the amount of light scattered and estimates the concentration of CCN per volume. Delene et al (1998) and Delene and Deshler (2000) have a more detailed description and analysis of this instrument operation.

The PCASP, FSSP300, and FSSP100 provide information on particle concentrations in size defined ranges. The instruments pass a beam of light with specific wavelength and intensity, through the air sample. The particles in the sample scatter the light beam with an intensity that depends on their size, shape and composition. The instrument measures the amount of light scattered with sensors. By knowing the intensity of scattered light and particles composition (water for FSSP100 and FSSP300) or sodium chloride (for PCASP), it is possible to infer the size range of particles to which they belong. A more detailed description of the operation and measurement uncertainty associated with the PCASP can be obtained on Strapp et al (1992).

2.1.2. Instruments location

The external instruments are installed in a pod on the wing and internal within the fuselage in the cabin. The location characterizes the way that each machine takes the air sample to analyze. The external instruments are in direct contact with the air all the time. Thus, the air sampling is instantaneous. But the internal instruments take the air sample by an air inlet and a section of hose that transports it to the device. In this case, the instrument uses a suction pump to enter the air. The path that the air sample covers must be considered when comparing the data from internal and external instruments. For instance, the CN counter readings show a 1 second delay compared with those obtained instantaneously from the PCASP and the FSSP.

2.2. Composition of atmospheric particles derived by their physical properties

To calculate the particles size we used Mie's theory. It says that the intensity of light scattered in all directions depends on the size and composition of the particle and the wavelength of incident light. Mie's theory considers spherical particles, so the optical counters used this assumption to obtain a value about the size of atmospheric particles.

A particle exposed to a beam of light will eliminate some of the energy that hits on it. This phenomenon of extinction is given by the combination of absorption and scattering of light. Water droplets do not absorb radiation and only scatters light. With this justification, we assume that the dispersion coefficient values are a good approximation to the extinction coefficient of particles.

The amount of light scattered by a particle depends on three combined effects: 1) reflection, 2) refraction and 3) diffraction. Refraction depends on the composition of the particle, while the diffraction depends on the wavelength of incident light, the size, and the shape of the particle. Optical counters use these concepts to calculate the size of particles in an air sample. With this information it is possible to characterize the particles counted in certain size ranges, obtaining the size distribution spectrum of particles.

The instruments are calibrated using refraction indexes for water (1.33), ammonium sulphate (1.48) or sodium chloride particles (1.54). Figure 1 shows the dispersion efficiency depending on the particles diameter with the three refraction indexes from Mie's theory.

Figure 1 shows the sensitivity of dispersion efficiency by varying the composition of particles. If we consider a particle with 0.4 μm diameter, the changes on its composition from water ($n = 1.33$) to sodium chloride ($n = 1.58$) is approximately 30 times. This indicates that the difference in the amount of light scattered by a cloud droplet and a NaCl particle increases by several orders of magnitude.

The coefficient of dispersion of a population of particles with size r_0 to r_n is defined by

$$\sigma_s = \int_{r_0}^{r_n} Q_s(\eta, r, \lambda) \pi r^2 n(r) dr \tag{1}$$

where Qs is the scattering efficiency, which is a function of refractive index (η) radius of the particle (r) and the wavelength of scattered light (λ). And $n(r)$ represents the concentration of a population of particles according to radius r.

$$\sigma_s = \sum_{i=1}^{n} Q_s(\eta, r_i, \lambda) \pi r_i^2 n(r_i) dr \tag{2}$$

With this equation and the data obtained by particle counters we calculated the particles dispersion coefficients for the vicinity of the clouds. We used 28 different refractive indexes, ranging from 1.33 to 1.60, resulting in a matrix of 28 dispersion coefficients for each data. The values were compared against the dispersion coefficients obtained directly from a nephelometer, inferring the approximate refractive index and a possible particles composition.

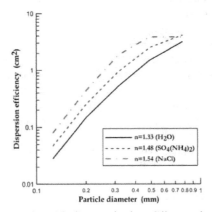

Figure 1. Dispersion efficiency and particle diameter for three different refractive indexes

2.3. Sampling

In EPIC 2001 project we did 19 flights to investigate the ocean-atmosphere interaction, and clouds and aerosol particles properties in the Eastern Pacific. Nine flights were conducted within the ITCZ. The flights were in the area between 8° - 12° North latitude, and 93° - 97° West longitude (figure 2). During the flights were searched and selected young convective and precipitation clouds.

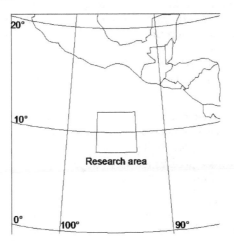

Figure 2. EPIC 2001 research area.

3. Particles processing by clouds

If a mass of moist air is forced to raise it creates a cloud, either by buoyancy promoted by surface heating or by mechanical means, such as climbing up a slope pushed by the wind. When the mass of air containing water vapor is cooled by adiabatic ascent, the vapor pressure increases until it reaches the complete saturation (RH = 100%). The increase in RH results in the formation of cloud droplets, because some particles act as CCN. The particles grow by molecular diffusion to the drop and form a solution. New droplets can interact with interstitial particles in the cloud or collide and coalesce with other droplets changing its composition. If the water droplet evaporates, then releases an aerosol particle with different mass and composition than the original particle. The particle is physical, chemical and hygroscopic different. According to the Köhler curve, a particle increasing its mass may trigger saturation values lower than a smaller particle (Pruppacher and Klett, 1997). In other words, the atmospheric particles that are processed by clouds acquire properties to become more efficient CCN. In addition, the size distribution of particles is modified and can directly influence the evolution of the cloud. The processes change the physical and chemical characteristics of particles, but their concentrations can be identified by analyzing the shapes of the distributions of sizes.

3.1. Particles sizes distribution

Atmospheric particles have different origins; some are from natural sources such as oceans, volcanoes, soil, pollen, forest fires, and so on. And human activities also generate particles that reach the atmosphere. Motor vehicles, power generation, industrial boilers and incineration of solid waste are some sources of anthropogenic particles. The diversity of emissions presents a wide range sizes, concentrations and compositions of particles. And their size range spans several orders of magnitude, ranging from nanometers to hundreds of micrometers. Similarly, the concentrations might be from 1.0^7 to 1.0^{-6} particles per cubic centimeter. In addition, the particles in the atmosphere undergo processes that transform their physical and chemical properties. Therefore the study of atmospheric particles is complicated. However, an important tool in particles analysis is the construction of size distribution graphs. These charts provide relevant information such as the nature of the particles (i.e., maritime, continental, urban, or rural zones), because each place has a kind of specific particle sources.

In the study we use particle size distributions from inside, outside and away from the clouds to identify and analyze possible changes of particles properties interacting with cloud droplets. Particle counters at the C130 aircraft provided the information. The instruments measure aerosol particles and cloud droplets in different size ranges and cover a wide range of sizes (~ 0.1 to 50 microns).

The PCASP dehydrates particles reducing the relative humidity below 30%, but the FSSP measure them at ambient relative humidity, so we calculate the dry particle diameter from both FSSP, based on the Tang's theory (1976) and Tang and Munkelwitz (1977), to obtain a size distribution of dry particles.

3.2. Interaction between particles and clouds

There are four main processes involved in the interaction particles-clouds:

a.   Vertical distribution

The classic physics model for developing convective clouds indicates that the aerosol particles are incorporated from the base of the cloud. Some particles form droplets that grow vertically while being transported by updrafts currents generated by latent heat during a phase change from vapor to liquid. A few drops reach the top of the cloud, where updrafts currents lose strength by neutral stability between the cloud and the environment. At this point, the interaction of clouds with dry air dilutes and evaporates drops. In this mixing and evaporation zone of droplets is where the particles, used as CCN to form cloud droplets, are released back into the upper top of the cloud reaching the high troposphere and in some cases of deep convection may lead them to the lower stratosphere. Some researchers have shown that this mechanism is the main transport of particles from the boundary layer to free troposphere (i.e., Flossmann, 1998).

b.   Mass incorporation into the drop by diffusion

A particle in a high relative humidity environment will grow by diffusion and condensation of vapor molecules producing a cloud droplet. The particle can be diluted to form a solution within the droplet. In that case, if the droplet is in an atmosphere of various gases that can be absorbed by the same specie (i.e., SO_2 in marine clouds) there is an increase in the mass concentration of solute within the droplet changing its physical properties (mass increase). A rapid change in pH also transform the chemical properties, resulting in the dissolution of species in a solution (Hegg and Hobbs, 1982; Leaitch, 1996; Leaitch et al, 1986, O'Dowd et al, 2000).

When a particle is in high relative humidity (~ 80%) environment, it becomes an effective site for oxidizing species in aqueous phase (Chameides and Stelson, 1992). For example, SO_2 dissolved in a particle can react with ozone and hydrogen peroxide. In an acid particle (H_2SO_4) with low pH, the oxidant is hydrogen peroxide, but for particles with high pH (i.e., $[NH_4]_2SO_4$), the oxidant will be ozone. The first reaction is more important in maritime areas, because SO_2 is abundant from dimethyl sulphide emissions produced by phytoplankton. O'Dowd et al, (2000) estimate that under mass incorporation conditions a particle can increase its size to double in about 400 seconds.

c.   Collision-coalescence of drops

When the droplets have certain size, they grow more efficiently by collision-coalescence. The collision and coalescence among cloud droplets is mainly governed by gravitational effects, so large droplets fall faster than small ones. This process produces a decrease in the concentration of drops, but form larger particles and evaporate the droplet mass. Each collision-coalescence between two original CCN becomes in to one drop, which has a mass equal to the sum of the two nuclei. If the original CCN have a different composition, it also changes the chemical composition of the resulting drop.

d.   Mechanical removal

The removal of particles by precipitation is a cleaning process from the atmosphere. This mechanism helps to maintain a balance between sources and sinks of particles. The precipitation removes mechanically particles by inertial collection and transportation of raindrops, and also removes the nuclei when the drops become rain. However, it depends on the size of the interstitial space of the particles, related to the size of the drop. Experimental studies of Chate et al, (2003) demonstrated that this mechanism is more efficient on particles in the range of coarse mode (> 1 μm). Other studies have also shown that removal by inertial collection and transportation only affects to a small percentage of particles that are on base of the cloud (Wang and Pruppacher, 1977).

3.3. Mechanisms of interaction between particles and clouds

The mechanisms that modify the properties of atmospheric particles are varied and have different efficiencies depending on the size of the particles. The size distribution and the area plots of particles in the study showed four patterns that may be associated with processes of interaction between particles and clouds:

1.   Vertical transport with mixing and dilution with minimum changes in the size
2.   Aqueous phase oxidation of aerosol precursors ($\leq$ 1 micron)
3.   Droplet coalescence (> 1 micron)
4.   Removal by precipitation

Charts on figure 3 illustrate the general features that are described as follows:

1.   Vertical transport with mixing and dilution

This cloud processing mechanism, discussed in detail by Flossman (1998), transports particles from cloud base to upper regions of the cloud where they eventually are detrained, either at cloud top edges or by mixing with ambient aerosols at detrainment level. PSD signatures take one of two forms. If RH at the point of measurement is higher than RH at cloud base, then PSD exhibits a tail at larger sizes exceeding that of the cloud base PSD (Fig. 3, pattern A1). Other studies have shown the correlation between RH and changes in particle size near cloud boundaries (e.g., Baumgardner and Clarke, 1998). This, or the particles have mixed with air close of to the same RH as at the cloud base so that the resulting PSD is one that has approximately the same shape as at cloud base, but with lower concentrations as a result of the dilution with ambient air (Fig. 3, pattern A2).

2.   Aqueous phase oxidation of aerosol precursors

In-cloud oxidation of dissolved species is a process that increases the mass of aerosol particles and may change their composition (Hegg and Hobbs, 1982; Leaitch, 1996; O'Dowd et al., 2000). The likely precursor gas in the EPIC research region is SO_2, which evolves from dimethyl sulfide produced by phytoplankton or from anthropogenic sources, as discussed below. The PSD pattern produced by this process will be indistinguishable from pattern A unless additional information is known about the aerosol chemistry. As discussed in section

2.4, measurements were not made for particle composition, but the average refractive index of particles could be estimated. A comparison of the average refractive index at cloud base with the near-cloud and far-cloud values at higher altitudes suggest changes in particle composition, as shown in Fig. 3 where the cloud base refractive index is near that of sea salt (1.54), while the near-cloud value at 2500 m is closer to that of ammonium sulfate (1.48). The observed differences are based on a technique that has a large amount of uncertainty and is used qualitatively in the present study as an indicator of composition change.

3.   Droplet coalescence

Coalescence decreases the number concentration of particles while shifting the mass to large sizes. Each coalescence event decreases the number of original CCN by one and the resulting mass is the sum of the two nuclei. If nuclei are of different composition, then this process also changes the chemistry of the particle contained in the resulting drop. The large particle mode, with a peak between 5 – 6 μm, seen in Fig. 3 indicates coalescence, since neither the cloud base nor far-cloud PSD have particles in this size range.

4.   Removal by precipitation

Precipitation removes particles mechanically by inertial or nucleation scavenging when cloud droplets become raindrops. Mechanical scavenging depends on the size of interstitial aerosol in relation to the raindrop size. Experimental results (Wang and Pruppacher, 1977) suggest that only a few percent of the interstitial and sub-cloud particles are removed by this mechanism and this is not considered as a major factor here. The majority of aerosols removed by precipitation will be those that are in cloud droplets growing by condensation and coalescence to precipitable sizes. Figure 3 illustrates this process where PSDs at the cloud base level are quite different depending upon whether the measurements were made at the actual cloud base or in the far-cloud air. The far-cloud PSD has particles of super-micron sizes, but such particles are noticeably missing at the cloud base. In this particular case, the cloud base measurements were made after the cloud had formed and the super-micron particles had been activated and grew quickly to droplet sizes that could coalesce and precipitate.

4. Analysis and discussion

Five flights and ten cloud systems were selected for analysis based on a visual evaluation of the records made with the forward- and side-looking video cameras on the aircraft. The criteria was that no other clouds could be seen within around 10 km on either side of a cloud line, such that far-cloud samples represent "ambient" aerosols, i.e., lacking any recently processed particles by clouds.

4.1. Time series to identify clouds

We studied the data from flights 7, 9, 12, 13 and 17. The flights were conducted in convective clouds by passing through the clouds at different levels (1000, 2500, 4200, and 6000 meters). Moreover, we passed through the cloud base (300 meters) and at surface level (30 meters).

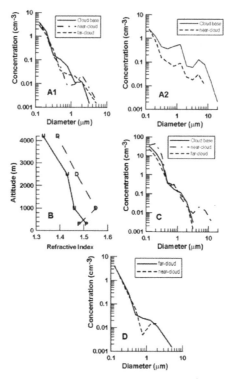

Figure 3. Cloud processing signatures

Furthermore, during the flights were conducted surveys of the atmosphere's vertical profile. This information is useful because surveys have different locations, but they are associated to the clouds recorded data.

We observed changes in the properties of atmospheric particles and the processes responsible. Also we recorded environmental and weather conditions in zones which the property changes occur more frequently, comparing both sides of the clouds.

The droplets concentration in a cloud is function of the particles number in the atmosphere, so any variation in the amount of particles will affect the cloud microphysics evolution. In marine areas the concentration of particles is about 100 per cm³, if there is a greater amount of particles is likely to pollution particles are present. The EPIC 2001 research area is located approximately 800 - 1000 km away from Mexico and Central America, allowing the transport of pollutants from the continent to the area when the prevailing winds are favourable. The opposite situation is also possible in maritime areas away from the coast. Wind patterns during flights 12 and 13 shows weather characteristics of maritime areas.

The cloud boundaries are identified by means of videotapes taken from the plane C-130 and the analysis of time series using reference measurements obtained by the FSSP100. The criterion was to consider the instrument's concentration records ≥ 1 cm^{-3} obtained within the cloud. In EPIC 2001, we identified 10 cloud systems that met the minimum information necessary for the study. Table 2 shows the location, date and time of each cloud systems. The data correspond to the average information of all transects made to the system.

Flight #	Date 2001	Cloud System	Time Period (UTC)	Location	Particle Source	300 m		Cloud base length (m)	Cloud base T (°C)	Cloud base wind speed (m s^{-1})	Cloud base RH (%)	Cloud top length (m)	Cloud top T (°C)	Cloud top wind speed (m s^{-1})	Cloud top RH (%)	4200 m	
						CN Conc (cm^{-3})	PCASP Conc (cm^{-3})									CN Conc. (cm^{-3})	PCASP Conc (cm^{-3})
7	sep-16	1	16:46-17:19	12.3°N, 93.7°W	HG	910	345	4800	25.60	1.4	79	***	***	***	***	***	***
7	sep-16	2	18:42-20:12	11.9°N, 95.2°W	HG	830	227	25740	25.04	2.5	80	13900	2.44	9.3	91	3370	447
9	sep-20	3	18:16-20:11	10.5°N, 95.9°W	MR	380	66	18093	24.18	1.3	85	500	3.63	0.3	65	320	32
9	sep-20	4	18:56-20:24	8.2°N, 95.8°W	MR	200	40	12860	23.63	1.5	83	2640	2.95	3.4	79	810	48
12	sep-28	5	17:03-18:12	9.3°N, 93.9°W	MR	460	138	13970	25.26	3	83	680	3.7	7.2	62	590	32
12	sep-28	6	19:14-20:20	11.9°N, 94.1°W	MR	420	143	4550	24.87	1.2	80	7040	3.74	2.7	80	818	14
13	sep-29	7	18:31-19:03	11.4°N, 94.6°W	MR	360	98	11440	24.06	1.6	84	2970	3.97	1.9	88	15574	50
13	sep-29	8	19:36-20:22	12.4°N, 94.9°W	MR	390	64	9460	24.15	1.5	84	8690	3.88	2.7	85	1119	98
17	06-oct	9	18:34-19:49	11.9°N, 93.9°W	HG	1900	696	36200	26.02	2.3	82	5060	3.92	5.8	81	2810	384
17	06-oct	10	20:51-21:36	11.8°N, 94.1°W	HG	1600	510	10560	26.53	0.9	68	2480	3.2	2.4	72	1080	350

Table 2. Characteristics of cloud system selected for the analysis

Flights 9, 12, and 13 were made on days with "maritime" (MR) aerosol background and when winds came from the southwest. "Higher" (HG) aerosols concentrations correspond to flights 7 and 17 with average concentrations significantly higher than the other three flights. Table 2 summarizes the time, location and type of cloud systems.

4.2. Cloud boundaries comparison

A convective cloud in its formative stage has a rapid vertical development. The ambient air surrounding the cloud incorporates into the cloud increasing its volume with the vertical expansion. This process is known as entrainment, so its mixing process is more efficient outboard in the cloud with sub-saturated air from the environment. The droplets evaporate and particles served as CCN are released back into the atmosphere. The mixing with ambient air and the evaporation of cloud droplets produces a cooling air parcel which generates a negative buoyancy force. This could result in shear zones, as the turbulence caused by this effect helps to mix ambient air with the cloud and evaporate more drops. To define the distance corresponding to the cloud border, we did a time series properties analysis of the particles in the clouds, to evaluate the zone of influence. Figure 4 shows the changes in properties of the particles in the vicinity of the cloud. The area of influence for the interaction of particles with the cloud covers a distance of approximately 500 m from the cloud's border. The FSSP100 drops concentration measurements correspond to the solid line, which marks the cloud's boundary. Each tick mark represents 110 m at the aircraft's average velocity of 110 m s^{-1}. The dotted lines represent the behavior of the average diameter of particles measured by the FSSP300 (dash) and FSSP100 (dash and dot). These values denote the limits of the transition region between ambient and cloud air. Diameter measurements show a significant increase of particles up to ~500 m from the cloud's boundary. That determines the representative area to evaluate the mechanisms that modify the properties of particles due to the interaction with the cloud.

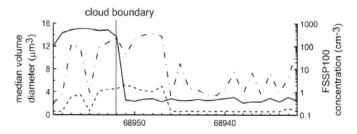

Figure 4. Medium volume diameters measured by FSSP300 (dashed line) and FSSP100 (dot-dash line) indicating a cloud boundary.

The particles properties on both sides of the cloud border must be different. The particles characteristics in the neighboring area of the cloud have a determining effect on the modification of some of their properties. Changes in humidity, altitude and environmental conditions, where the samples are collected, affect the particles concentration and size. So, in order to compare both sides of the cloud we analyze data from the clouds vicinity (500 m from the border of the cloud) to obtain average values. Figure 5 shows the average concentrations of vertical profiles measured by the PSCASP (> 0.1 microns) on the cloud borders (cases identified by U, D or P). A third profile, corresponding to a far area from the cloud (average 500 to 1500 m from the border (dotted line), is used to compare the profiles against those close to the cloud.

In conditions with and without pollution, particle profiles measured in the vertical gradient has two patterns. One shows a constant value or a decrease with height. The second shows an increase in concentration up to 2500 m and then decreases to lower values than those measured at the base. It is possible to observe cases where the particle concentration profiles are similar to the concentration profile far away from the cloud. This is due to the presence of dilution processes, where particle concentrations in the vicinity of the cloud are lower than in remote areas. When the concentration of particles near the cloud is greater than the environment it is possible that there is an increase of particle size < 0.1 µm to ranges that is not possible to detect. In summary, measures of concentration of particles increased with height indicating that the smaller particles (< 0.1 µm) at the base of the cloud grow to detectable size ranges for the instrument during transport through the cloud. On the other hand, when the concentrations decrease with height means a dilution effect.

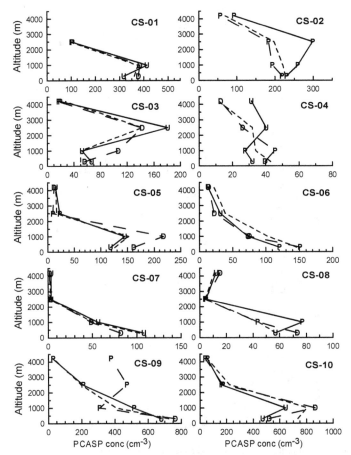

Figure 5. Vertical profiles of particle concentrations measures for the 10 cloud cases. The far-cloud profiles (short dashed) and near-cloud profiles on each side of the cloud (solid and long dashed lines).

Figure 6 shows concentrations measured with FSSP100 (> 2 μm) in the vicinity of the cloud (solid and dotted lines) and away from the cloud (dotted line). In the region where there are drops of cloud measurements, we used them to detect giant particles (> 1 micron) coming from the ocean. In most cases, the concentration values for particles greater than 1 micron near the cloud have a maximum at 1000 m high. It happens perhaps because smaller size particles increased their volume within the cloud.

Sometimes it is possible to observe when the particle concentrations decrease with height that there is an increase in the concentrations of larger diameters. The combination of patterns is indicative of changes in particles by mixing and dilution on the total concentration dominated by small particles. Also, a simultaneous increase in the concentrations of particles suggests that it there is a change in particles size.

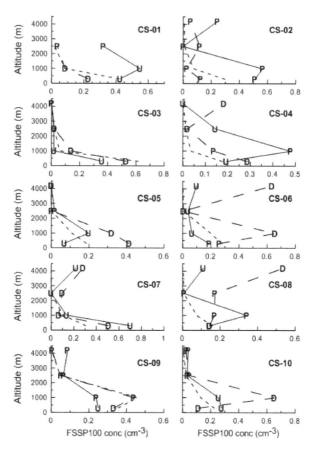

Figure 6. The vertical profiles of concentrations measured with the FSSP-100.

4.3. Processes classification and evaluation

There are different processes at each side of the cloud and in each level of each clouds system. We used as reference the processes signal description and their property. This subjective classification is based on an examination of vertical profiles, shown in Figs. 4, 5, and 6 and on the shapes of PSDs associated with the near-cloud region at each level. The 300 m passes were only evaluated for evidence of removal by precipitation (pattern D). Figure 7 summarizes the frequency of each category. There were 49 classifications made on MR days and 35 during HG days. A classification could not be made in 10% of the MR cases and 5% of the HG ones. The largest fraction of the observations was classified as pattern A (40% and 60% of MR and HG cases, respectively). Pattern C matched only 4% of the MR observations and none during HG days. Patterns B and D were equally represented during both MR and HG days.

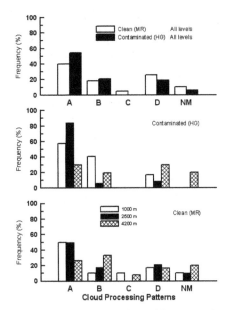

Figure 7. Percentages of cloud processing signatures, derived from vertical profiles of particle
properties and evaluation of particle size distributions, for the MR cases (bottom panel), HG cases
(middle), and combined cases.

There were no consistent trends with altitude between MR and HG days. Particles at 1000 m
and 2500 m altitude were predominantly like pattern A, in both the MR and HG cases. In
MR cases, there were slightly more type B than type A particles at 4200 m. In HG cases, the
4200 m particles were evenly distributed between patterns A and D.

4.4. Particle composition estimation

The refractive index was used to estimate the particles composition at the ends of the clouds.
We calculated and average refractive indices at 500 m from the borders of the cloud. Figure
8 shows the results. The refractive index profile for the area between 1000 and 1500 m away
from the border of the cloud is also shown. This area is considered free from the influence of
particle processing by clouds.

We consider three refractive indexes as benchmarks. The refraction indexes of water (1.33),
ammonium sulfate (1.48), and sodium chloride (1.54). Another factor we take into account is
the size distribution and composition of particles as a function of their height in normal
conditions without the influence of pollution. The type and composition of particles depend
on local sources. Thus, in sea areas the main sources are the production of salt particles from
the surface of the oceans that are caused by wind friction and breaking waves. These
particles are mainly in the lower parts of the troposphere near its source. In the upper
troposphere, main sources of particle are the conversion of gas to particle and deposition by

clouds (Hobbs, 1993) which produce small particles (< 0.1 μm). In marine areas the particles are composed of sulfate, because they are formed by the condensation of SO_2. Particles ranged between 0.1 to 1 microns are composed of sulfate (Hobbs, 1993). Based on the sources we expected to find high ammonium sulfate in the lower troposphere.

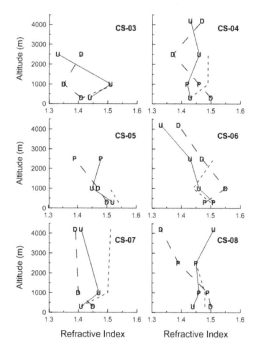

Figure 8. Derived refractive indexes for far- and near-cloud vertical profiles for the maritime (MR) day cases.

In estimating the composition of particles from optical counters measurements, the environmental conditions can strongly influence the outcome. Figure 9 shows the comparison between the dispersion coefficients obtained with a nephelometer and the dispersion coefficient calculated from the particles size distribution. Data were collected on transects at 2500 m without pollution (Figure 9 right) and with pollution from the continent (Figure 9 left). The reference line 1:1 is used to compare dispersion coefficients calculated and from nephelometer, assuming three different compositions.

Figure 9 (left) shows that polluted cases at the same height had calculated coefficients higher than those measured with instrumentation. This may be due to anthropogenic particles that absorb and scatter light. So, the measured values will be lower than calculated, since it is not taken into account the particles absorption. Moreover, no polluted cases have a better correlation because marine particles do not absorb light. This feature allows us to use this technique to more easily estimate the composition of particles in cases without contamination.

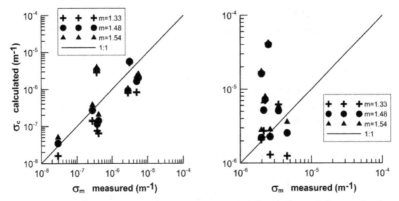

Figure 9. Scattering coefficients compared against those measured with the nephelometer for three
refractive indexes: 1.33, 1.48, and 1.54, for MR days (a) and high aerosol concentrations (HG) days (b).

5. Effects of atmospheric particles

Atmospheric particles play an important role in the planet radiative budget. Their effects on
radiative forcing by absorbing solar radiation backscattered or as facilitating clouds
formation are important objects of study.

The uncertainty of particle-radiation interactions is still very large. For example, some
particles with sulfate or organic carbon cool the atmosphere. On the other hand, black
carbon particles warm it, because they absorb visible light and convert it into thermal
energy (IPCC, 2000). This strange balance increases the uncertainty of the magnitude of their
effects on the atmosphere.

5.1. Direct effects of particles processed by clouds

The cloud process and modify particles size and composition. Larger particles scatter more
sunlight and increase the extinction of light. This impact on the radiative balance can be
estimated with the optical depth, since the extinction of particles is expressed:

$$\tau(\lambda) = \int_0^\infty \sigma_e(\lambda, z) dz \qquad (3)$$

Where τ is the optical thickness and σ_e is the particle's extinction coefficient. In our case, we
assume that the particle's composition is mainly sodium chloride (sea salt) and sulfates. The
particle does not absorb visible light, so its extinction and dispersion coefficients are equal.
Thus, the above equation becomes:

$$\tau = \sum_{n=30}^{4200} \sigma_{s(n)} \Delta z \qquad (4)$$

To calculate the optical depth, we use the size distribution spectra for the following heights:
30, 300, 1000, 2500 and 4200 m on both sides of the cloud, as well as the average between
1000 and 1500 m away from the cloud.

Table 3 shows the optical depth on both sides of the cloud (upwind and downwind) and data away from the cloud that are used as reference atmosphere without the influence of processed particles. The values are higher near than far away from the cloud. Figure 10 shows the optical depth near and far from the cloud. The particle optical depth near the cloud is 10 times higher than distant to the cloud. System 7 has a ratio about 1:1 indicating that the physical and optical properties of particles near and far from the cloud do not exhibit noticeable differences. The data suggest that particles near to the cloud in system 7 have not been yet processed or it could be a very young cloud with no mixing with ambient air.

cloud	τ (U)	τ (D)	τ (far)
1	0.125	0.021	0.027
2	0.158	0.373	0.018
3	0.116	0.124	0.013
4	0.152	0.196	0.017
5	0.192	0.118	0.010
6	0.247	0.316	0.084
7	0.171	0.231	0.211
8	0.272	0.306	0.005
9	0.322	0.355	0.036
10	0.196	0.613	0.041

Table 3. Optical depth on both sides of the cloud

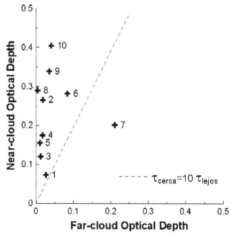

Figure 10. Optical depths calculated for particle size distributions at 30, 300, 1000, 2500, and 4200 m, near-cloud and far-cloud cases. These optical depths are compared for the ten cases (labeled by their number).

The decrease in the amount of radiation reaching Earth's surface can increase the optical depth, suggesting that cloud particles processed promote a local cooling. Indeed, satellite images of cloud cannot detect the optical depth because it is very large.

5.2. Indirect effects of particles processed by clouds

Five flights and ten cloud systems were selected for analysis based on a visual evaluation of the records made with the forward- and side-looking video cameras on the aircraft. The criteria was that no other clouds could be seen in 10 km on either side of a cloud line, such that far-cloud samples represent "ambient" aerosols, i.e., lacking any recently processed particles by clouds. Flights were also classified by aerosol type. Figure 11 shows frequency distributions of CN and PCASP measured concentrations at ≤ 300 m for those five days. Flights 9, 12, and 13 were made on days with "maritime" (MR) aerosol background and when winds came from the southwest. "Higher" (HG) aerosols concentrations correspond to flights 7 and 17 with average concentrations significantly higher than the other three flights. Table 2 summarizes the time, location and type of cloud systems.

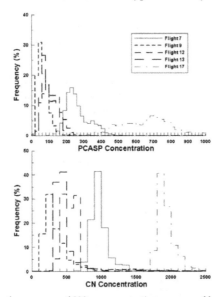

Figure 11. The frequency of occurrence of 300 m concentrations measured by the PCASP (top panel) and CN (bottom) are shown here for the five days used in the case studies.

The cloud albedo depends on the concentration of droplets (Twomey, 1974). One way to estimate, with good approximation, the changes in albedo (A) is using the Meador and Weaver (1980) equation:

$$A = \frac{(1-g)\tau}{1+(1-g)\tau} \tag{5}$$

Where g is the asymmetry factor, which is the average cosine of scattering angle. For the scattering clouds by sunlight g = 0.85 (Hobbs, 1993), we can simplify the equation 5.3 to:

$$A = \frac{\tau}{\tau+6.7} \tag{6}$$

The cloud optical depth (τ) of h that contains a concentration of droplets n (r), with radius r is given by:

$$\tau = \pi h \int_0^\infty Q_e r^2 n(r) dr \tag{7}$$

Where Qe is the extinction efficiency factor for the wavelengths of visible radiation ($\lambda = 400 - 700$ nm).

We calculated the single scattering albedo employing equations 5.4 and 5.5 for each concentration of drops at 6000 m, using a layer of cloud thickness h = 100 m, based on the distance of a datum to another along the horizontal axis (data per second). The 6000 m level is considered the top of the ice-free clouds. Previous figure shows the histograms of the single scattering albedo calculated inside the cloud for different environmental conditions. The albedo on HG episodes ranged between 0.8 – 0.9, while in MR days ranged between 0.6 – 0.8. The results agree with those obtained theoretically by Lohmann et al (2000) stating that anthropogenic pollution causes a diminution in the effective radius of cloud droplets and an increase in the albedo of the cloud.

There is a relationship between the maximum concentrations registered by both FSSP100 and PCASP, indicating a higher concentration of cloud droplets in the episodes with anthropogenic influence and resulting in a diminution in droplet size, because a bigger amount of CCN compete for moisture in the air. Last figure shows the different droplets average diameters in pollution-free days (~14 microns) and polluted days (~10 microns).

Analysis shows the indirect effect of the particles in the formation of convective clouds. During episodes of anthropogenic contamination, the concentration of droplets in the cloud increases and their size decrease, thus causing low rainfall. These phenomena will increase the albedo of the cloud, because it depends on the concentration of drops (Twomy, 1974).

Table 4 shows the values of optical depth (τ) and albedo (A) of the clouds studied. The highest values of albedo were presented in systems 1 and 9, corresponding to days with pollution.

Cloud	τ	Albedo
1	66.46	0.63
2	27.35	0.46
3	31.75	0.42
4	29.38	0.52
5	34.04	0.49
6	65.48	0.58
7	37.29	0.53
8	43.28	0.52
9	165.90	0.73
10	40.18	0.54

Table 4. Optical depth (τ) and albedo (A) of clouds studied

6. Conclusions

The physical and optical properties analysis of atmospheric particles is focused on the observation of several processes involved in convective clouds and their environment. We have studied cloud systems on Mexico's Pacific ITCZ. The research flights were conducted during September and October 2001. The data obtained point to some relevant cases marked by the weather and cloud characteristics. The analysis and evaluation of information allows us to reach the following conclusions.

We identify the most important interaction processes between particles and clouds, which can cause changes in the size and composition of atmospheric particles: a) diluting the concentration of particles with minimal changes in size, b) increasing atmospheric concentration of submicron particles (≤ 1 μm), c) increasing the concentration of atmospheric supermicron particles (> 1 μm) d) removal of supermicron particles. The analysis of particles and clouds interaction shows that the most common contact mechanisms were: a) vertical transportation with mixing and dilution, which occurred in 44% of the MR days and 55% on HG episodes b) oxidation of aqueous phase particles are present in 20% and 24% days MR and HG events, respectively, c) coalescence of droplets occurred in 18% and 15% days MR and HG, respectively.

The particles change their optical properties and the way they interact with solar radiation and clouds. Particles that are processed in the vicinity of the cloud increase the optical depth. The growth comes in quantities up to 10 times larger than the value recorded in distant particles. Therefore, variations in the optical properties of particles affect directly the radiative balance and influence in local climate.

The cloud observations were classified into two categories: typical values of maritime areas with prevailing westerly winds and low concentrations of cloud condensation nuclei (conc. < 500 particles per cm^3) and values influenced by anthropogenic pollution (conc. < 1800 particles/cm^3).

Increasing the concentration of particles in a place influenced by a pollution source also enlarged the number of CCN. Data analysis shows a good correlation between the concentration of CCN at cloud base and the concentration of droplets inside the cloud ($r^2 = 0.92$), which explains the clouds albedo augmentation on days with influenced by anthropogenic pollution.

Future work considers the application of detailed microphysics models to evaluate the different processes of interaction of particles and clouds. Thus, also intends to use these models to analyze the effect of these particles processed in the dynamics of the cloud as well as the influence on processes like rain.

Author details

J.C. Jiménez-Escalona
ESIME U. Ticomán, Instituto Politécnico Nacional, Gustavo A. Madero, Mexico City, Mexico

O. Peralta
CCA, Universidad Nacional Autónoma de México, Ciudad Universitaria, Mexico City, Mexico

7. References

Albrecht B.A., 1989, Aerosol, cloud microphysics, and fractional cloudines, Science, 262, 226-229.

Alfonso, L. and G.B. Raga, 2002: Estimating the impact of natural and anthropogenic emissions on cloud chemistry. Part I: Sulfur cycle. Atmospheric Research, 62, 33-55.

Baumgardner D., Cooper, W.A., Radke, L.F., 1996, The interaction of aerosols with developing maritime and continental cumulus clouds, 12th Int. Conf. on Clouds and Prec. Zurich, 308-311.

Baumgardner, D., Clarke, A., 1998, Changes in aerosol properties with relative humidity in the remote southern hemisphere marine boundary layer. J. Geophys. Res., Vol.103, No. D13, 16,525-16,534.

Chameides, W.L., Stelson, A.W., 1992, Aqueous phase chemical processes in deliquescent sea-salt aerosol: a mechanism that couples the atmospheric cycles of S and sea-salt, J. Geophys. Res. 97, 20565 – 20580.

Charlson R.J., Schwartz S. E., Hales J.M., Cess R.D., Coakley J.A., Hansen Jr J. E., Hofmann D.J., 1992, Climate forcing by anthropogenic aerosols, Science, Vol. 255, 423–430.

Chate D. M., Rao P.S.P., Naik M.S., Momin G. A., Safai P. D., Ali K., 2003, Scavenging of aerosols and their chemical species by rain, Atmos. Environ, 37, 2477 – 2484.

Coakley J. A., and Grams G., 1976, Relative influence of visible and infrared optical propierties of a stratospheric aerosol layer on the global climate, J. Appl. Meteorol., 15, 679 – 691.

DeFelice T. P., Saxena V. K., 1994, On the variation of cloud condensation nuclei in association with cloud systems at a mountain-top location, Atmos. Res., 31, 13-39.

DeFlelice, T.P., Cheng, R. J. 1998, On the phenomenon of nuclei enhancement during the evaporative stage of a cloud, Atmos. Res., 47-48, 15-40.

Delene D.J., Deshler T., Wechsler P., Vali G., 1998, A ballon-borne cloud condensation nuclei counter, J. Geophys. Res. 103 (D8), 8927–8934.

Delene D.J. and Deshler T., 2000, Calibration of photometric cloud condensation nucleus counter designed for deployment on a balloon package, J. Atmos. Oceanic Tech, 17(4):, 459-467.

Emanuel, K. A., 1982, Inertial inestability and mesoscale convective systems. Part II: Symmetric CISK in a baroclinic flow. J. Atmos. Sci. 39, 1080-1097.

Flossmann, A.I., 1998, interaction of Aerosol Particles and Clouds, J. Atmos. Sci., Vol 55,879-887.

Hansen J. E., et al., 1980, Climatic effects of atmospheric aerosols, Ann. New York Acad. Sci., 338, 575 – 587.

Hegg D. A., Hobbs, P.V., and L. F. Radke, 1980, Observations of the modification of cloud condensation nuclei in wave clouds, J. Rech. Atmos, 14, 217-222.

Hegg D. A., Hobbs, P.V., 1982. Measurements of sulphate production en natural clouds. Atmos. Environ., 2663 – 2668.

Hobbs, P.V. 1993: Aerosols-Cloud Interactions, Aerosol-Clouds-Climate Interactions, Academic Press. pp. 33-73

Holton J. R., 1992, An introduction to dynamic meteorology, Academic Press, 507 pp.

Hudson, J. G., 1989: An instantaneous CCN spectrometer. J. Atmos. Ocean. Technol.,6, 1055-1065.

Jiménez-Escalona J.C. and Peralta O, 2010. Processing of aerosol particles in convective cumulus cloud: a case study in the Mexican east pacific, Advances in Atmospheric Sciences, Vol. 27, No. 6, pp 1331 – 1343.

Leaitch, W.R., Strapp, J.W., Wiebe, H.A., Isaac, G.A., 1986. In: Pruppacher, H.R., Semonin, R.G., Slinn, W.G.N. (Eds.), Precipitation Scavenging, Dry deposition, and Resuspension. Elsevier, pp. 53 - 59.

Leaitch, W.R., 1996. Observations pertaining to the effect of chemical transformation in cloud on the antropogenic aerosol size distribution. Aerosol Sci. Technol. 25, 157 – 173.

Lohmann U., Tselioudis G., and Tyler C., 2000, Why is the cloud albedo-particle size relationship different in optically thick and optically thin clouds?, Geophys. Res. Lett, 27, No 8, 1099 – 1102.

Meador, W.E. and Weaver, W.R., 1980, Two-Stream Aproximations to Radiative Tranfer in Planetary Atmospheres: A Unified Description of Existing Method and a New Improvement, J. Atmos. Sci, Vol 37, No. 3, 630-643

Mie, G., 1908: Beitrage zur optik trüber medien speziell kolloidaler metallösungen. Ann. Phys. 25, 377-445.

Naoki K., Hobbs P. V., Isjizaka Y., Quian G. W., 2001, Aerosol properties around marine tropical cumulus clouds, J. Geophys. Res. 106, D13, 14435–14445

O' Dowd, C.D., Lowe, J. A., Smith, M.H., 2000, The effect of clouds on aerosol growth in the rural atmosphere, Atmos. Res. 54, 201 –221.

Pollack J. B., et al., 1981, Radiative propierties of the background stratospheric aerosols and implications for perturbed conditions, Geophys. Res. Lett., 8, 26 – 28.

Pruppacher H. R. and Klett J. D., 1997: Microphysics of Clouds and Precipitation, Kluwer Academic Publisher., 954 pp.

Raga, G.B. and P.R. Jonas, 1993 a: Microphysical and radiative properties of small cumulus clouds over the sea. Quart. J. Royal Meteor. Soc., 119, 1399-1417.

Raga, G.B. and P. R. Jonas, 1993 b: On the link between cloud-top radiative properties and sub-cloud aerosol concentrations. Quart. J. Royal Meteor. Soc., 119, 1419-1425.

Rosenfeld, D., 1999: TRMM Observed first direct evidence of smoke from forest fires inhibiting rainfall. Geophys. Res. Letters, 26, 3105-3108.

Squires, P., 1958. The microstructure and colloidal stability of warm clouds. Tellus 10, 256-271.

Strapp J. W., Leaitch W. R., and Liu P. S. K., 1992, Hydrated and dried aerosol-size-distribution measurements from the particle measuring systems FSSP-300 probe and the deiced PCASP-100X probe, J. Atmos. Ocean. Technol.,Vol 9, No 5, 548-555.

Tang I. N., 1976, Phase tranformation and growth of aerosol particles composed of mixed salts, J. Aerosol Sci., Vol 7, 361 – 371.

Tang I. N., Munkelwitz H. R., 1977, Aerosol growth studies-III; Ammonium Bisulfate Aerosols in a moist atmosphere, J. Aerosol Sci., Vol 8, 321 – 330.

Twomey, S., 1974: Pollution and the planetary albedo, Atmos. Environ, 8, 1251-1256.

Twomey, S. 1991, Aerosols Clouds and Radiation, Atmos. Environ., 25A, 2435-2442.

Wallace J. M. and Hobbs P. V., 1977, Atmospheric Science, An introductory survey, 467 pp.

Wang, P. K. and H. R. Pruppacher, 1977: An experimental determination of the efficiency with which aerosol particles are collected by water drops in subsaturated air, J. Atmos. Sci., 34, 1664–1669.

Weller, B., B. Albrecht, S. Esbensen, C. Eriksen, A. Kumar, R. Mechoso, D. Raymond, D. Rogers, D. Rudnick, 1999: A science and implementation plan for EPIC: An eastern Pacific investigation of climate processes in the coupled ocean-atmosphere system. [Available online from http://wwwatmos.washintong.edu/gcg/EPIC/].

Chapter 8

Characteristics of Low-Molecular Weight Carboxylic Acids in PM2.5 and PM10 Ambient Aerosols From Tanzania

Stelyus L. Mkoma, Gisele O. da Rocha,
José D.S. da Silva and Jailson B. de Andrade

Additional information is available at the end of the chapter

1. Introduction

Carboxylic acids together with water-soluble inorganic ions are an important group of water-soluble organic compounds in the atmospheric aerosols (Jacobson et al., 2000, Bourotte et al., 2007). They are highlighted because they account for substantial portion of atmospheric aerosols, and potentially control chemical and physical properties of the particles. Consequently, they may have direct and indirect effects on the earth's radiation balance by scattering incoming solar radiation, which counteracts the global warming (IPCC, 2007). More attention has been paid to carboxylic acids due to their potential to modify the hygroscopic properties of atmospheric particles, including cloud condensation nuclei activity and hence to change global radiation balance (Kerminen, 2001; Peng et al., 2001). Major water-soluble inorganic ions are associated with atmospheric visibility degradation, adverse human health effects, and acidity of precipitation (Dockery & Pope, 1996; IPCC, 2007).

Among the organic acids, low molecular weight carboxylic acids such as acetic, oxalic and malonic are generally most abundant in the atmospheric aerosols. Carboxylic acids in variable concentrations have been reported in various environments including rural and urban atmosphere (Kawamura & Sakaguchi, 1999; Kerminen et al., 2000; Nicolas et al., 2009) and have different source origin, including biomass burning, fossil fuel combustion (Kawamura 1987; Narukawa et al., 1999), sea spray, traffic and industrial emissions and photochemical oxidation of precursors from anthropogenic and biogenic origin (Kawamura & Sakaguchi, 1999; Limbeck & Puxbaum, 1999; Kumar et al., 2001; Chakraborty & Gupta, 2010). Other sources for carboxylic acids in the marine atmosphere include in-cloud and heterogeneous formations (Warneck, 2003).

Chemical composition of PM2.5 and even that of PM10 aerosols is important to gain insights into sources and of their toxicity and to evaluate effectiveness of abatement strategies for relevant emission sectors. Particulate matter (PM) with aerodynamic diameter less than 2.5 μm (PM2.5) exhibited stronger relation with health than those with aerodynamic diameter less than 10 μm (PM10), but other studies have reported a strong potential of PM10 to human health (Salma et al., 2002; Kappos et al., 2004). Most studies on low molecular weight carboxylic acids and their related compounds (Limbeck et al., 2001; Limon-Sanchez et al., 2002; Kawamura & Yasui, 2005) and major ions (Harrison et al., 2004; Karthikeyan & Balasubramanian, 2006; Mariani et al., 2007; Kundu et al., 2010; Mkoma et al., 2010) have extensively been reported.

In Africa similar aerosols measurements especially of organic components are missing. Therefore, a full scenario of air quality is far from being revealed because some pollutants including carboxylic acids have not been measured. The knowledge of elucidating chemical composition, levels, and source profiles of aerosols in the Tanzania atmosphere remains a challenge and is needed for both scientific and policy reasons. The continuous changes in socioeconomic and political environments in Tanzania result in changes in development, particularly in transport, industry, energy, and construction sectors. This chapter reports for the first time in Tanzania, composition of low molecular weight carboxylic acids in PM2.5 and PM10 aerosol samples collected from a rural background atmosphere in Morogoro. An insight of characteristics of water-soluble inorganic ions is also discussed in this chapter.

2. Experimental

2.1. Aerosol sampling site

Aerosol samples were collected at a rural site in Morogoro (300,000 inhabitants) between 26 April and 10 May 2011. This site is located at about 200 km west of the Indian Ocean and the city of Dar es Salaam, a business capital in Tanzania (Fig. 1). The samples were collected at Solomon Mahlangu Campus of Sokoine University of Agriculture (06°47'41"S, 37°37'44"E, altitude 504 m a.s.l.). This site is located about 5 km from Morogoro central area and major road systems and possible aerosol sources include biomass burning, agriculture, livestock and soil dust. Approximately 70% of this area is covered by vegetation and about 15% with pasture field. Conversely, tropical savannah is the most important land cover in large part of the sampling site.

2.2. Aerosol collection

Two samplers were used in parallel to collect aerosol particles: a "Gent" PM2.5 and PM10 filter holder each with two quartz fibre filters (Whatman QM-A) in series. Quartz fibre filters can adsorb volatile organic compounds (VOCs) causing positive artifacts when measuring PM and particulate OC. On the other hand, semi-volatile organic compounds (SVOCs) in aerosols may partially evaporate during sampling resulting in negative artifacts (Turpin et al., 2000; Mader et al., 2003; Hitzenberger et al., 2004). The quartz fibre filters were pre-fired at 550 °C during 24 h before use. Samplers operated at a flow rate of 17 L/min and were mounted on grass survey at SMC synoptic station approximately 2.7 m above ground level. The sampling was carried out approximately in 24 h intervals and exchange of filters during sampling periods was done at 7:30 am. A total of 11 sets of actual filed samples and 2 blanks were collected for each sampler and used in this chapter. After sampling the exposed filters were folded in half face to face, placed in polyethylene plastic bags and kept frozen at -4 °C during storage and transported cool to the laboratory of research and development in chemistry (LPQ) at the Institute of chemistry, Federal University of Bahia (UFBA). The samples were stored in a freezer at −20 °C prior to analysis. All the procedures were strictly quality-controlled to avoid any possible contamination of the samples.

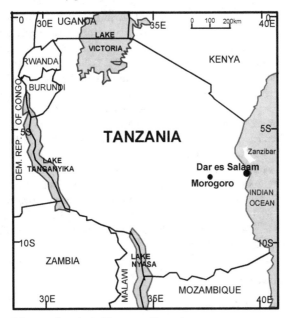

Figure 1. Location of the sampling site in Morogoro, Tanzania

During the sampling period meteorological data were collected from the site. The daily winds were predominantly south-easterly with an average speed of 6.8 m/s. Average temperature was 26.8 °C and average relative humidity was 73%. The recorded maximum temperature and relative humidity were 29.8 °C and 79.5%, while minimum values were 23.7 °C and 63.5%, respectively. During the campaigns 5 days hand rainfall of a total 19.9 mm.

2.3. Aerosol analyses

For particulate mass measurements, the filter samples were weighed before and after sampling with an analytical microbalance balance Mettler Toledo MX5 (reading precision 1 µg). Before weighing, the filters were conditioned in a chamber equipped with hydro-thermometer clock at a temperature of 20 °C and relative humidity of 40% for 48 h and the weightings were done under these conditions.

For determination of carboxylic acids and water-soluble ions one-half of 12.88 cm² portions punched from of each PTFE filter was extracted using 5 ml Milli-Q ultrapure water (resistivity of 18.2 MΩcm, Barnstead International, USA) in a shaker tubes Model AT56 (Fanem Ltd, Sao Paulo, Brazil) for 5 minutes, followed by filtering through Polytetrafluoroethylene (PTFE) filter (0.45 μm pore size, Sartorius Stedim, Germany). The concentrations of aqueous extracts were determined by Dionex ion chromatography ICS 1100 and ICS 2100 for acids/anions and cations respectively which was equipped with an auto sampler (Dionex ICS Series AS-DV). An analytical column AS16 (3 x 50 mm) with AG16 guard column (3 x 50 mm) and CSRS-300 I (2 mm) suppressor in ion-exchange mode was used to determine carboxylates (monocarboxylates: formate and acetate; dicarboxylates: oxalate, malonate, succinate, and maleate; ketocarboxylate: pyruvate) and water-soluble anions (chloride Cl⁻, nitrate NO_3^- and sulphate SO_4^{2-}). The eluent gradient programme was sweeping from 6.0 to 8.0 mmol/L KOH in 35 minutes under flow rate of 0.38 μL/min, except for acetic acid which was determined in another run, reducing injection time to avoid overlap of peaks. For determination of water-soluble cations (NH_4^+, Na^+, K^+, Mg^{2+} and Ca^{2+}) an analytical column CS16 and Guard column CG16 (both 3 x 50 mm) and CSRS-I (2 mm) suppressor in a chemical mode were used. An eluent of 17.5 mmol/L H_2SO_4 was used at flow rate of 0.35 μL/min. The injection volume was 25 μL for all detection. Peak identification was confirmed based on a match of ion chromatograph retention times and standard samples. Limit of detection determined as mean equal to 3 times standard deviation of the field blank value corresponded to a range of 0.008 to 0.017 ng/L for carboxylates, 0.008 to 0.023 ng/L for anions and 0.021 to 0.083 ng/L for cations. Limits of quantification were between 0.026 and 0.058 ng/L for carboxylates, 0.028 and 0.078 ng/L for anions and 0.063 and 0.252 ng/L for cations.

3. Results and discussion

3.1. Concentrations of PM mass

Mean PM mass concentrations and associated standard deviations and ranges as derived from the two low-volume samplers are shown in Table 1. The results showed that mean mass concentration of PM2.5 and PM10 aerosols during the campaign were 13±3.5 μg/m³ and 16±2.3 μg/m³, respectively. The percentages of PM2.5 mass in PM10 size fraction (Fig. 2) found to range from 44–99% with a mean of 83±29%. These results indicate that most of PM mass was in PM2.5 size fraction. High PM2.5/PM10 ratios for PM mass indicate that there is small contribution from soil dust, which is known to be mostly associated with PM10 aerosols. Currently in Tanzania, the ambient air quality standard limit values for inhalable particulate matter are 60 to 90 μg/m³ for PM10 (TBS, 2006). The mean concentrations for PM10 mass at our site in Morogoro were below these average limit values. In addition, the current data sets were in line with levels reported in our previous studies (Mkoma et al., 2009a,b; Mkoma et al., 2010). Nevertheless, when compared PM mass data from our rural site in Tanzania are in line with few available other data sets for rural sites in Southern Africa (Nyanganyura et al., 2007). They are also comparable to or lower to other sites in Europe and Asia (Van Dingenen et al., 2004; Gu et al., 2010; Maenhaut et al., 2011; Ram & Sarin, 2011).

3.2. Concentrations of carboxylates ions

Table 1 present mean total concentrations and range of carboxylates (TCAs) which were 23.7±6.5 ng/m^3 (range: 13.3-36.5 ng/m^3) in PM2.5 and 36.4±12 ng/m^3 (range: 10.7-58.2 ng/m^3) in PM10 aerosols. Oxalate and malonate were most abundant carboxylates in PM2.5 accounting for 32.5% and 31.85% of total carboxylates, respectively, whereas in PM10 acetate was most abundant accounted for 62.5% of total carboxylates followed by oxalate which accounted for 32.6% of total carboxylates. Other studies have also reported oxalates to be most abundant carboxylate in aerosol samples (Mochida et al., 2003; Warneck, 2003). Pyruvate was also found in substantial amount and formate the least abundant counting on average 3% of total carboxylates in each of the aerosol fractions. Succinate and malonate were below detection limit in PM2.5 and PM10 aerosols, respectively. The total carboxylates accounted for 0.18% to total PM2.5 mass and 0.22% to PM10 mass. In comparison with other studies, the mean concentrations of all measured carboxylates in Tanzania were lower to those reported in urban and rural sites around the world (Souza et al., 1999; Kerminen et al., 2000; Yao et al., 2003; Kawamura & Yasui, 2005).

3.3. Water-soluble inorganic ions and ratios

Chemical characteristics of water-soluble inorganic ions and their relative abundances in PM2.5 and PM10 aerosols are also shown in Table 1. In both aerosol fractions, water-soluble Mg^{2+} was the most important cation and SO_4^{2-} the main anionic species. On average Mg^{2+} accounted for 44.4% of total water-soluble ions in PM2.5 and 24.7% in PM10 whereas SO_4^{2-} accounted for 22.8% and 35.2% of total ions in PM2.5 and PM10, respectively. High levels of crustal element Mg^{2+} together with Ca^{2+} are essentially attributable to soil/mineral dust dispersal. As to reasonable NH_4^+ levels (8% of total ions) in PM2.5, this may be due to presence of ammonia gas from biomass burning especially during smoldering combustion (Andreae & Merlet, 2001) and from agricultural activities in particular cattle raising (Street et al., 2003; Stone et al., 2010). Water-soluble K^+, a good indicator for biomass burning, was second most abundant cation in PM2.5 accounted for 10.6% of total water-soluble ions.

For SO_4^{2-} the higher levels could be attributed to its efficient formation by in-cloud processing of SO_2 (Yao et al., 2003) and from secondary formation processes (Allen et al., 2004). As to low NO_3^- levels, this is likely due to the fact that the site is rural with little or no traffic and undoubtedly there are less anthropogenic emissions of precursor gas NO_x. Also as to low concentrations of Na^+ which is mainly derived from sea-salt, this is presumably due to long distance (about 200 km) from the Indian Ocean to our sampling site. The observed levels for water-soluble ions are comparable with those reported in our previous work in Morogoro (Mkoma et al., 2009a; Mkoma et al., 2010). It appears that the levels of

SO_4^{2-}, NO_3^-, and NH_4^+ in PM10 fractions are substantially lower in Tanzania than at European rural sites (Putaud et al., 2004) and Asia (Aggarwal & Kawamura, 2009; Pavuluri et al., 2011).

Species	PM2.5					PM10				
	Mean	SD	Min.	Max.	Rel. Ab.	Mean	SD	Min.	Max.	Rel. Ab.
PM mass ($\mu g/m^3$)	13.3	3.5	8.2	19.5	-	16.2	2.3	12.5	20.5	-
Carboxylates ions (ng/m^3)										
Formate, FA	0.71	0.30	0.38	1.27	3.00	1.2	0.6	0.5	2.6	3.4
Acetate, Ac	5.4	2.2	0.4	7.8	22.9	22.7	3.3	16.3	27.5	62.5
Oxalate, Oxa	7.7	2.7	4.7	13.1	32.5	11.8	7.2	4.5	31.0	32.6
Malonate, Mal	7.5	3.9	4.8	18.3	31.8	-	-	-	-	-
Succinate, Suc	-	-	-	-	-	1.6	2.4	0.2	6.5	4.5
Pyruvate, Pyr	2.4	0.7	1.5	3.8	9.9	2.2	0.7	0.9	3.3	6.1
Total carboxylate	23.7	6.5	13.3	36.5	-	36.4	12.0	10.7	58.2	-
Water-soluble ions (ng/m^3)										
NH_4^+	37.8	11.7	21.0	66.5	8.4	26.2	12.5	10.7	54.0	4.1
NO_3^-	4.5	1.6	1.2	8.0	1.0	25.1	12.4	10.5	52.0	3.9
SO_4^{2-}	102	27	69.1	160	22.8	237	125	38.8	487	36.6
Cl^-	2.3	0.4	1.6	3.1	0.5	9.6	5.7	3.3	20.2	1.5
Na^+	15.6	2.5	13.2	21.5	3.5	98.5	42.0	41.1	174.8	15.2
K^+	47.5	24.1	22.0	97.0	10.6	37.6	28.2	6.3	88.5	5.8
Mg^{2+}	199	33.7	155	265	44.4	158	69	16.4	228	24.5
Ca^{2+}	39.3	8.5	28.1	58.8	8.8	56.2	42.9	1.0	107	8.7
Total ions	448	88	348	585	-	646	214	256	1108	-

Rel. Ab. = Relative abundances

Table 1. Mean concentrations, ranges and relative abundances (%) of carboxylates and water-soluble inorganic ions in PM2.5 and PM10 aerosols from Morogoro.

To determine the impact of marine sources on chemical composition of aerosol particles in PM2.5 and PM10 fractions, sea-salt ratios were calculated for each inorganic ion using Na as a reference element, assuming all Na to be of marine origin. The ratios for Cl^-/Na^+, SO_4^{2-}/Na^+, K^+/Na^+, Mg^{2+}/Na^+ and Ca^{2+}/Na^+ in PM2.5 were 0.15 (0.25), 6.48 (1.81), 2.95 (0.04), 12.97 (0.04), 2.44 (0.12), respectively. The corresponding values in PM10 were 0.10 (0.25), 2.51 (1.81), 0.43 (0.04), 1.68 (0.04), 0.75 (0.12), respectively. Values in brackets represent average ratios for each ion in sea-water (Brewer, 1975). Larger ratios of ions indicate incorporation of non-marine constituents in aerosols. As to low mean Cl^-/Na^+ ratios than sea-water ratio indicates that a minor fraction of Na^+ may be contributed from other sources such as mineral dust. But also low ratio could be due to modifications of sea-salt fraction by non-marine constituents. Chloride loss may be explained by

heterogeneous reaction of airborne sea-salt with acidic gases and aerosol species
(Millero, 2006).

3.4. Time series of PM mass and selected aerosol species

Time series of PM mass, selected acids and ions species in PM2.5 and PM10 fractions as a
function of sampling time are shown in Figs. 2 and 3. Nss-SO$_4{}^{2-}$ in Fig. 2 was hereby
obtained by subtracting sea-salt contribution from measured SO$_4{}^{2-}$ data. Sea-salt
contribution of SO$_4{}^{2-}$ was obtained as 0.252Na$^+$, whereby Na$^+$ is the measured concentration
of Na$^+$ and 0.252 is SO$_4{}^{2-}$/Na$^+$ ratio in the bulk seawater composition given by Riley and
Chester (1971). As can be observed in Fig. 2, selected species in both size fractions showed
no clear trends that can be noted but showed slightly variation during sampling period
especially for PM10 aerosols. The observed behaviour of the species could be resulted from
variations in sources strengths and meteorological conditions, such as mixing height.
Additionally, high relative humidity (mean: 73%) during the campaign could serve as
removal mechanisms hence lead to a daily variation in carboxylates levels.

In this study, oxalate concentrations were high to a factor of 10 than those of formate during
the sampling period and in both PM2.5 and PM10 aerosol particles. These results indicate
that formate was mainly from photochemical oxidation, while oxalates might have other
sources besides photochemical oxidation. On the other hand, the concentrations of acetate
were high than those of oxalate. Acetic acid in the atmosphere has been reported to be
produced by oxidations of longer-chain dicarboxylic acids (Kawamura et al., 1996).
Therefore, the observed acetate levels suggest that longer-chain dicarboxylic acids were
possibly available at our site. Unfortunately, no data for high molecular weight dicarboxylic
acids measured for the campaign.

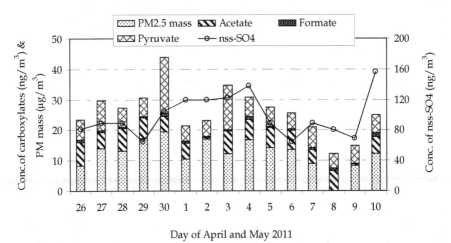

Day of April and May 2011

Figure 2. Time series of PM mass, carboxylates and nss-SO$_4{}^{2-}$ in PM2.5 fraction at Morogoro

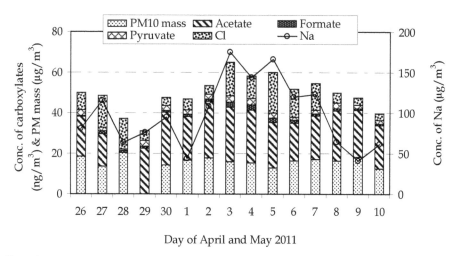

Day of April and May 2011

Figure 3. Time series of PM mass, carboxylates and Na⁺ and Cl⁻ in PM10 fraction at Morogoro

3.5. PM2.5 to PM10 ratios

The average PM2.5 (fine) to PM10 (coarse) percentage ratios and associated standard deviations for PM mass, carboxylates and various water-soluble inorganic ions are shown in Fig. 4. The ratios were calculated on the basis of data for PM2.5 and PM10 samples taken in parallel and then averaged over all samples from the sampling period. The mean fine to coarse ratios for all species with exception of those for acetate, Na^+, Ca^{2+}, NO_3^- and Cl^- were predominantly associated with fine fraction (for more than 55%). High fine/coarse ratio for PM mass may be due to a less contribution from soil dust, which is known to be mostly associated with coarse particles. For carboxylates, high ratios (even larger than 70%) are considered to be attributed from secondary organic aerosols (SOA), biomass burning activities and high temperature (average: 26.8 °C during sampling period). Concentrations of oxalate in PM2.5 showed strong correlations ($r^2 = 0.70$) with those in PM10 aerosols. The slope of linear regression equations (PM2.5 = 2.48 x PM10) indicated that oxalate was mainly present in fine fraction during sampling period. On the other hand, different size distributions between oxalate and acetate could be related to their different physical characteristics. Acetate in PM2.5 fraction could easily volatilize (more volatile than oxalate) to gas phase, part of which could be absorbed on PM10 particles.

For water-soluble inorganic ions, as expected, sea-salt elements (Na, Cl) and indicator element for crustal matter (Ca) were predominantly (for more than 62%) associated with PM10 aerosols. NH_4^+ and nss-SO_4^{2-} were mainly present in the fine aerosols suggesting that these species originated from high temperature sources and/or gas-to-particle conversion. The nss-SO_4^{2-} is due to oxidation of SO_2, which is predominantly from anthropogenic origin (e.g. biomass burning). A well-known indicator for biomass burning, K was associated with the fine particles (about 100%) suggesting vegetative

emissions and crustal source could be an important source for K⁺ aerosols at this site with
small impact from biomass burning activities.

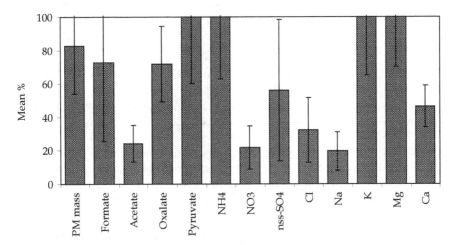

Figure 4. Mean contribution (%) of PM2.5 aerosols mass and selected aerosol species in PM10 fraction
during the campaign in Morogoro.

3.6. Sources of carboxylates and water-soluble ions in aerosols

3.6.1. Correlation analysis

Correlation coefficients of PM mass, carboxylates and source indicators, shown in Table 2
were performed in order to understand their possible sources and formation mechanisms.
The selected source indicators include K⁺ for biomass burning and vegetation emissions,
Na⁺ and Cl⁻ for sea spray or waste burning, and SO_4^{2-} for secondary formation of different
mechanisms. Temperature and wind speed have been used as additional parameters to
illustrate the atmospheric behaviours of carboxylic acids. In this study K⁺ had good
correlation with formate (r^2=53) and moderate to poorly correlations with other
carboxylates in PM10. This indicates that formate could be originated from biomass
and/or waste burning emissions but other carboxylates are considered to have other
important sources than biomass burning. It can also be observed from Table 2 that there
were possible similar sources for formate and other carboxylates (oxalate, Succinate and
pyruvate) as verified by good correlation between them in PM2.5 and PM10 aerosols.
Pyruvate shows good correlation with acetate (r^2 = 54) and succinate (r^2=51) in PM10
aerosols. These indicate a feature of photochemical decomposition of succinic acid (Yao et
al., 2002).

Sea-salt derived aerosols have been reported to have particle with aerodynamic diameter
between 1 and 5 μm (Kerminen et al., 2000). In this study, we found Na⁺ has pronounced

amount in PM10 aerosol and to slightly extent Cl⁻, suggesting sea spray could be one of the contributing sources of the aerosol components at the site. But even though Na⁺ correlate well (r²=73) with Cl⁻, the calculated Cl⁻ to Na⁺ mass ratio to sea-salt component in aerosol particles had mean value between 0.15 (PM2.5) and 0.10 (PM10). This suggest that continental contributions was most important than marine contribution since Cl⁻/Na⁺ ratio in marine aerosols varies between 1.0 and 1.7 (Chesselet et al., 1972).

Sulphate has been used as reference to investigate major formation routes of carboxylic acids (Yu et al., 2005). As shown in Table 2, formate and pyruvate showed good correlation with SO_4^{2-} in PM2.5, suggesting that in-cloud and heterogeneous formations play an important role in the formation of carboxylates. On the other hand, poor correlation of malonic with SO_4^{2-} in PM2.5 suggests that possibly the acid is volatile at ambient temperatures (Peng et al., 2001). Acetate and oxalate showed poor correlations with SO_4^{2-} in both aerosol fractions, indicating that they mainly originated from primary emissions sources and/or the atmospheric processes different from those of SO_4^{2-}. This is contrary to what have been observed in other studies that in-cloud and heterogeneous formations can yield a good correlation between oxalate and SO_4^{2-} (Yu et al., 2005).

Wind speed poorly correlated with most carboxylates except with acetate (r²=51) in PM10 aerosols. This indicates that in addition to secondary formation carboxylates were mainly generated from local sources, while acetate might be related to long range aerosols transport to the sampling site. It should also be noted that primary emissions are major sources of precursors for most carboxylic acids (Kawamura & Yasui, 2005).

Species	FA	Ac	Oxa	Mal	Suc	Pyr	SO_4^{2-}	NO_3^-	Cl^-	Na^+	K^+	Ca^{2+}	WS
FA		0.26	0.10	0.19	–	**0.91**	**0.60**	0.31	0.10	–	0.38	–	-0.23
Ac	0.42		0.21	0.11	–	0.28	-0.04	0.45	0.27	–	-0.09	–	-0.21
Oxa	**0.77**	0.29		0.31	–	0.11	0.34	0.23	0.17	–	0.30	–	-0.07
Mal	–	–	–		–	0.17	0.17	**0.51**	0.33	–	0.13	–	0.08
Suc	**0.57**	0.42	-0.11	–		–	–	–	–	–	–	–	–
Pyr	0.48	**0.54**	0.13	–	**0.51**		**0.52**	0.21	-0.01	–	0.31	–	-0.20
SO_4^{2-}	–	0.26	0.30	–	0.31	-0.10		-0.16	-0.06	–	**0.93**	–	0.06
NO_3^-	**0.88**	0.17	**0.74**	–	**0.55**	0.30	0.11		0.78	–	-0.41	–	-0.03
Cl^-	**0.55**	-0.29	**0.56**	–	0.28	-0.11	0.05	**0.81**		–	-0.26	–	-0.12
Na^+	**0.81**	0.18	**0.74**	–	0.39	0.36	0.23	**0.87**	**0.73**		–	–	–
K^+	**0.53**	0.35	0.38	–	0.45	0.45	0.22	0.39	0.18	0.45		–	-0.02
Ca^{2+}	**0.65**	0.23	**0.52**	–	0.30	0.32	0.25	**0.58**	0.32	**0.58**	**0.51**		–
WS	-0.10	**0.51**	0.25	–	-0.44	-0.23	0.28	-0.06	-0.08	-0.19	0.11	0.26	

WS = Wind speed

Table 2. Correlation coefficients for PM mass, carboxylates, selected ions and wind speed in PM2.5 (upper diagonal triangle) and PM10 fractions (lower diagonal triangle) in Morogoro. Correlation coefficients that are larger than 0.50 are indicated in bold.

3.6.2. Concentrations ratios

The ratio of acetic to formic acid has been used as good indicator of contributions of primary (high ratio) and secondary sources (low ratio) to carboxylic acids (Talbot et al., 1988, 1990; Grosjean, 1992). As can be seen in Table 3, low acetate/formate ratios for both PM2.5 and PM10 aerosols particles indicate that secondary formation was an important contributing source of carboxylates at our site. This suggestion is supported by the fact that high mean average temperature during sampling period (mean: 26.8 °C) might be controlling factor in determining the contribution of primary and secondary sources to these carboxylates. However, there are various types of the atmospheric reactions forming carboxylic acids (e.g. formic, acetic, and oxalic) in urban and near urban atmospheres, which include oxidation of unsaturated fatty acids originating from cooking activities, ozone oxidation of olefins emitted from vehicular exhausts (Scheff & Wadden, 1993) and oxidation of aromatic hydrocarbons (Kawamura & Ikushima, 1993).

The ratio of oxalic acid to total dicarboxylic acid (for this study oxalic, malonic, succinic acids) can be used to evaluate aging process of organic aerosols (Kawamura & Sakaguchi, 1999), because diacid such as oxalic acid can be produced by oxidations of longer-chain dicarboxylic acids (Kawamura et al., 1996). In this study oxalate to total dicarboxylates ratios show low values in both aerosol fractions, indicating that aerosols emitted from various sources and transported to this site were less and equally aged. Since there relative humidity was high during the campaign (up to 73% on average), it is supposed that oxalate was also produced in aqueous phase. Aqueous phase chemistry in aerosol and/or cloud droplets is important in production of oxalic acid (Warneck, 2003). On the other hand, mean ratio of oxalate to K^+ in PM2.5 aerosols was 0.19±0.08, somewhat close to or higher than those reported range (0.03–0.1) for flaming and smoldering phases in burning plumes (Yamasoe et al., 2000). This suggests that carboxylates might be originated from biogenic sources with contribution from biomass burning emissions.

Ratio	PM2.5		PM10	
	Mean±Stdev	Range	Mean±Stdev	Range
Acetate/Formate	0.24±0.31	0.07–1.07	0.05±0.03	0.03–0.10
Oxalate/Total dicarboxylates	0.33±0.08	0.23–0.49	0.33±0.17	0.15–0.83
Oxalate/K^+	0.19±0.08	0.07–0.36	–	–

Table 3. Mean ratios and ranges for carboxylates and K^+ in PM2.5 and PM10 aerosols in Morogoro.

4. Conclusion

PM2.5 and PM10 aerosols samples were collected from a rural site Morogoro, Tanzania and analysed for low molecular weight carboxylates and water-soluble inorganic ions. Oxalate and malonate were dominant species in PM2.5 while acetate was most prominent species in

PM10 aerosols followed by oxalate. Of the ionic components, SO_4^{2-}, K^+, and Mg^{2+} in PM2.5 and SO_4^{2-}, Na^+, and Mg^{2+} in PM10 made lager contribution to total water-soluble inorganic aerosol mass. Various ratios and correlations between carboxylates and ions used for possible source identification suggest that primary emissions, secondary formation, and to a slightly extent sea spray and biomass burning could be the sources for the aerosols at this site. The ratio of acetate to formate was used to distinguish primary and secondary sources of these carboxylates and was found to be close to reported value for secondary reactions, indicating dominance of secondary sources. Substantial concentration of carboxylates and water-soluble ions observed in the Morogoro atmosphere suggest that it was urgent to study the characteristics and sources of these species to better understand their roles in the Tanzania environment. However, more work is needed to determine longer-chain (high) molecular weight carboxylic acids and related organic compounds and their seasonal variations in other urban and rural sites in Tanzania.

Author details

Stelyus L. Mkoma
Faculty of Science, Sokoine University of Agriculture, Tanzania

Gisele O. da Rocha, José D.S. da Silva and Jailson B. de Andrade
Instituto de Química, Universidade Federal da Bahia, Brazil

Acknowledgement

The authors are grateful to funding from Directorate of Research and Postgraduate Studies, Sokoine University of Agriculture (SUA)-Tanzania; Conselho Nacional de Desenvolvimento Científico e Tecnológico (CNPq)-Brazil; and Instituto Nacional de Ciência e Tecnologia, de Energia e Ambiente (INCT-E&A), Instituto de Química, Universidade Federal da Bahia (UFBA)-Brazil. We also wish to acknowledge Mr. Filbert T. Sogomba (SUA) for help in logistics and collection of aerosol samples and Cibele Cristina de Araújo Soares (UFBA) for PM mass measurements.

5. References

Aggarwal, S.G. & Kawamura, K. (2009). Carbonaceous and inorganic composition in long-range transported aerosols over northern Japan: Implication for aging of water-soluble organic fraction. *Atmospheric Environment*, Vol. 43, No. 16, (May 2009), pp. 2532–2540, ISSN: 1352-2310

Allen, A.G., Cardoso, A.A. & da Rocha, G.O. (2004). Influence of sugar cane burning on aerosol soluble ion composition in Southeastern Brazil. *Atmospheric Environment*, Vol. 38, No. 30, (September 2004), pp. 5025-5038, ISSN: 1352-2310

Andreae, M.O. & Merlet, P. (2001). Emission of trace gases and aerosols from biomass burning. *Global Biogeochemical Cycles,* Vol. 15, No. 4, (December 2001), pp. 955–966, ISSN: 1944–9224

Bourotte, C., Curi-Amarante, A.-P., Forti, M.-C., Luiz, A., Pereira, A.,. Braga, A.L. & Lotufo, P.A. (2007). Association between ionic composition of fine and coarse aerosol soluble fraction and peak expiratory flow of asthmatic patients in São Paulo city (Brazil). *Atmospheric Environment,* Vol. 41, No. 10, (March 2007), pp. 2036-2048, ISSN: 1352-2310

Brewer, P.G. (1975). Minor elements in sea water, In: *Chemical Oceanography, Vol. 1,* 2nd edition, J.P Riley and G. Skirrow (Ed.): 415-496, Academic Press, ISBN 0-12-588601-2, London, UK

Chakraborty, A. & Gupta, T. (2010). Chemical Characterization and Source Apportionment of Submicron (PM1) Aerosols in Kanpur Region, India. *Aerosol and Air Quality Research,* Vol. 10, No. 4, (December 2010), pp. 433–445, ISSN: 2071-1409

Chesselet, R., Morelli, M. & Buat-Menard, P. (1972). Some aspects of the geochemistry of marine aerosols. In: Dyrssen, D., Jagner, D. (Eds.), *The Changing Chemistry of the Oceans,* Proceedings of Nobel Symposium 20, Wiley-Interscience, New York, p. 92.

Dockery, D. & Pope, A. (1996). Epidemiology of acute health effects: Summary of time-series studies. In: *Particles in Our Air: Concentration and Health Effects,* J.D. Spengler, and R. Wilson, (Eds.): 123-147, Harvard University Press, ISBN 0-674-24077-4, Cambridge, MA

Grosjean, D. (1992). Formic and acetic acids: emissions, atmospheric formation and dry deposition at two southern California locations. *Atmospheric Environment,* Vol. 26A, No. 18, (December 1992), pp. 3279–3286, ISSN: 1352-2310

Gu, J., Bai, Z., Liu, A., Wu, L., Xie, Y., Li, W., Dong, H. & Zhang, X. (2010). Characterization of Atmospheric Organic Carbon and Element Carbon of PM2.5 and PM10 at Tianjin, China. *Aerosol Aerosol and Air Quality Research,* Vol. 10, No. 2, (April), pp. 167–176, ISSN: 2071-1409

Harrison, R.M., Jones, A.M. & Lawrence, R.G. (2004). Major component composition of PM10 and PM2.5 from roadside and urban background sites. *Atmospheric Environment,* Vol. 38, No. 27, (September 2004), pp. 4531, ISSN: 1352-2310

Hitzenberger, R., Berner, A., Galambos, Z., Maenhaut, W., Cafmeyer, J., Schwarz, J., Miiller, K., Spindler, G., Wieprecht, W., Acker, K., Hillamo, R. & Makela, T. (2004). Intercomparison of methods to measure the mass concentration of the atmospheric aerosol during INTERCOMP2000-influence of instrumentation and size cuts. *Atmospheric Environment,* Vol. 38, No. 38, (December 2004), pp. 6467–6476, ISSN: 1352-2310

Intergovernmental Panel on Climate Change (IPCC), (2007). IPCC fourth assessment report (2007), Contribution of working group I. Cambridge University Press, London, p. 996

Jacobson, M.C., Hanson, H.C., Noone, K.J. & Charlson, R.J. (2000). Organic atmospheric aerosols: review and state of the science. *Reviews of Geophysics*, Vol. 38, No. 2, (May 2000), pp. 267–294, ISSN: 1944–9208

Kappos, A.D., Bruckmann, P., Eikmann, T., Englert, N., Heinrich, U., Höppe, P., Koch, E., Krause, G.H.M., Kreyling, W.G., Rauchfuss, K., Rombout, P., Klemp, V.S., Thiel, W.R. & Wichmann, H.E. (2004). Health effects of particles in ambient air. *International Journal of Hygiene and Environmental Health*, Vol. 207, No. 4, (August 2004), pp. 399-407, ISSN: 1438-4639

Karthikeyan, S. & Balasubramanian, R. (2006). Determination of water-soluble inorganic and organic species in atmospheric fine particulate matter. *Microchemical Journal*, Vol. 82, No. 1, (January 2006), pp. 49 – 55, ISSN: 0026-265X

Kawamura, K. & Ikushima, K. (1993). Seasonal changes in the distribution of dicarboxylic acids in the urban atmosphere. *Environmental Science and Technology*, Vol. 27, No. 10, (September 1993), pp. 2227–2233, ISSN: 1520-5851

Kawamura, K., Kasukabe, H. & Barrie, L. (1996). Source and reaction pathways of dicarboxylic acids, ketoacids and dicarbonyls in Arctic aerosols: one year of observations. *Atmospheric Environment*, Vol. 30, No. 10-11, (May 1996), pp. 1709–1722, ISSN: 1352-2310

Kawamura, K. & Sakaguchi, F. (1999). Molecular distribution of water soluble carboxylic acids in marine aerosols over the Pacific Ocean including tropics. *Journal of Geophysical Research*, Vol. 104, No. D3, (February 1999), pp. 3501–3509, ISSN: 2156–2202

Kerminen, V.-M. (2001). Relative roles of secondary sulfate and organics in atmospheric cloud condensation nuclei production. *Journal of Geophysical Research*, Vol. 106, No. D15, (August 1999), pp. 17321–17333, ISSN: 2156–2202

Kerminen, V.-M., Ojanen, C., Pakkanen, T., Hillamo, R., Aurela, M. & Merilaien, J. (2000). Low-molecular-weight dicarboxylic acids in an urban and rural atmosphere. *Journal of Aerosol Science*, Vol. 31, No. 3, (March 2000), pp. 349–362, ISSN: 0021-8502

Kawamura, K. & Yasui, O. (2005). Diurnal changes in the distribution of dicarboxylic acids,ketocarboxylic acids and dicarbonyls in the urban Tokyo atmosphere. *Atmospheric Environment*, Vol. 39, No. 10, (March 2005), pp. 1945–1960, ISSN: 1352-2310

Kumar, A.V., Patil, R.S. & Nambi, K.S.V. (2001). Source Apportionment of Ambient Particulate Matter at Two Traffic Junctions in Mumbai, India. *Atmospheric Environment*, Vol. 35, No. 25, (September 2001), pp. 4245–4251, ISSN: 1352-2310

Kundu, S., Kawamura, K., Andreae, T.W., Hoffer, A. & Andreae, M.O. (2010). Diurnal variation in the water-soluble inorganic ions, organic carbon and isotopic compositions of total carbon and nitrogen in biomass burning aerosols from the LBA-SMOCC campaign in Rondônia, Brazil. *Journal of Aerosol Science*, Vol. 41, No. 1, (January 2010), pp. 118-133, ISSN: 0021-8502

Limbeck, A. & Puxbaum, H. (1999). Organic acids in continental background aerosols. *Atmospheric Environment*, Vol. 33, No. 12, (June 1999), pp. 1847–1852, ISSN: 1352-2310

Limbeck, A., Puxbaum, H., Otter, L. & Scholes, M.C. (2001). Semi-volatile behavior of dicarboxylic acids and other polar organic species at a rural background site (Nylsvley, RSA). *Atmospheric Environment*, Vol. 35, No. 10, (April 2001), pp. 1853–1862, ISSN: 1352-2310

Limon-Sanchez, M.T., Arriaga-Colina, J.L., Escalona-Segura, S. & Ruíz-Suárez, L.G. (2002). Observations of formic and acetic acids at three sites of Mexico City. *Science of the Total Environment*, Vol. 287, No. 3, (March 2002), pp. 203–212, ISSN: 0048-9697

Mader, B.T., Schauer, J.J., Seinfeld, J.H., Flagan, R.C., Yu, J.Z., Yang, H., Lim, H.J., Turpin, B.J., Deminter, J.T., Heidemann, G., Bae, M.S., Quinn, P., Bates, T., Eatough, D.J., Huebert, B.J., Bertram, T. & Howell, S. (2003). Sampling methods used for the collection of particle-phase organic and elemental carbon during ACE-Asia. *Atmospheric Environment*, Vol. 37, No. 11, (April 2003), pp. 1435–1449, ISSN: 1352-2310

Maenhaut, W., Nava, S., Lucarelli, F., Wang, W., Chi, X. & Kulmala, M. (2011). Chemical composition, impact from biomass burning, and mass closure for PM2.5 and PM10 aerosols at Hyytiälä, Finland, in summer 2007. *X-Ray Spectrometry*, Vol. 40, No. 3, (May/June 2011), pp. 168–171, ISSN: 1097-4539

Mariani, R.L. & de Mello, W.Z. (2007). PM2.5–10, PM2.5 and associated water-soluble inorganic species at a coastal urban site in the metropolitan region of Rio de Janeiro. *Atmospheric Environment*, Vol. 41, No. 13, (April 2007), pp. 2887–2892, ISSN: 1352-2310

Millero, F.J. (2006). Chemical Oceanography (3rd ed.): 496, Taylor and Francis, CRC Press, Boca Raton, ISBN 0849322804, FL

Mkoma, S.L., Wang, W., Maenhaut, W. & Tungaraza, C.T. (2010). Seasonal Variation of Atmospheric Composition of Water-Soluble Inorganic Species at Rural Background Site in Tanzania, East Africa. *Ethiopian Journal of Environmental Studies and Management*, Vol. 3, No. 2, (June 2010), pp. 27-38, ISSN: 1998-0507

Mkoma, S.L., Maenhaut, W., Chi, X., Wang, W. & Raes, N. (2009b). Chemical composition and mass closure for PM10 aerosols during the 2005 dry season at a rural site in Morogoro, Tanzania. *X-Ray Spectrom*etry, Vol. 38, No. 4, (July/August 2009), pp. 293–300, ISSN: 1097-4539

Mkoma, S.L., Chi, X., Maenhaut, W., Wang, W. & Raes, N. (2009a). Characterisation of PM10 Atmospheric Aerosols for the Wet Season 2005 at Two Sites in East Africa. *Atmospheric Environment*, Vol. 43, No. 3, (January 2009), pp. 631-639, ISSN: 1352-2310

Mochida, M., Kawabata, A., Kawamura, K., Hatsushika, H. & Yamazaki, K. (2003). Seasonal variation and origins of dicarboxylic acids in the marine atmosphere over the western North Pacific. *Journal of Geophysical Research*, Vol. 108, No. D6, (March 2003), pp. 4193–4203, ISSN: 2156-2202

Narukawa, M., Kawamura, K., Takeuchi, N. & Nakajima, T. (1999). Distribution of dicarboxylic acids and carbon isopotic compositions in aerosols from 1997 Indonesian forest fires. *Geophysical Research Letters,* Vol. 26, No. 20, (October 1999) pp. 3101–3104, ISSN: 1944–8007

Nicolas, J.F., Galindo, N., Yubero, E., Pastor, C., Esclapez, R. & Crespo, J. (2009). Aerosol Inorganic Ions in a Semiarid Region on the Southeastern Spanish Mediterranean Coast. *Water, Air and Soil Pollution,* Vol. 201, No. 1-4, (July 2009), pp. 149–159, ISSN: 1573-2932

Nyanganyura, D., Maenhaut, W., Mathuthu, M., Makarau, A. & Meixner, F.X. (2007). The chemical composition of tropospheric aerosols and their contributing sources to a continental background site in northern Zimbabwe from 1994 to 2000. *Atmospheric Environment,* Vol. 41, No. 12, (April 2007), pp. 2644-2659, ISSN: 1352-2310

Pavuluri, C.M., Kawamura, K., Aggarwal, S.G. & Swaminathan, T. (2011). Characteristics, seasonality and sources of carbonaceous and ionic components in the tropical aerosols from Indian region. *Atmospheric Chemistry and Phys*ics, Vol. 11, No. 15, (August 2011), pp. 8215–8230, ISSN: 1680-7316

Peng, C., Chan, M.N. & Chan, C.K. (2001). The hygroscopic properties of dicarboxylic and multifunctional acids: measurements and UNIFAC predictions. *Environmental Science and Technology,* Vol. 35, No. 22, (November 2001), pp. 4495–4501, ISSN: 1520-5851

Putaud, J.-P., Raes, F., Van Dingenen, R., Baltensperger, U., Brüggemann, E., Facchini, M.-C., Decesari, S., Fuzzi, S., Gehrig, R., Hüglin, C., Laj, P., Lorbeer, G., Maenhaut, W., Mihalopoulos, N., Müller, K., Querol, X., Rodriguez, S., Schneider, J., Spindler, G., ten Brink, H., Tørseth, K. & Wiedensohler, A. (2004). A European aerosol phenomenology-2: chemical characteristics of particulate matter at kerbside, urban, rural and background sites in Europe. *Atmospheric Environment,* Vol. 38, No. 16, (May 2004), pp. 2579-2595, ISSN: 1352-2310

Ram, K. & Sarin, M.M. (2011). Day and night variability of EC, OC, WSOC and inorganic ions in urban environment of Indo-Gangetic Plain: Implications to secondary aerosol formation. *Atmospheric Environment,* Vol. 45, No. 2, (January 2011), pp. 460–468, ISSN: 1352-2310

Riley, J.P. & Chester, R. (1971). Introduction to Marine Chemistry: 465, Academic Press, ISBN 0125887507, London

Salma, I., Balashazy, I., Winkler-Heil, R., Hofmann, W. & Zaray, G. (2002). Effect of particle mass size distribution on the deposition of aerosols in the human respiratory system. *Journal of Aerosol Science,* Vol. 33, No. 1, (January 2002), pp. 119-132, ISSN: 0021-8502

Scheff, P. & Wadden, R.A. (1993). Receptor modeling of volatile organic compounds: 1. Emission inventory and validation. *Environmental Science and Technology,* Vol. 27, No. 4, (April 1993), pp. 617–625, ISSN: 1520-5851

Souza, S.R., Vasconcellos, P.C. & Carvalho, L.R.F. (1999). Low molecular weight carboxylic acids in an urban atmosphere: winter measurements in São Paulo City,

Brazil. *Atmospheric Environment*, Vol. 33, No. 16, (July 1999), pp. 2563–2574, ISSN: 1352-2310

Stone, E.A., Schauer, J.J., Pradhan, B.B., Dangol, P.M., Habib, G., Venkataraman, C. & Ramanathan, V. (2010). Characterization of emissions from South Asian biofuels and application to source apportionment of carbonaceous aerosol in the Himalayas. *Journal of Geophysical Research*, Vol. 115, No. D06301, doi:10.1029/2009JD011881 (March 2010), ISSN: 2156–2202

Streets, D.G., Yarber, K.F., Woo, J.-H. & Carmichael, G.R. (2003). Biomass burning in Asia: Annual and seasonal estimates and atmospheric emissions. *Global Biogeochemical Cycles*, Vol. 17, No. 4, (October 2003), pp. 1099–1118, ISSN: 1944–9224

Talbot, R.W., Andreae, M.O., Berresheim, H., Jacob, D.J. & Beecher, K.M. (1990). Sources and sinks of formic, acetic and pyruvic acids over central Amazonia: 2. Wet deposition. *Journal of Geophysical Research*, Vol. 95, No. D10, (January 1990), pp. 16799–16811, ISSN: 2156–2202

Tanzania Bureau of Standards, TBS (2006). *National Environmental Standards Compendium EMDC 6(1733)*. TZS 845: 2006 Air Quality Specification. p. 74.

Turpin, B.J., Saxena, P. & Andrews, E. (2000). Measuring and simulating particulate organics in the atmosphere: problems and prospects. *Atmospheric Environment*, Vol. 34, No. 18, (July 2000), pp. 2983–3013, ISSN: 1352-2310

Van Dingenen, R., Raes, F., Putaud, J.-P., Baltensperger, U., Charron, A., Facchini, M.-C., Decesari, S., Fuzzi, S., Gehrig, R., Hansson, H.-C., Harrison, R.M., Hüglin, C., Jones, A.M., Laj, P., Lorbeer, G., Maenhaut, W., Palmgren, F., Querol, X., Rodriguez, S., Schneider, J., ten Brink, H., Tunved, P., Tørseth, K., Wehner, B., Weingartner, E., Wiedensohler, A. & Wåhlin, P. (2004). A European aerosol phenomenology-1: physical characteristics of particulate matter at kerbside, urban, rural and background sites in Europe. *Atmospheric Environment*, Vol. 38, No. 16, (May), pp. 2561-2577, ISSN: 1352-2310

Warneck, P. (2003). In-cloud chemistry opens pathway to the formation of oxalic acid in the marine atmosphere. *Atmospheric Environment*, Vol. 37, No. 17, (June 2003), pp. 2423–2427, ISSN: 1352-2310

Yamasoe, M.A., Artaxo, P., Miguel, A.H. & Allen, A.G. (2000). Chemical composition of aerosol particles from direct emissions of vegetation fires in the Amazon Basin: water-soluble species and trace elements. *Atmospheric Environment*, Vol. 34, No. 10, (July 2000), pp. 1641–1653, ISSN: 1352-2310

Yao, X., Fang, M. & Chan, C.K. (2002). Size distributions and formation of dicarboxylic acids in atmospheric particles. *Atmospheric Environment*, Vol. 36, No. 13, (May 2002), pp. 2099–2107, ISSN: 1352-2310

Yao, X., Fang, M., Chan, C.K. & Hu, M. (2003). Formation and size distribution characteristics of ionic species in atmospheric particulate matter in Beijing, China: (2) dicarboxylic acids. *Atmospheric Environment*, Vol. 37, No. 21, (July 2003), pp. 3001–3007, ISSN: 1352-2310

Yu, J., Huang, X., Xu, J. & Hu, M. (2005). When aerosol sulfate goes up, so does oxalate: implication for the formation mechanisms of oxalate. *Environmental Science and Technology*, Vol. 39, No. 1, (January 2005), pp. 128–133, ISSN: 1520-5851

Impacts of Biomass Burning in the Atmosphere of the Southeastern Region of Brazil Using Remote Sensing Systems

F. J. S. Lopes, G. L. Mariano, E. Landulfo and E. V. C. Mariano

Additional information is available at the end of the chapter

1. Introduction

Usually, both for the scientific community and to the general public, there is a tendency to associate air pollution with large urban centers (mainly coming from motor vehicles or factory chimneys). However, large areas, especially tropical regions, live with another source of pollution: the biomass burning. According to the Intergovernmental Panel on Climate Change (IPCC) report, biomass burning is the major source of air pollution and is considered an important environmental problem with several impacts on local, regional and global levels [1]. Biomass burning includes burning of forests, grasslands, and croplands. Large quantities of gases and materials, besides trace elements, are emitted into the atmosphere by this action. This can affect both the regional and global climate through the interaction with solar radiation and the chemical and physical processes in the atmosphere. A large amount of these burning points occurred in the southern part of the Amazon basin during the dry season and the product of these emissions can be transported to some cities in the southeast of the country, a highly polluted region, and with cities with serious air pollution problems at the urban environment. Moreover, with the growing demand for biofuels in Brazil, the cultivation of sugarcane has been expanding considerably in southeastern Brazil, being a strong contribution to poor air quality in the region due to the burning of that culture, aiming to facilitate its harvest.

A very useful tool in studies of the effects of burning in the atmosphere is the Lidar (Light Detection and Ranging) technique, which gives vertical profiles of aerosols and allows the monitoring of the temporal evolution of the atmospheric structure, as well as to obtain values of backscattering coefficient. This technique is characterized by high spatial and temporal resolution, allowing the measurement of small concentrations of different gases (mainly water vapor), aerosols and local meteorological parameters such as wind direction and temperature, depending, however, on the type of Lidar system and the wavelength used.

A Lidar system operates on the same physical principle of Radar, but using a laser beam as emission source; the detection components are composed by a telescope and an optical analyzer system. In the case of Lidar, a light pulse is directed into the atmosphere. The light beam interacts with the atmospheric compounds and is scattered in all directions by particles and molecules. A portion of the light is absorbed and other portion is scattered back towards the Lidar system. The backscatter light is collected by a telescope and focused upon a photo detector capable to measure the signal intensity as function of the distance from the system.

Lidar systems, both aboard of satellites and ground-based, have been used in synergy with sunphotometer systems to detect and track the transport of aerosol from the MidWestern portion of the Brazilian territory to the Southeastern region, mainly the São Paulo city, where a ground-based elastic backscatter Lidar system is installed [2]. Here, some results obtained using a methodology developed to detect biomass burning aerosol loaded in the atmosphere at specific locations of the Brazilian territory will be presented, mainly in the Midwestern and Northern portion, considered to be one of the biggest producers of biomass burning in South America due to the vast quantities of forest, pasture and plantations, and track the transport trajectories of the aerosol plumes into the areas located at the Southeastern portion of Brazil, in the city of São Paulo, where it is possible to identify such plumes using the AERONET sunphotometer and the elastic backscatter Lidar system. Initially, Ångström Exponent (AE) and Aerosol Optical Depth (AOD) values retrieved from the AERONET sunphotometer network and the MODIS satellite, combined with AOD and Lidar ratio (LR) products from the CALIOP measurements are used in order to identify biomass burning aerosols loaded in the atmosphere in the MidWestern and Northern portion of the Brazilian territory. Attenuated backscatter profile images and aerosol optical properties values from CALIOP are used to monitor and track such aerosol plumes to the Southeastern region. If the transportation of biomass burning to São Paulo city is confirmed using the HYSPLIT (Hybrid Single Particle Lagrangian Integrated Trajectory Model) backtrajectories, the backscatter Lidar and the sunphotometer data are used to analyze the presence of those plumes at the São Paulo atmosphere. Such results can be a strong indication that the city of São Paulo, one of the most polluted cities in the world, is affected not only by the presence of aerosol from local sources but also by aerosols produced in remote sources.

2. Instrumentation

This study has the aim of analyzing the biomass burning aerosol optical properties using data from the CALIOP sensor installed on board the satellite CALIPSO and MODIS sensor aboard the TERRA and AQUA satellite. In addition, ground-based data measurements such as an elastic backscatter Lidar system and the AERONET sun photometer data will be used. In this section some details of each instrument will be presented.

2.1. Moderate Resolution Imaging Spectroradiometer (MODIS) Sensor

Through the EOS (Earth Observing System) program, initiated in 1980 with the main objective to allow continuous observations for a period of at least 15 years of global changes, various sensors have been launched, and among them the MODIS in 1999 (Moderate Resolution Imaging Spectroradiometer) is highlighted.

The MODIS sensor is on board the polar orbiting satellites TERRA and AQUA launched in 1999 and 2002, respectively. The sensor was the first designed to obtain global observations of aerosols with moderate resolution (between 250m and 1000m depending on the wavelength used).

The AQUA satellite is part of the so-called A-Train constellation [3], which also contains the Aura satellite (launched in 2004) to study the atmospheric chemistry and dynamics with emphasis on the sensor for ozone monitoring OMI - Ozone Monitoring Instrument, PARASOL (2004) to study the water and carbon cycle, CALIPSO (2006) to study the profile of aerosol and clouds and CloudSat (2006) to study the clouds.

MODIS has 36 spectral bands between 0.4 and 14.5 μm, allowing the generation of several products related to aerosol, such as aerosol optical depth over the ocean and land with a resolution of 10x10 km (at nadir), and the size and type distribution over oceans and type of aerosol over the continent. General and operational characteristics of the sensor can be found in Barnes et al. (1998) [4].

The MODIS aerosol data analyzed consist of the aerosol product level 3, MOD08. This level of data is generated daily for the entire globe offering several properties related to aerosols such as optical depth over ocean/continent and Ångström Exponent over the continent. The spatial resolution of the data is 1.0° for level 3. In this chapter, data from the AOD MODIS sensor (AQUA satellite) between 2003 and 2010 were used.

Any given set of data, ordered from the lowest to the highest value, have a central value which is called a median. Likewise, it is possible to think of the values dividing the set of data in four equal parts, that is, the quartiles, Q1, Q2 and Q3, with Q2 being equal to the median. The values dividing the set in 10 equal parts are the deciles, and the values which divide the set in 100 equal parts are the percentiles. A percentile is the value of a data set below which a certain amount of the observations are. For instance, the 90th percentile represents the number below which 90% of the data is found. The 50th percentile is equivalent to the median [5].

The AOD is a dimensionless coefficient, indicating the amount and efficiency of solar radiation extinction by optically active material for a given wavelength. The optical depth may be defined as an attenuation coefficient of a beam of light which undergoes scattering or absorption during its passage through any given element [6].

Higher values of optical depth lead to lower values for the optical transmittance of the air column, with the intensity of solar radiation on the surface also being smaller. Consequently, the temporal and spatial evolution of this radiation depends on the atmospheric optical depth, which will depend on local factors, since these cause variations in how the solar radiation in the electromagnetic spectrum and the direction of propagation is distributed.

The AOD can be divided into some components, due to Rayleigh and Mie scattering, as well as the absorption by atmospheric particles. Therefore, the optical depth is a measure of transparency, being defined as the fraction of radiation (or light) that is scattered or absorbed in a path. An easy example is that of a fog. The fog between an observer and an object immediately in front of him has an optical depth tending to zero. If the object is moving away from the observer's, the optical depth will increase until it reaches a large value where the object is no longer visible.

The optical depth indicates the amount of absorbing and scattering optically active material found in the path of a beam of radiation. It is defined as the integral over the optical path of the product of the total quantity of molecules present in the medium and the cross section of extinction for each wavelength. The optical depth is expressed by:

$$\tau_\lambda = \int \sigma_\lambda N(x) dx \tag{1}$$

Where σ_λ is the extinction cross section, dx the path of integration and N(x) the volume number density of optically active atoms or molecules [particles cm^{-2}] [6]. It expresses the amount of light removed from a beam by scattering or absorption during its path in any means. Being I_0 the intensity of the radiation source and I the intensity observed after a certain path, the optical depth may be defined by the following equation:

$$\frac{I}{I_0} = e^{-\tau} \tag{2}$$

In the atmospheric sciences, the atmospheric optical depth is commonly referred as the vertical path from the surface of the earth or the altitude of the observer into space. Since it regards the vertical path, the optical depth of a sloping path is $\tau' = \mu\tau$, which is called the air mass factor, which to that atmosphere is usually defined as $\mu = 1/cos\Theta$ where Θ is a zenith angle corresponding to a certain path. Thus:

$$\frac{I}{I_0} = e^{-\mu\tau} \tag{3}$$

2.2. CALIPSO satellite

Ground-based systems are very useful for monitoring local and regional aerosol properties in the atmosphere, playing an important role in the estimation of the aerosol influence on the radiation budget. However, in order to cover the cloud and aerosol vertical distribution in a global range, NASA and CNES agencies has launched a satellite with a Lidar system as a main operational instrument on board [7]. The CALIPSO satellite was launched in April 2006, and since then is part of the NASA's A-Train satellite constellation as mentioned before. CALIPSO is flying in a 705 km sun-synchronous polar orbit with an equator-crossing time of about 13:30 in local time, covering the whole globe in a repeat cycle of 16 days [8]. The CALIPSO payload consists of three co-aligned nadir-pointing instruments designed to operate autonomously and continuously. Two of them are passive sensors and can provide a view of the atmosphere surrounding the Lidar curtain, namely, a wide field camera with a spatial resolution of 125 m for pixels and a three-channel infrared imaging radiometer instrument at each of the two wavelengths [9]. The lasers are Q-Switched to provide a pulse length of about 20 ns. The receiver subsystems measure the attenuated backscattering signal intensity at 1064 nm and the two orthogonal polarization components at 532 nm. From the backscattering signal measured by the receiver system, the CALIOP data products are assembled and separated in two categories: Level 1 and 2 products. Level 1 products are composed of calibrated and geolocated profiles of the attenuated backscatter returned signal, and are separated in three types, the total attenuated backscatter profile at 1064 nm, total attenuated backscatter profile (the sum of parallel and perpendicular signals) and the perpendicular backscatter signal, both at 532 nm [8, 10]. The level 2 products are derived from level 1 products and are classified in three types: layer products, profile products and vertical feature mask (VFM). Layer products provide the optical properties of aerosol and clouds integrated or averaged in the layers detected in the atmosphere. The profile products provide retrieved backscatter and extinction profiles within the detected layers. The VFM provide information of the cloud and aerosol location, and also their types. The level 2 data of aerosol layers provide information about values of the AOD, LR, and information from the heights of top and bottom of the layers detected by the CALIOP sensor.

2.3. AERONET sunphotometer

AERONET is a global network of optical monitoring of atmospheric aerosols, maintained by NASA and expanded throughout various research institutions around the world. This network has more than 200 measuring points, 22 in South America. The sunphotometer system from the AERONET network is a remote sensing instrument very useful not only to work in synergy with Lidar but also to retrieve several optical properties from aerosol loaded in the atmosphere.

The CIMEL 318A spectral radiometer is a solar-powered weather-hardy robotically pointed Sun and sky instrument. This instrument is installed on the roof of the Physics Department at the University of São Paulo (USP). The CIMEL photometer performs measurements of the AOD at several wavelengths in the visible and the near-infrared spectral region to enable the assessment also of the Ångström Exponent [11]. The principle of operation of this system is to acquire aureole and sky radiances observations using a great number of scattering angles from the Sun, through a constant aerosol profile to retrieve the aerosol size distribution, the phase function and the AOD. For this study, the channels used are centered at 340, 440, 500, 670, 870, 940 and 1020 nm, with a 1.2 μm full FOV (field of view) angle. The measurements are taken pointed directly to the Sun (four sequences) or to the sky (five sequences) in nine different pre-programmed sequences [11]. The inversion of the solar radiances measured by the CIMEL sunphotometer to retrieve the aerosol optical depth values is based on the Beer-Bouguer-Lambert equation, assuming that the contribution of multiple scattering within the small FOV of the sunphotometer is negligible. The aerosol optical depth at 532 nm was determined by the relation:

$$\frac{\tau_{532}^{aer}}{\tau_{500}^{aer}} = \left(\frac{532}{500}\right)^{-a} \tag{4}$$

Where the Ångström Exponent [12] was derived from the measured optical depth in the blue and red channels (440nm and 670 nm):

$$a = -\frac{\log\left(\frac{\tau_{440}^{aer}}{\tau_{670}^{aer}}\right)}{\log\left(\frac{440}{670}\right)} \tag{5}$$

The AE is also an indirect mean to retrieve the particle size distribution [13] and its possible composition [14, 15]. Concerning the uncertainty, the major source of error would be in the calibration procedure, which is proportional to the associated uncertainty of the AOD at a given wavelength [16].

2.4. Elastic backscatter Lidar system (MSP-Lidar)

A ground-based elastic backscatter Lidar system is in operation since 2001 in the Environmental Laser Applications Laboratory at the Centre for Laser and Applications (CLA) at the Nuclear Energy Research Institute (IPEN). The MASP and instrument locations in this area are depicted in Figure 1 (point L).

The Lidar technique is based on the emission of a collimated laser beam through the atmosphere and on the detection of the backscattered laser light by the suspended atmospheric aerosols and molecules. A backscattering Lidar can thus provide information on the Planetary Boundary Layer (PBL) mixed depth, entrainment zones and convective cell structure, aerosol distribution, clear air layering, cloud-top altitudes, cloud statistics, atmospheric transport processes and other inferences of air motion [17-20]. The Lidar system is a single-wavelength backscatter system pointing vertically to the zenith and operating in the coaxial mode. The light source is based on a commercial Nd:YAG laser (Brilliant by Quantel SA) operating at the second harmonic frequency (SHF), namely at 532 nm, with a fixed repetition rate of 20 Hz. The average emitted power can be selected up to values as high as 3.3 W. The emitted laser pulses have a divergence of less than 0.5 mrad. A 30 cm diameter telescope (focal length f=1.5 m) is used to collect the backscattered laser light. The telescope's FOV is variable (1–2 mrad) by using a small diaphragm. Lidar is currently used with a fixed FOV of the order of 1 mrad, which according to geometrical calculations [21] permits a full overlap between the telescope FOV and the laser beam at heights around 300 m above the Lidar system. This FOV value, in accordance with the detection electronics, permits the probing of the atmosphere up to the free troposphere (12-15 km asl). The backscattered laser radiation is then sent to a photomultiplier tube (PMT) coupled to a narrowband (1 nm full width at half maximum - FWHM) interference filter to assure the reduction of the solar background during daytime operation and to improve the signal-to-noise ratio (SNR) at altitudes greater than 3 km. The PMT output signal is recorded by a Transient Recorder in both analog and photon counting modes. Data are averaged between 2 and 5 min and then summed up over a period of about 1 h, with a typical spatial resolution of 15 m, which corresponds to a 100 ns temporal resolution.

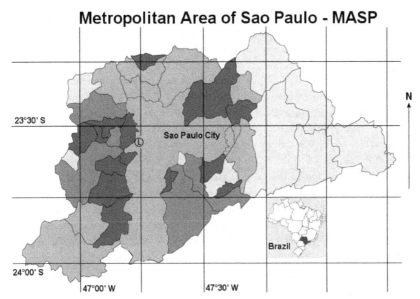

Figure 1. Metropolitan Area of São Paulo (MASP) map with the sampling locations indicated: (1) São Paulo University Campus (USP), site of the Lidar and AERONET sunphotometer, in the point L.

In the present stage, the inversion of the Lidar profile is based on the solution of the basic Lidar equation taking into account the atmospheric solar background radiation correction [22]. The Lidar equation is presented according to equation (6), where $P(\lambda, z)$ is the Lidar signal received from a distance z at the wavelength λ, P_0 is the emitted laser power, ξ is all Lidar system parameters, $\beta_m(\lambda, z)$ and $\beta_{aer}(\lambda, z)$ are the atmospheric volume backscatter coefficients for the molecular and aerosol contribution, respectively; and $\alpha_{aer}(\lambda, z)$ is the volume extinction coefficient at range z.

$$P(\lambda, z) = P_o \xi \left[\frac{\beta_m(\lambda,z)+\beta_{aer}(\lambda,z)}{z^2}\right] \times exp\left[-2\int_0^z \alpha(\lambda, z')dz'\right] \tag{6}$$

The retrieval of the aerosol optical properties is based on the measurements of the aerosol backscatter coefficient β_{aer} at 532 nm, up to an altitude of 5 to 8 km. The determination of the vertical profile of the aerosol backscatter and extinction coefficients relies on the Lidar inversion technique following a modified Klett's algorithm [23, 24] under the assumption of the single scattering approximation. One has, however, to bear in mind that this inversion technique is an ill-posed problem in the mathematical sense, leading to errors as large as 30% when applied [22]. To make the Lidar equation solvable it is necessary to establish a relation between $\alpha(\lambda, z)$ and $\beta(\lambda, z)$. This is achieved assuming the backscatter-to-extinction ratio (LR) as:

$$LR = \frac{\alpha_{aer}(\lambda,z)}{\beta_{aer}(\lambda,z)} \tag{7}$$

However, it is known that the LR depends on several physical-chemical parameters inherent to the aerosols being inspected, such as the aerosol refractive index, size and shape distribution of the aerosol particles [25]. To derive the appropriate "correct" values of the vertical profile of aerosol backscatter coefficient in the lower troposphere an iterative inversion approach was used (by "tuning" the LR values), based on the intercomparison of the AOD values derived by the Lidar and a collocated AERONET sunphotometer [26], assuming the absence of stratospheric aerosols and that the PBL is homogeneously mixed between ground and the altitude of 300 m, where the Lidar overlap factor is close to 1. Once the correct values of the vertical profile of the aerosol backscatter coefficient were derived (when the difference between the AODs derived by sunphotometer and Lidar was less than 10%), the Klett's inversion technique was reapplied, using the appropriate LR values, to retrieve the final values of the vertical profiles of the backscatter and extinction coefficient at 532 nm. The vertical profiles of pressure and temperature measured by radiosoundings launched twice a day, at 12 UTC and 00 UTC in a distance about 10 km from the place where the MSP-Lidar system is located, are used in order to obtain the molecular contribution based in the Bucholtz's approach [27].

3. Analysis of the active fires in the state of São Paulo

The fire detection method employed by INPE (National Institute for Space Research) is a digital non supervised clustering algorithm which selects pixels as burning if the AVHRR radiometric temperature exceeds 46°C [28], using the images of several satellites, which operate in the band between 3.7 and 4.1 µm, selecting the pixels (resolution elements) with higher temperature. Data from the morning overpass are used to identify fire pixels. Fire counts identified by INPE are provided in a weekly base in grid format. They consist of a 7-day sum at 0.5-degree increments from 17°N to 40°S and 85°W to 34.5°W [29]. Data of active fires obtained through the NOAA-12 satellite were used, for the period between January, 1999 and July, 2007 at the studied region (pixels located over the state of São Paulo).

Through the spatial distribution of fires (Figure 2) obtained by the NOAA-12 satellite, it is possible to notice that the regions with the highest number of outbreaks of fires during the selected period coincides with sugarcane crops, especially during the winter and spring, periods with the largest number of outbreaks. The central-northern, central, central-eastern and central-western regions of the São Paulo State are highlighted as the areas where there is a high rate of outbreaks due to the presence of the sugarcane culture.

The results of Figure 2 and Table 1 indicate that the winter was the season with the highest number of fire outbreaks during the period, when approximately 15,000 outbreaks were recorded, followed closely by spring with approximately 12,000 outbreaks, while in summer only 2000 spots were registered. During the harvest period, which occurs from May to November, the plantation areas are burnt a few hours before the manual cutting, and this period coincides with the dry season in Southeastern Brazil. From December to April the wet pattern prevails and there are only few activities for burning around the state of São Paulo. The number of fire outbreaks tends to increase from the month of March, achieving its maximum in the trimester July-August-September. The peak of the burning season (August) coincides with the least amount of precipitation in the region, while the reverse is also found, burning minimum in the months with maximum precipitation (December and January).

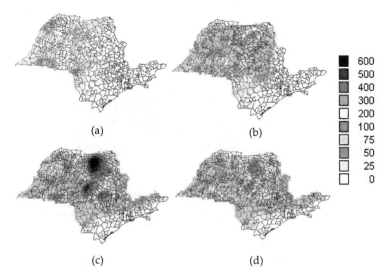

Figure 2. Outbreaks of fire during the period from January, 1999 to July, 2007: (a) summer; (b) autumn; (c) winter and (d) spring.

The great quantity of active fires during the dry season, combined with low humidity in the region, can cause several health problems in the people living near the plantations and burnings of sugarcane. The association between internal and external exposure to the biomass smoke and its effects on health has been reported in some areas of Asia and India [30, 31]. Carbon deposition in the lungs occurred consistently in patients exposed to biomass burning [32]. Unlike most regions where external biomass burning is sporadic, the biomass burning in the region of São Paulo is a common and scheduled activity, due to the areas with sugarcane crops.

The seasonal and interannual variation of the active fires in the State of São Paulo can be analyzed in Table 1. Through this table it can be noticed that the highest value of active fires occurred in 1999 followed by a decrease until 2001. From this year on, a slightly constant tendency of decrease in numbers of forest fires was noticed, except for the two last measured years (2005 and 2006). It is important to highlight that the measurements and monitoring in 2007 only goes until July, before the burning maximum period – August, September and October (as seen on Figure 2), probably due to technical problems with the satellite acquisitions. Besides the annual active total fires, the seasonal variation of the number of fires for each year can also be observed in the Table 1, where it can be clearly seen that the maximum active fires seasons are winter and spring, followed by autumn and summer in almost every analyzed year, except for the years of 2000 and 2004, where the total active fires detected were greater in spring and summer than in winter and autumn, respectively. For the year of 2004, it was observed that the rain rates for these two periods were normal according to the climatology, without an obvious reason for that variation. It can also be seen the difference between the number of fires during the maximum fires season (winter) and the minimum (summer), reaching a value 3 times higher at some periods.

Year/ Season	Summer	Autumn	Winter	Spring	Total
1999	57	924	2633	2401	6090
2000	598	451	2202	1288	4489
2001	78	277	1658	1026	3091
2002	120	670	1657	1416	3829
2003	97	519	1759	1065	3420
2004	74	387	1118	1484	3171
2005	215	680	1912	1498	4174
2006	53	933	1889	1490	4398
2007	108	602	306	-	983
Total	993	5443	15134	11668	33645

Table 1. Active fires measured by the NOAA-12 satellite for the State of São Paulo – Brazil

4. MODIS aerosol optical depth

Figure 3 shows the maximum values of AOD for the southeastern region of Brazil from 2003-2010 (latitude between 14°S and 25°S and longitude between 54.1°W and 38.1°W). This region is known for having main urban centers, especially the metropolitan area of São Paulo and Rio de Janeiro, with large amounts of industries. In addition, the city of Cubatão is located at this region (50 km from MASP), in the state of São Paulo, known in the 1980s and 1990s as one of the most polluted cities in the world [33].

Despite the known major sources of pollution in the southeastern region of Brazil, it is possible to notice high maximum AOD values only during spring and winter. Except for these periods, and for one day in 2003 and other in 2010, both at summer, the maximum AOD over the Southeastern region was always below 1.0. The average of the maximum AOD between 2003 and 2010 was 0.492±0.505, indicating the high variability of the maximum AOD over the region.

The 95th percentile calculated using data on maximum daily AOD for the southeastern region of Brazil was 1.143. This means that 95% of the daily data (MOD08) from the MODIS sensor (from a total of 2921 measurements spanning from 2003 to 2010) have lower values than 1.143. At first, this value can be considered low, but aspects of the sensor used must be taken into account, such as a spatial resolution of 1 km and that the

maximum concentration of aerosols tends to be found near the surface, hindering the use
of remote sensing for this task.

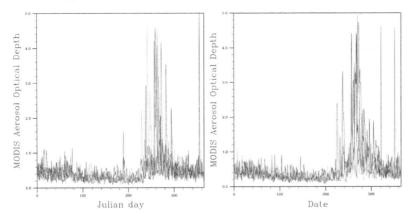

Figure 3. MODIS Maximum Aerosol Optical Depth over southeast of Brazil (a) 2003-2006 and (b) 2007-
2010. Years: (a) 2003 (orange); 2004 (blue); 2005 (green); 2006 (red) and (b) 2007 (orange); 2008 (blue);
2009 (green); 2010 (red)

Both the Amazon and Midwestern Brazil are great sources of aerosols into the atmosphere,
and the maximum daily value of AOD found on the Southeastern region can be compared
to these areas. In all of these regions the largest occurrences of elevated AOD are during the
winter and spring in the Southern Hemisphere. The occurrence of higher AOD levels during
this period for the Northern and Central region is basically due to the highest amount of
biomass burning in the region (dry season). In Brazil, biomass burning in the Amazon
region occurs with greater intensity during the dry season (July - October) and primarily
affects the ecosystems of forest, pasture and Cerrado [34]. The small amount of precipitation
and the increased atmospheric stability at this period contributes to the persistence of
aerosols in the atmosphere, as well as a higher transport of biomass burning aerosols from
the Amazon region (Section 6).

Analyzing the 146 days with maximum AOD higher than the 95th percentile (1.143) found
between 2003-2010 it is observed that approximately 82% of the cases occurred during the
spring in the Southeastern hemisphere (Figure 4). When considering the period
corresponding for spring and winter, the same has 144 days with values above the 95th
percentile (99% approximately). The emissions from these fires have significant impacts on
the concentration of gases and aerosols. The results presented here emphasize the
importance of monitoring the aerosols for the period between June and October. The values
recorded during summer and fall may be related, among other reasons, with burning events
in the Southeastern region and transport of biomass burning aerosols from distant regions,
such as the Amazon and MidWestern region of Brazil regions, Bolivia, the Northern portion
of Argentina and the North region of South America. Another possible explanation for the
values found are the urban aerosols emitted into the studied area.

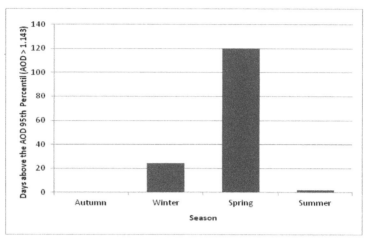

Figure 4. Days above the AOD 95th Percentile (maximum AOD by MODIS > 1.143) over southeast of Brazil

5. Aerosol detection in the burning region

As mentioned before, some regions in the Brazilian territory suffers from intense anthropogenic biomass burning activities, such as forest and sugarcane plantation fires, during the so-called dry season. High concentrations of biomass burning aerosol particles, produced mainly in the Amazon basin and the Brazilian Mid-Western region, can be detected due to these fire activities in the tropical forest, savanna and pasture [35]. In this context, data from AERONET sunphotometer, MODIS sensor and CALIPSO satellite were employed in order to detect possible sources of biomass burning aerosols and map its transportation from Mid-Western to the Southeastern region in Brazil.

Initially the image data measured by the MODIS sensor aboard the Terra satellite were used to identify possible signs of smoke from biomass burning. In the image retrieved on 21 August 2007, presented in Figure 5, it can be clearly seen the presence of a dense smoke layer in the region South and Southeast of Campo Grande (CG - Lat: 20°26'16" Long: 54°32'16"). The trajectories generated using the HYSPLIT model indicate the transport of air masses from the Midwestern region of Brazil to the Southeast, where the MSP-Lidar system and the AERONET sunphotometer are installed. Figure 5 also shows that the CALIOP sensor aboard the CALIPSO satellite overpass the region near of Campo Grande during the nighttime of 21 August 2007, as can be seen by the descending trajectory (in the left of Figure 5). Furthermore, the sensor made measurements during daytime on the same day, however, in a region between the cities of Campo Grande and São Paulo, which will be helpful in the monitoring processes of the aerosol transport from the region of CG to MASP.

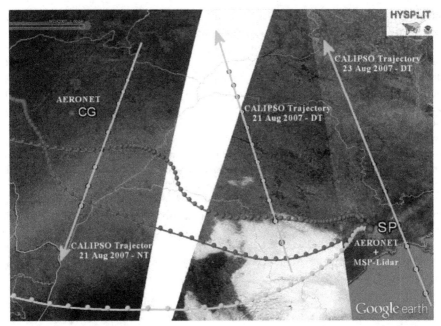

Figure 5. Image of the MODIS sensor aboard the Terra satellite showing a dense smoke layer in the South and Southwest region of Mato Grosso State (Campo Grande) on 21 August 2007. The green arrows show the nighttime (descending) and daytime (ascending) trajectories of the CALIPSO satellite for the same day. The trajectories generated by the HYSPLIT model show that aerosol masses were transported from the Midwestern region of Brazilian territory to the Southeastern region where the MSP-Lidar system and AERONET sunphotometer are installed.

After identifying the presence of a dense smoke layer through the MODIS image, some optical properties of the atmospheric aerosol retrieved by the AERONET sunphotometer installed in CG site were analyzed. The AOD and AE products can provide information about the absorption and extinction characteristics and the size distribution of the aerosol in the atmosphere. In general, high values of AOD are associated to high extinction (absorption) of radiation, and in the same reasoning, high values of AE are associated to the fine mode size distribution of aerosols. These two characteristics are considered a signature of biomass burning aerosol. Figure 6 shows the scatter plot of Ångström Exponent as a function of the Aerosol Optical Depth for the month of August 2007 and case previously selected at Campo Grande (points in red), the grey filled lines are the AOD and AE median values for all data and the dashed lines are the respective median standard deviation. The AE values below the horizontal line corresponding to the median, AE=1.46±0.09, are an indication that most of the aerosol load is in the coarse mode size distribution. Those values above the median line belong to the fine mode size type of aerosols. The AE indicates a small sized particle distribution similar to those found in the presence of biomass burning aerosols as shown by [36]. Making the same interpretation to the AOD median vertical line,

AOD = 0.198±0.078, the values in the left side of the AOD median corresponding to low extinction and absorption of radiation aerosol types, and the right side is related to the high extinction and absorption radiation types. The median values were assumed as a better statistical evaluator since it was found using a skewness test that the AOD vs. AE distribution is rather asymmetrical, and instead of the standard deviation it has been used the median standard deviation as the measure of the variability or dispersion of the data set, according to [02]. The scatterplot in Figure 6 was divided into four distinctive sectors, I, II, III and IV. Each of them represents different types and sizes of aerosol according to the AOD and AE values. This study should be focusing mainly on regions II and IV, which correspond to fine mode size distribution and high absorption and extinction aerosol types, and coarse mode and high absorption and extinction aerosol types, respectively. In sector II there is a strong indication of the predominance of biomass burning aerosols in the atmosphere as the large AE corresponds to small sized particles and the large AOD for a strong absorbing type of aerosols. The same reasoning can be applied to sector IV, although the values below the AE median values can be associated to particle growth during the long-range transport [37]. As can be seen in Figure 6, the most AOD and AE values of aerosol measured during 21 August 2007 (red points) are inside the region II, representing biomass burning aerosol types.

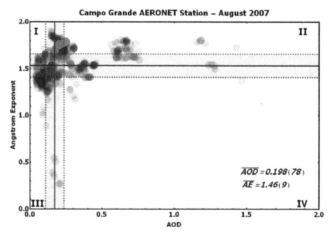

Figure 6. Scatterplot of the Ångström Exponent versus Aerosol Optical Depth at 532 nm for the case study selected (21 August 2007) and for the complete period of the August 2007. The four distinctive sectors represent different types and sizes of aerosol according to the AOD and AE values. The region II represents aerosols with biomass burning products characteristics, with high values of AOD and AE. The calculated median and median standard deviation values are AOD = 0.198±0.078 and AE = 1.46±0.09, respectively.

On 21 August 2007, the CALIPSO satellite overpasses the region near the AERONET site in CG during the nighttime (descending trajectory), with the closest approach to the AERONET site of 126 km (horizontal distance) at 05:00 UTC approximately. During the daytime on the same day the satellite made measurements in an area between Campo Grande and São Paulo, as can be seen in Figure 5. The AOD and LR retrieved by the CALIOP sensor were obtained using the Level 2 Aerosol Layer products Version 3.0 for both trajectories during 21 August 2007. Figure 7 shows the AOD (left panel) and LR values (right panel) according to the latitude for the nighttime trajectory (21 Aug 2007 – NT in figure 5).

In the left panel of figure 7 it can be observed the values of AOD retrieved by CALIOP sensor spanning from 0.65 to 0.30. Such values are higher compared with those measured by the AERONET sunphotometer installed at CG site, represented by the red points in Figure 6. The high values of AOD measured by CALIOP sensor are a strong indication that highly absorbent aerosols are present in the atmosphere; in addition, it is a strong indication that these particles are from biomass burning aerosols [38]. Such evidence is confirmed analyzing the Lidar Ratio values, which can indicate the most likely type of aerosol detected, as shown at the right panel of Figure 7. The CALIOP sensor signed LR values of 70 sr. According to [39, 40], the LR value of 70 sr corresponds to the aerosol type from biomass burning or continental polluted air. However, biomass burning aerosol types differ according to their detection altitude. Generally, such layers are detected above the PBL, localized approximately between 3 to 5 km and more [41,42]. In the case of the CALIOP nighttime measures on 21 August 2007 in the region of the CG's AERONET site, the detected layers correspondent to the AOD and LR values presented in Figure 7 were localized between 2.5 and 4 km, which is Above the PBL, demonstrating that the detected layers are mostly loaded by biomass burning aerosols. The 532 nm Total Attenuated Backscatter profile presented in Figure 8 shows an intense backscatter signal from an aerosol layer at high altitude, localized above the PBL approximately between 2.5 and 4 km, which is the biomass burning layer detected near the AERONET sunphotometer site at CG pointed by the dashed line in red. Such aerosol type is confirmed by the Vertical Feature Mask (VFM) product presented in Figure 9, which identifies the subtype of each aerosol in the atmosphere [43].

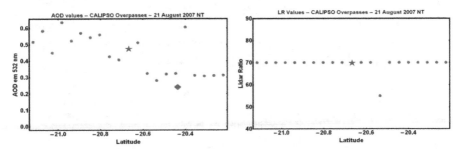

Figure 7. AOD and LR distribution as function of latitude (21 Aug 2007 – NT trajectory in figure 5) measured by the CALIOP sensor on 21 August 2007 during the nighttime. The star marks the point of the satellite closest approach to the AERONET site installed at CG. The red triangle represents the value of AOD measured by the AERONET sunphotometer during the closest approximation time.

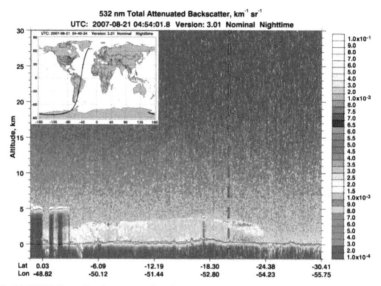

Figure 8. CALIOP 532 nm Total Attenuated Backscatter profile along with the orbit track in the graphic embedded in the upper left. The dashed line in red represents the closest approach to the region of Campo Grande on 21 August 2007 around 05:00 (UTC). It can be noticed an intense aerosol layer detached above the PBL, localized approximately between 2.5 and 4.0 km of altitude.

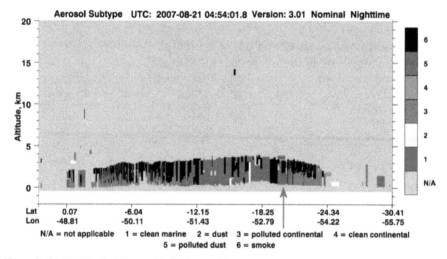

Figure 9. CALIOP Vertical Feature Mask of aerosol layers measured in the region of the CG's AERONET sunphotometer on 21 August 2007 around 05:00 (UTC). It can be seen an aerosol layer above the PBL localized approximately between 2.5 and 4.0 km altitude and classified as biomass burning aerosol type (6 = smoke). The layer immediately below, within the PBL, is classified as a mixture of dust and pollution (5 = polluted dust aerosol type).

6. Transport and detection of biomass burning aerosols

For the days when the presence of aerosols from biomass burning were detected by the sensors onboard CALIPSO and TERRA satellites, and also by the sunphotometer on CG site, air mass trajectories were generated using the HYSPLIT model, that is a complete system for computing simple trajectories to complex dispersion and deposition simulations using either puff or particle approaches [44]. The HYSPLIT trajectories are computed based on the Global Data Assimilation System (GDAS), an operational system from the National Weather Service of the National Centers for Environmental Prediction (NCEP) and it was used to investigate if the air mass parcels in the Midwestern have been dislocated towards the Southeastern region where the MSP-Lidar system is installed. The purpose to use this trajectory model is to constrain the direction of the air masses to improve the correlation between the optical properties (i.e., AOD and Lidar ratio) measured by two different instruments spatially separated, i.e., CALIOP sensor and ground-based AERONET sunphotometer and MSP-Lidar system. According to Figure 5, the HYSPLIT trajectories show that there were transportation of biomass burning aerosols generated in the Midwestern region of Brazil, i.e. Campo Grande region, to the Southeastern, where the MSP-Lidar system is installed, as well as an AERONET sunphotometer system. The same analysis presented in the previous section was performed to the daytime measurement made by the CALIOP sensor in its trajectory in the region between the cities of CG and SP (Figure 5), using the AOD and LR retrieved using the Level 2 Aerosol Layer products Version 3.0. Figure 10 shows the AOD (left panel) and LR values (right panel) as function of the latitude for the daytime trajectory (21 Aug 2007 – DT in Figure 5). It can be seen that in this region the CALIOP sensor obtained not so high AOD values, being around 0.10. The LR values varied around 40-55 sr, representing the dust aerosol type or a mixture of dust and pollution (polluted dust) [39,40]. In the left panel of Figure 10 low values of AOD can be observed, around 0.10. Such values are lower compared to those measured by the AERONET sunphotometer installed at CG site, represented by the red points.

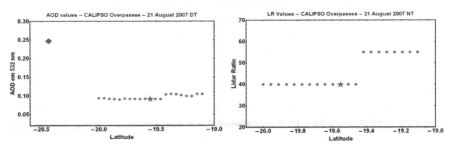

Figure 10. AOD and LR distribution as function of latitude (21 Aug 2007 – DT trajectory in figure 5) measured by the CALIOP sensor on 21 August 2007 during the daytime. The star marks the point of the satellite closest approach to the AERONET site installed at CG. The red triangle represents the value of AOD measured by the AERONET sunphotometer during the closest approach time.

Two possible reasons can explain the detection of different values of AOD and LR from those values retrieved on the nighttime measurements (Figure 7). The first one is that the detected aerosols have their optical properties changed due to long-range transport, mixing with other aerosol types or absorbing moisture from the atmosphere, undergoing aging processes. The low value of the Lidar ratio (40 to 55 sr) also indicates the aging processes of the aerosols; such lower values may be linked to an increase of particle size by moisture absorption and/or a reduction of the light absorption capability of the particles [37]. However, the second reason can be the fact of the transported aerosol masses only reaches the region measured by the CALIOP sensor on 22 August 2007, a day after the CALIPSO satellite overpasses.

7. Aerosol detection over the metropolitan region of São Paulo

In order to track the biomass burning plume detected near the AERONET sunphotometer site in Campo Grande, backtrajectories were generated using the HYSPLIT model. Those backtrajectories show that the aerosol plumes were transported from the Midwestern region, near Campo Grande, to the Southeastern of the Brazilian territory, more precisely to the city of São Paulo, where there are two remote sensing equipments installed, the AERONET sunphotometer and an elastic backscatter 532 nm Lidar system these backtrajectories, presented in Figure 5 and Figure 11, it is possible to observe that aerosol air masses trajectories start at different altitude levels (3.5, 4.0 and 4.5 km AGL) on 21 August 2007 and move towards the Southeastern part of Brazil, reaching the MASP on 23 August 2007 at an altitude also around 3.5 and 4.5 km. The altitudes of the starting trajectories in CG region are at the same range altitude of the aerosol layers detected by the CALIOP sensor signed as biomass burning according to Figure 8.

On 23 August 2007, measurements carried out with the MSP-Lidar system, between 12:40 and 22:12 UTC, show the MASP atmosphere heavily loaded by aerosols, in addition, in Figure 12 the presence of a thin aerosol layer above the PBL can be seen from 3.5 to almost 5 km of altitude. Such vertical profile measured using MSP-Lidar system together with the backtrajectories from HYSPLIT model give a strong indication that this thin layer detached from the PBL is composed by biomass burning aerosols from the Midwestern region of Brazil.

Backscatter profile analyses were performed applying the Klett inversion method together with the AOD values retrieved by the AERONET sunphotometer. The 532 nm backscatter profile, presented in Figure 13, was calculated for the period between 17:00 to 18:00 (UTC), which corresponds to the CALIPSO closest approach to the MASP. The backscatter maximum for this measurement period was observed between 0.5 and around 2.0 km altitude range. This aerosol layer showed a quite stable maximum of 0.0033 $km^{-1}sr^{-1}$ during the period of measurements and represents the aerosol trapped in the PBL layer. The Lidar ratio in this period presents a constant value of 41 sr, calculated between 0.5

and 5.0 km range altitude, which represent the dust aerosol type [45]. Such values of LR are not consistent with those values for biomass burning; however, it should be kept in mind that the Klett's inversion method only provide a constant LR value for the whole atmosphere column; in addition, the MASP is considered one of the most polluted cities in the world, having several different aerosol types loaded in its atmosphere, which can turn the retrieval of the LR value and the confidence of the aerosol type classification very difficult tasks. Furthermore, the HYSPLIT backtrajectories leads to a strong indication that the detached layer between 3.5 and 5 km is the biomass burning aerosol transported from the Midwestern region of Brazil. Backscatter profile analyses performed in posterior time period (around 20:00 UTC) presents LR values of 60 sr, consistent to the biomass burning aerosol type according to [45].

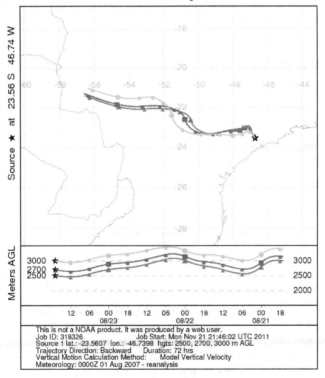

Figure 11. Air mass trajectories generated by the HYSPLIT model. Such trajectories identify the transport of aerosols from the Midwestern region of Brazilian territory on 21 August 2007 for the Southeast region of Brazil, reaching the MASP on 23 August 2007.

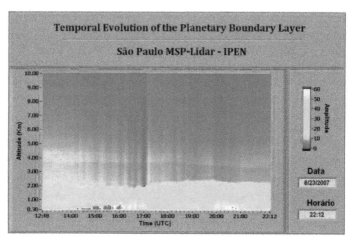

Figure 12. Range corrected backscatter profile at 532 nm for measurements carried out on 23 August 2007 by the MSP-Lidar system between 12:48 to 22:12 (UTC). It can be seen a thin layer of aerosol between 3.5 and 5 km transported from the Midwestern region of the Brazilian territory.

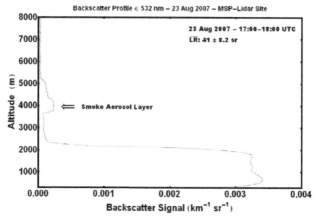

Figure 13. Backscatter coefficient profile at 532 nm measured on 23 August 2007 by the MSP-Lidar system between 17:00 to 18:00 (UTC).

On the same day the AERONET sunphotometer installed at MASP obtained AOD values spanning from 0.20 to 0.38, indicating the detection of a high amount of absorbent aerosols loaded in the atmosphere of the MASP. Figure 14 shows the AOD and LR values obtained by the CALIOP sensor aboard the CALIPSO satellite on 23 August 2007, which overpasses the region close to MASP with closest approach of about 80 km. The CALIOP AOD values for this case are spanning from 0.08 to 0.14, which is quite low to the values retrieved by AERONET. Such difference between AOD retrieved by both instruments might have occurred due to the fact that the sunphotometer is installed within the MASP, which is considered one of the most polluted megacities of the world, with high quantity and variety of aerosols in its atmosphere. On the other hand, the trajectory of the CALIPSO satellite took place at about 80 km of horizontal distance to the east direction of the MASP, as can be checked in Figure 5. In this case, the CALIPSO satellite must have measured different aerosol layers with different optical properties from those measured by the AERONET sunphotometer. Furthermore, the values of LR are practically constant with values of 55 sr, representing aerosols of dust or polluted dust types. Taking into account the closest distance between CALIPSO trajectory and the MSP-Lidar system site, these CALIOP LR values can be considered in agreement with that retrieved by the MSP-Lidar system, which obtained values spanning from 33 to 66 sr in the whole time period (Figure 12).

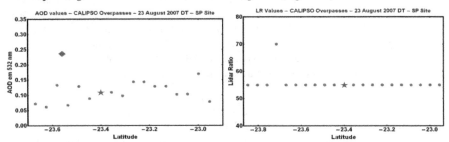

Figure 14. AOD and LR distribution as function of latitude (DT trajectory in Figure 5) measured by the CALIOP sensor on 23 August 2007 during the daytime. The star marks the point of the satellite closest approach to the AERONET site installed at MASP. The red triangle in the left panel represents the value of AOD measured by the AERONET sunphotometer during the closest approach time.

8. Conclusions

In this study, an extensive biomass burning event observed at two different sites in Brazilian territory and track the aerosol transportation from Midwestern to Southeastern region of Brazil have been described. The synergy of MODIS and CALIOP sensor, both aboard of satellites, the AERONET sunphotometer and the elastic backscatter Lidar system measurements allowed to estimate key optical characteristics of the aerosols observed during this event. These results are important for the radiative forcing study in the Brazilian territory and for the study of the effects of certain types of aerosol in the air quality of the megacities such as São Paulo, due to aerosols produced in the MASP as well as those transported from distant regions.

Spatial distribution of fires in the São Paulo State obtained by the NOAA-12 satellite indicates that the winter is the season with the highest number of fire outbreaks. In addition, during the harvest period, which occurs from May to November, the plantation areas are burnt a few hours before the manual cutting, resulting in large quantities of aerosols being emitted into the atmosphere. AOD measurements from MODIS sensor indicate high maximum values during spring and winter (Southeastern dry season). Average of the maximum AOD between 2003 and 2010 was 0.492±0.505, indicating the high variability of the maximum AOD over this region. As in the Brazilian Southeastern region, the large occurrence of higher AOD levels during the dry season for the Midwestern region is basically due to the highest amount of biomass burning loaded in the atmosphere.

One day backtrajectories calculated for 21 August 2007 using the HYSPLIT model indicated that at 3.5, 4.0 and 4.5 km altitude air masses were advected from regions with intensive biomass burning activities in the Midwestern portion of the Brazilian territory. For the biomass burning episode under investigation in the Midwestern region on 21 August 2007, the AOD and LR at 532 nm was calculated using measurements from the CALIOP sensor and the AERONET sunphotometer installed at Campo Grande. The AOD retrieved by CALIOP sensor spanning from 0.30 to 0.65 and the LR was almost constant at 70 sr. The AERONET provides on the same day mean values of AOD and AE of 0.2 and 1.67, respectively, both values higher than the median in the August period (AOD = 0.198±0.078 and AE = 1.46±0.09). Such values correspond to fine mode size distribution and high absorption and extinction aerosol types, which is a strong indication of the predominance of biomass burning aerosols in the atmosphere. MODIS data image also provide a strong indication of biomass burning loaded in the atmosphere in the Midwestern region.

Daytime CALIPSO trajectory on 21 August 2007 together with the HYSPLIT backtrajectories provide evidences that air masses of biomass burning aerosol were transported towards the Southeastern region of Brazilian territory. However, AOD value of 0.10 and LR value of 40-55 sr indicate that those biomass burning aerosol layers have been undergoing an aging process, absorbing moisture from the atmosphere or even mixing with other aerosol types.

Air mass trajectories generated by the HYSPLIT model showed that these biomass burning aerosols were transported from the CG region towards the Metropolitan Area of São Paulo, reaching MASP on 23 August 2007 at an altitude range of 3.5 to 4.5 km. The backscatter coefficient profile retrieved by the MSP-Lidar system showed an atmosphere heavily loaded of aerosol trapped within the PBL and also a distinguishable aerosol layer between 3.5 and 5 km of altitude. LR values retrieved by the MSP-Lidar, reaching maximum values of 66 sr, are in agreement with LR values of 55 sr signed by the CALIOP sensor. This LR value of 55 sr indicates the presence of polluted dust aerosol type instead of biomass burning type, though. However, it needs to be taken into account that the closest approach of the CALIPSO satellite was 80 km eastern of the MSP-Lidar site, which can lead to the fact that both instruments were not probing the same air mass parcels. Furthermore, the LR values retrieved from MSP-Lidar system in the whole period of time indicate the presence of several amount of aerosols present in the atmosphere of São Paulo, which can disguise the individual effect of the optical properties from biomass burning aerosols, which may have had their optical properties altered due to transport and aging processes [37].

This case study shows that the synergy and combination of analysis using several remote sensing instruments, whether passive or active, results in a better understanding of the aerosol optical properties in the atmosphere. Furthermore, these results show that aerosols produced in different regions can be transported by long-range distances. In this sense, megacities such as the Metropolitan Area of São Paulo, which have a large number of local aerosol sources is subject to the influence of aerosol pollution produced by remote sources.

Author details

G. L. Mariano*
Federal University of Pelotas, UFPEL, Pelotas, Brasil

F. J. S. Lopes
São Paulo University, USP, São Paulo, Brasil

F. J. S. Lopes and E. Landulfo
Nuclear and Energy Research Institute, IPEN, São Paulo, Brasil

E. V. C. Mariano
National Institute for Space Research, INPE, São José dos Campos, Brasil

Acknowledgement

The authors wish to acknowledge all the CALIPSO team for data obtained from the NASA Langley Research Center Atmospheric. They also gratefully acknowledge the team of AERONET sunphotometer network, CPTEC/INPE for the active fires and the NOAA Air Resources Laboratory for the provision of the HYSPLIT transport and dispersion model and READY website used in this publication. The first author wishes to acknowledge the financial support of Fundacao para o Amparo da Pesquisa do Estado de Sao Paulo FAPESP under the project number 2011/14365-5. The secong author wish to acknowledge the Research Foundation of the State of Rio Grande do Sul and the post graduate program in Meteorology at the Federal University of Pelotas.

9. References

[1] Forster, P., Ramaswamy, V., Artaxo, P., et al. (2007): Changes in Atmospheric Constituents and in Radiative Forcing, in: Climate Change 2007: The Physical Science Basis. Contribution of Working Group I to the Fourth Assessment Report of the Intergovernmental Panel on Climate Change, edited by: Solomon, S., Qin, D., Manning, M., Chen, Z., Marquis, M., Averyt, K. B., Tignor, M., and Miller, H. L., Cambridge University Press, Cambridge, UK and New York, NY, USA.

[2] Landulfo, E., Lopes, F.J.S. (2009) Initial approach in biomass burning aerosol transport tracking with CALIPSO and MODIS satellites, sunphotometer and a backscatter Lidar system in Brazil. (2009) Proceedings of SPIE — The International Society for Optical Engineering 7479: 747905.

[3] Stephens, G.L., D.G. Vane, R.J. Boain, G.G. Mace, K. Sassen, Z. Wang, A.J. Illingworth, E.J. O'Connor, W.B. Rossow, S.L. Durden, S.D. Miller, R.T. Austin, A. Benedetti, C. Mitrescu, and CloudSat Science Team (2002) The CloudSat mission and the A-Train: A new dimension of space-based observations of clouds and precipitation. Bull. Amer. Meteorol. Soc. 83: 1771-1790.

* Corresponding Author

[4] Barnes, W.L.; Tpagano, T.S.; Salomonson. V.V. (1998) Prelaunch characteristics of the Moderate Resolution Imaging Spectroradiometer (MODIS) on EOS-AM1. IEEE Trans. Geosci. Remote Sensing, 36: 1088–1100.

[5] Spiegel, M.R.; Stephens, L.J. (2009) Estatística. Porto Alegre: Bookman, 597p

[6] Coulson, K. L. (1975) Solar and terrestrial radiation: methods and measurements. New York: Academic Press. 479 p.

[7] Winker, D., W. H. Hunt, and M. J. McGill (2007), Initial performance assessment of CALIOP, Geophys. Res. Lett., 34, L19803, doi: 10.1029/2007GL030135.

[8] Winker, D. M., Vaughan, M.A., Omar, A., Hu, Y., Powell, K.A., Liu, Z., Hunt, W.Z., Young, S.A. (2009), Overview of the CALIPSO mission and CALIOP data processing algorithms, J. Atmos. Oceanic Technol., 26, 2310, doi:10.1175/2009JTECHA1231.1.

[9] Hunt, W. H., D. M. Winker, M. A. Vaughan, K. A. Powell, P. L. Lucker, and C. Weimer (2009), CALIPSO Lidar Description and Performance Assessment, J. Atmos. Oceanic Technol., 26, 1214, doi:10.1175/2009JTECHA1223.1

[10] Hostetler, C. A., Z. Liu, J. Reagan, M. A. Vaughan, D. M. Winker, M. Osborn, W. H. Hunt, K. A. Powell, and C. Trepte (2006), CALIOP Algorithm Theoretical Basis Document - Calibration and level 1 data products. Release 1.0, PC-SCI-201 Part 1, NASA Langley Research Center, Hampton, Virginia, USA, Available: http://www.calipso.larc.nasa.gov/resources/project documentation.php. Accessed 2012 May 07.

[11] Holben, B. N., Eck, T. F., Slutsker, I., Tanré, D., Buis, J. P., Setzer, A., Vermote, E., Reagan, J. A., Kaufman, Y. J., Nakajima, T., Lavenu, F., Jankowiak, I., and Smirnov, A. (1998) AERONET a federal instrument network and data archive for aerosol characterization, Remote Sensing of Environment 66: 1–16.

[12] Ångström, A. (1964) The parameters of atmospheric turbidity. Tellus, 16, 64–75.

[13] Junge, C. (1963) Air chemistry and radioactivity. New York: Academic Press Inc., 382 pages.

[14] Deepak, A., Gerber, H.E. (1983) Aerosols and their climate effects. Series Report 55, International Council of Scientific Unions and WMO, Switzerland.

[15] D'Almeida, G.A., P. Koepke, and E. P. Shettle (1991) Atmospheric aerosols, global climatology and radiative characteristics. Deepak Publ. 561p.

[16] Hamonou, E., Chazette, P., Balis, D., Dulac, F., Schneider, X., Galani, E., Ancellet, G., Papayannis, A. (1999), Characterization of the vertical structure of Saharan dust export to the Mediterranean basin. Journal of Geophysical Research. 104: 22257–22270.

[17] Ferrare, R.A., Schols, J.L., Eloranta, E.W. (1991). Lidar observations of banded convection during BLX83. Journal of Applied Meteorology. 3: 312–326.

[18] Mel, S.H., Spinhirne, J.D., Chou, S.H., Palm, S.P. (1985) Lidar observations of vertically organized convection in the planetary boundary layer over the ocean. Journal of Climate and Applied Meteorology. 24: 806–821.

[19] Crum, T.D., Stull, R.B., Eloranta, E. (1987) Coincident Lidar and aircraft observations of entrainment into thermals and mixed layer. Journal of Climate and Applied Meteorology. 26: 774–788.

[20] Balis, D., Papayannis, A., Galani, E., Marenco, F., Santacesaria, V., Hamonou, E., Chazette, P., Ziomas, I., Zerefos, C. (2000) Tropospheric LIDAR aerosol measurements and sun photometric observations at Thessaloniki, Greece. Atmospheric Environment. 34: 925-932.

[21] Chourdakis, G., Papayannis, A., Porteneuve, J. (2002) Analysis of the receiver response for a non-coaxial Lidar system with fiber-optic output. Applied Optics. 41: 2715-2723.

[22] Papayannis, A., Chourdakis, G. (2002) The EOLE Project. A multi-wavelength laser remote sensing (Lidar) system for ozone and aerosol measurements in the troposphere and the lower stratosphere. Part ii: aerosol measurements over Athens, Greece. International Journal of Remote Sensing. 23: 179–196.

[23] Klett, J. D. (1981) Stable analytical inversion solution for processing Lidar returns. Appl. Opt. 20: 211-220.

[24] Klett, J. D. (1985) Lidar inversion with variable backscatter/extinction ratios, Appl. Opt. 24: 1638–1643.

[25] Liou, K. N. (2002) An Introduction to Atmospheric Radiation, 2nd ed. Academic Press, California. 583p.

[26] Marenco, F., V. Santacesaria, A. F. Bais, D. Balis, A. di Sarra, A. Papayannis, and C. Zerefos (2002), Optical properties of tropospheric aerosols determined by Lidar and spectrophotometric measurements (Photochemical Activity and Solar Ultraviolet Radiation campaign), Appl. Optics, 36, 6875-6886, doi:10.1364/AO.36.006875.

[27] Bucholtz, A. (1995) Rayleigh-scattering calculations for the terrestrial atmosphere. Applied Optics. 34: 2765-2773, doi: 10.1364/AO.34.002765.

[28] Setzer, A. W.; Pereira, M. C. (1991) Amazonia biomass burnings in 1987 and an estimate of their tropospheric emissions. Ambio, 20: n. 1.

[29] Feltz, J. M., Prins, E.M., Setzer, A.W. (2001) A comparison of the GOES-8 ABBA and INPE AVHRR fire products for South America from 1995-2000. 11th conference on Satellite Meteorology and Oceanography, Bulletin of the American Meteorological Society 83: 1645–1648.

[30] Behera, D., Jindal, S.K., Malhotra, H.S. (1994) Ventilatory function in nonsmoking rural Indian women using different cooking fuels. Respiration 61: 89–92.

[31] Phonboon, K., Paisarn-Uchapong, O., Kanatharana, P., Agsorn, S. (1999) Smoke episodes emissions characterization and assessment of health risks related downwind air quality case study, Thailand. In: WHO Health Guidelines for Vegetation Fire Events. Geneva: World Health Organization: 334–358.

[32] Bruce, N., Perez-Padilla, R., Albalak, R. (2000) Indoor air pollution in developing countries: a major environmental and public health challenge. Bulletin WHO 78: 1078–1092.

[33] CETESB - Relatório de Qualidade do Ar no Estado de São Paulo (2009) Secretaria Do Meio Ambiente, Série Relatórios, São Paulo. Available: www.cetesb.sp.gov.br. Accessed: 2012 Feb 14

[34] Artaxo, P., Castanho, A.D.A. (1998). Aerosol concentrations and source apportionment in the urban area of São Paulo, Brazil. IAEA. International Atomic Energy Agency. TEC DOC series.

[35] Freitas, S. R. , Longo, K. M., Dias, M. A. F. S., Dias, P. L. S., Chatfield, R., Prins, E., Artaxo, P., Grell, G. A., F. S. Recuero, (2005) Monitoring the transport of biomass burning emissions in South America. Environmental Fluid Mechanics. 5: 135–167.

[36] Kaskaoutis, D., Kambezidis, H., Hatzianastassiou, N., Kosmopoulos, P., Badarinath, K., (2007). Aerosol climatology: on the discrimination of aerosol types over four AERONET sites. Atmospheric Chemistry and Physics Discussions 7, 6357–6411.

[37] Müller, D., Ansmann, A., Mattis, I., Tesche, M., Wandinger, U., Althausen, D., Pisani, G., (2007). Aerosol-type-dependent lidar ratios observed with raman lidar. Journal of Geophysical Reasearch 112, D16202. doi:10.1029/2006JD00829.

[38] Amiridis, V., Balis, D.S., Giannakaki, E., Stohl, A., Kazadzis, S., Koukouli, M.E., Zanis, P. (2009) Optical characteristics of biomass burning aerosols over southeastern Europe determined from UV–Raman Lidar measurements. Atmospheric Chemistry and Physics. 9: 2431–2440. doi:10.5194/acp-9-2431-2009.

[39] Omar, A. H., J. G. Won, D. M. Winker, S. C. Yoon, O. Dubovik, M. P. McCormick (2005) Development of global aerosol models using cluster analysis of Aerosol Robotic Network (AERONET) measurements. Journal of Geophysical Research. 110: D10S14, doi:10.1029/2004JD004874.

[40] Omar, A. H., D. M. Winker, C. Kittaka, M. A. Vaughan, Z. Liu, Y. Hu, C. R. Trepte, R. R. Rogers, R. A. Ferrare, K.-P. Lee, R. E. Kuehn, C. A. Hostetler (2009) The CALIPSO automated aerosol classification and Lidar Ratio Selection Algorithm. Journal of Atmospheric and Oceanic Technology. 26: doi:10.1175/2009JTECHA1231.1.

[41] Freitas, S. R., K. M. Longo, J. Trentmann and D. Latham (2010)Technical Note: Sensitivity of 1-D smoke plume rise models to the inclusion of environmental wind drag. Atmospheric Chemistry and Physics. 10: 585–594. doi:10.5194/acp-10-585-2010.

[42] Liu, Z., A. H. Omar, Y. Hu, M. A. Vaughan, D. M. Winker (2006). CALIOP Algorithm Theoretical Basis Document - Scene Classification Algorithms. Release 1.0, PC-SCI-202 Part 3, 2005, NASA Langley Research Center, Hampton, Virginia, USA, Available: http://www.calipso.larc.nasa.gov/resources/project documentation.php. Accessed 2012 May 07.

[43] Vaughan, M. A., D. M. Winker, and K. A. Powell (2006) CALIOP Algorithm Theoretical Basis Document - Part 2: Feature detection and layer properties algorithms. Release 1.01, PC-SCI-202 Part 2, NASA Langley Research Center, Hampton, Virginia, USA, Available http://www.calipso.larc.nasa.gov/resources/project documentation.php]. Accessed 2012 May 07.

[44] Draxler, R., G. Hess (1998). Description of the Hysplit 4 modeling system, NOAA Tech. Memo. ERL ARL-224, NOAA Air Resources Laboratory, Silver Spring, MD, USA, Available: http://ready.arl.noaa.gov/HYSPLIT.php. Accessed 2012 May 07.

[45] Cattrall, C., Reagan, J., Thome, K., Dubovik, O. (2005). Variability of aerosol and spectral Lidar and backscatter and extinction ratios of key aerosol types derived from selected Aerosol Robotic Network locations. Journal of Geophysical Research. 110: D10S11. doi:10.1029/2004JD005124

Permissions

The contributors of this book come from diverse backgrounds, making this book a truly international effort. This book will bring forth new frontiers with its revolutionizing research information and detailed analysis of the nascent developments around the world.

We would like to thank Hayder Abdul-Razzak, for lending his expertise to make the book truly unique. He has played a crucial role in the development of this book. Without his invaluable contribution this book wouldn't have been possible. He has made vital efforts to compile up to date information on the varied aspects of this subject to make this book a valuable addition to the collection of many professionals and students.

This book was conceptualized with the vision of imparting up-to-date information and advanced data in this field. To ensure the same, a matchless editorial board was set up. Every individual on the board went through rigorous rounds of assessment to prove their worth. After which they invested a large part of their time researching and compiling the most relevant data for our readers. Conferences and sessions were held from time to time between the editorial board and the contributing authors to present the data in the most comprehensible form. The editorial team has worked tirelessly to provide valuable and valid information to help people across the globe.

Every chapter published in this book has been scrutinized by our experts. Their significance has been extensively debated. The topics covered herein carry significant findings which will fuel the growth of the discipline. They may even be implemented as practical applications or may be referred to as a beginning point for another development. Chapters in this book were first published by InTech; hereby published with permission under the Creative Commons Attribution License or equivalent.

The editorial board has been involved in producing this book since its inception. They have spent rigorous hours researching and exploring the diverse topics which have resulted in the successful publishing of this book. They have passed on their knowledge of decades through this book. To expedite this challenging task, the publisher supported the team at every step. A small team of assistant editors was also appointed to further simplify the editing procedure and attain best results for the readers.

Our editorial team has been hand-picked from every corner of the world. Their multi-ethnicity adds dynamic inputs to the discussions which result in innovative outcomes. These outcomes are then further discussed with the researchers and contributors who give their valuable feedback and opinion regarding the same. The feedback is then collaborated with the researches and they are edited in a comprehensive manner to aid the understanding of the subject.

Apart from the editorial board, the designing team has also invested a significant amount of their time in understanding the subject and creating the most relevant covers. They scrutinized every image to scout for the most suitable representation of the subject and create an appropriate cover for the book.

The publishing team has been involved in this book since its early stages. They were actively engaged in every process, be it collecting the data, connecting with the contributors or procuring relevant information. The team has been an ardent support to the editorial, designing and production team. Their endless efforts to recruit the best for this project, has resulted in the accomplishment of this book. They are a veteran in the field of academics and their pool of knowledge is as vast as their experience in printing. Their expertise and guidance has proved useful at every step. Their uncompromising quality standards have made this book an exceptional effort. Their encouragement from time to time has been an inspiration for everyone.

The publisher and the editorial board hope that this book will prove to be a valuable piece of knowledge for researchers, students, practitioners and scholars across the globe.

List of Contributors

Inna Plakhina
Oboukhov Institute of Atmospheric Physics, RussianAcademy of Science, Russia

A.K. Srivastava
Indian Institute of Tropical Meteorology (Branch), Prof. Ramnath Vij Marg, New Delhi, India

Sagnik Dey
Centre for Atmospheric Sciences, Indian Institute of Technology Delhi, New Delhi, India

S.N. Tripathi
Department of Civil Engineering and Centre for Environmental Science and Engineering, Indian Institute of Technology Kanpur, Kanpur, India

Tomomi Hoshiko
Department of Urban Engineering, Graduate School of Engineering, The University of Tokyo, Tokyo, Japan

Kazuo Yamamoto and Fumiyuki Nakajima
Environmental Science Center, The University of Tokyo, Tokyo, Japan

Tassanee Prueksasit
Department of Environmental Science, Faculty of Science, Chulalongkorn University, Bangkok, Thailand

Sanat Kumar Das
Department of Atmospheric Sciences, National Taiwan University, Taiwan

S.V. Sunil Kumar and K. Parameswaran
Space Physics Laboratory, Vikram Sarabhai Space Centre, Thiruvananthapuram, India

Bijoy V. Thampi
Laboratoire de Météorologie Dynamique, IPSL, Place Jussieu, Paris, France

Alireza Rashki and C.J.deW. Rautenbach
Department of Geography, Geoinformatics and Meteorology, Faculty of Natural and Agricultural Sciences, University of Pretoria, Pretoria, South Africa
Department of Drylands and Desert Management, Faculty of Natural Resources and Environment, Ferdowsi University of Mashhad, Mashhad, Iran

Dimitris Kaskaoutis
Research and Technology Development Centre, Sharda University, Greater Noida, India

Patrick Eriksson
Department of Geology, Faculty of Natural and Agricultural Sciences, University of Pretoria, Pretoria, South Africa

J.C. Jiménez-Escalona
ESIME U. Ticomán, Instituto Politécnico Nacional, Gustavo A. Madero, Mexico City, Mexico

O. Peralta
CCA, Universidad Nacional Autónoma de México, Ciudad Universitaria, Mexico City, Mexico

Stelyus L. Mkoma
Faculty of Science, Sokoine University of Agriculture, Tanzania

Gisele O. da Rocha, José D.S. da Silva and Jailson B. de Andrade
Instituto de Química, Universidade Federal da Bahia, Brazil

G. L. Mariano
Federal University of Pelotas, UFPEL, Pelotas, Brasil

F. J. S. Lopes
São Paulo University, USP, São Paulo, Brasil

F. J. S. Lopes and E. Landulfo
Nuclear and Energy Research Institute, IPEN, São Paulo, Brasil

E. V. C. Mariano
National Institute for Space Research, INPE, São José dos Campos, Brasil

Printed in the USA
CPSIA information can be obtained
at www.ICGtesting.com
JSHW011452221024
72173JS00005B/1048